Repairing and Extending
Doors and Windows

**VAN NOSTRAND REINHOLD'S
BUILDING RENOVATION AND RESTORATION SERIES**

Repairing and Extending Weather Barriers   ISBN 0-442-20611-9

Repairing and Extending Finishes, Part I   ISBN 0-442-20612-7

Repairing and Extending Finishes, Part II   ISBN 0-442-20613-5

Repairing and Extending Nonstructural Metals   ISBN 0-442-20615-1

Repairing and Extending Doors and Windows   ISBN 0-442-20618-6

Repairing, Extending, and Cleaning Brick and Block   ISBN 0-442-20619-4

Repairing, Extending, and Cleaning Stone   ISBN 0-442-20620-8

Repairing and Extending Wood   ISBN 0-442-20621-6

Building Renovation and Restoration Series

Repairing and Extending Doors and Windows

METAL

WOOD

ENTRANCES

STORE FRONTS

CURTAIN WALLS

GLAZING

H. Leslie Simmons, AIA, CSI

VNR VAN NOSTRAND REINHOLD
New York

Copyright © 1991 by H. Leslie Simmons

Library of Congress Catalog Card Number 90-13029
ISBN 0-442-20618-6

All rights reserved. No part of this work covered by the copyright hereon may be reproduced or used in any form or by any means—graphic, electronic, or mechanical, including photocopying, recording, taping, or information storage and retrieval systems—without written permission of the publisher.

Printed in the United States of America

Van Nostrand Reinhold
115 Fifth Avenue
New York, New York 10003

Chapman and Hall
2-6 Boundary Row
London SE1 8HN, England

Thomas Nelson Australia
102 Dodds Street
South Melbourne 3205, Victoria, Australia

Nelson Canada
1120 Birchmount Road
Scarborough, Ontario M1K 5G4, Canada

16 15 14 13 12 11 10 9 8 7 6 5 4 3 2 1

Library of Congress Cataloging-in-Publication Data

Simmons, H. Leslie.
    Repairing and extending doors and windows : metal, wood, entrances, store fronts, curtain walls, glazing / H. Leslie Simmons.
       p.   cm. — (Building renovation and restoration series)
    Includes bibliographical references (p. 364) and index.
    ISBN 0-442-20618-6
    1. Doors. 2. Windows. 3. Buildings—Repair and reconstruction.
I. Title. II. Series: Simmons, H. Leslie. Building renovation and restoration series.
TH2270.S55 1991
690'.182—dc20
                                              90-13029
                                                  CIP

Contents

Series Foreword	xv
Preface	xix
Acknowledgments	xxv

CHAPTER 1 Introduction 1

1.1	What This Book Covers	2
1.2	Failure Types and Conditions	3
1.3	What to Do in an Emergency	3
1.4	**Professional Help**	4
1.4.1	Help for Building Owners and Managers	5
1.4.2	Help for Architects and Engineers	15
1.4.3	Help for General Building Contractors	17
1.5	**Prework On-Site Examination**	18
1.5.1	The Owner	18

1.5.2	Architects and Engineers	18
1.5.3	Building Contractors	19
1.6	**Demolition and Removal of Existing Construction**	**20**
1.6.1	Controls	20
1.6.2	Protection of Persons and Property to Remain or Be Reused	21
1.6.3	Performing Demolition Work	22
1.6.4	Disposition of Removed Materials	24
1.7	**General Requirements for Alterations**	**25**
1.7.1	Controls	25
1.7.2	Materials	26
1.7.3	Making Alterations, Patches, and Repairs	27

CHAPTER 2 Support Systems 29

2.1	Excess Structure Movement	30
2.2	Failed Steel or Concrete Structures	31
2.3	Failed Wood Structure or Wood Wall or Partition Framing	32
2.4	Failed Metal Wall or Partition Framing	35
2.5	Failed Concrete, Stone, or Masonry Walls or Partitions	37
2.6	Failed Other Building Elements	39
2.7	Where to Get More Information	40

CHAPTER 3 Wood and Metal Materials and Finishes 43

3.1	**Wood**	**43**
3.1.1	Wood Door and Panel Materials and Their Quality	44
3.1.2	Wood Window and Sliding-Glass-Door Materials	45
3.1.3	Wood Frame and Trim Materials	45
3.1.4	Preservative Treatment	46
3.2	**Carbon Steel and Stainless Steel**	**47**
3.2.1	Heat Treatment	48
3.2.2	Carbon-Steel and Stainless Steel Products	48
3.3	**Aluminum**	**49**
3.3.1	Alloy Designations	50
3.3.2	Temper Designations	50
3.3.3	Heat Treatment	50
3.3.4	Aluminum Products	51
3.4	**Copper Alloys**	**52**
3.4.1	Alloy Designations	53
3.4.2	Copper Alloy Products	53
3.5	**Other Metals**	**54**
3.6	**General Requirements for Finishes on Metals**	**54**

3.7	**Mechanical and Chemical Cleaning of and Finishes for Steel**	**55**
3.7.1	Mechanical Finishes for and Cleaning of Carbon Steel	55
3.7.2	Chemical Finishes for and Cleaning of Carbon Steel	56
3.7.3	Mechanical Finishes for and Cleaning of Stainless Steel	56
3.7.4	Chemical Finishes for and Cleaning of Stainless Steel	57
3.8	**Mechanical and Chemical Cleaning of and Finishes for Aluminum**	**57**
3.8.1	Mechanical Cleaning of and Finishes for Aluminum	58
3.8.2	Chemical Cleaning of and Finishes for Aluminum	59
3.8.3	Anodic Coatings	59
3.9	**Mechanical and Chemical Cleaning of and Finishes for Copper Alloys**	**61**
3.9.1	Mechanical Finishes for and Cleaning of Copper Alloys	61
3.9.2	Chemical Cleaning of and Finishes for Copper Alloys	62
3.10	**Inorganic Coatings on Metal**	**63**
3.10.1	Metallic Coatings	63
3.10.2	Vitreous Coatings	67
3.10.3	Laminated Coatings	68
3.11	**Organic Coatings**	**69**
3.11.1	Standards	69
3.11.2	Manufacturers and Products	70
3.11.3	Opaque Organic Coating Composition	72
3.11.4	Transparent Organic Coating Composition	72
3.11.5	Organic Coating Systems	75
3.11.6	Miscellaneous Materials	77
3.11.7	Special Paints	77
3.11.8	Powder Coatings	77
3.11.9	Film Laminate	77
3.11.10	Preparation of Surfaces to Receive Organic Coatings	78
3.11.11	Factory-Applied Organic Coatings	79
3.11.12	Shop Primer-Coat Application for Field-Applied Paint	80
3.11.13	Preparing the Site for Field Application of Organic Coatings	81
3.11.14	Field Touch-up of Shop Coats	82
3.11.15	Applying Organic Coatings in the Field	82
3.12	**Why Wood and Metal Materials and Finishes Fail**	**86**
3.12.1	Bad Materials	86
3.12.2	Selecting Inappropriate Materials	88
3.12.3	Selecting Inappropriate Finishes	89
3.12.4	Improper Preparation for Application of Finishes	91
3.12.5	Improper Finish Application	95
3.12.6	Failure of the Immediate Substrate	99
3.12.7	Failure to Protect Materials and Finishes	106
3.12.8	Failure to Properly Maintain Applied Finishes	107
3.12.9	Natural Aging	108

3.13	**Cleaning and Repairing of Woods and Metals and Their Finishes**	**110**
3.13.1	General Requirements	110
3.13.2	Repairing Specific Types of Damage	111
3.13.3	Cleaning Existing Wood and Metal Finishes That Do Not Need Repair	115
3.14	**Finishes That Cannot Be Refinished in the Field**	**119**
3.15	**Refinishing Existing Wood and Metal in the Field**	**119**
3.15.1	Materials and Manufacturers	121
3.15.2	Organic Coating Systems	122
3.15.3	Preparation for Applying Organic Coatings	122
3.15.4	Applying Paint	135
3.15.5	Applying Other Organic Coatings	136
3.16	**Where to Get More Information**	**137**

CHAPTER 4 Metal Doors, Frames, and Store Fronts — 147

4.1	**General Requirements for Metal Doors and Frames**	**148**
4.1.1	Standards	148
4.1.2	Controls	149
4.1.3	Materials	150
4.2	**Design and Fabrication of Carbon-Steel Doors and Flush Panels**	**152**
4.2.1	Carbon-Steel Door Types, Styles, and Designs	152
4.2.2	Carbon-Steel Panels	156
4.3	**Design and Fabrication of Carbon-Steel Frames**	**156**
4.3.1	Carbon-Steel Frame Types	159
4.4	**Design and Fabrication of Formed Stainless Steel and Copper-Alloy Swinging Doors and Frames**	**160**
4.5	**Design and Fabrication of Extruded Aluminum Swinging Doors, Frames, and Store Fronts**	**161**
4.5.1	Design and Fabrication of Extruded Aluminum Swinging Doors	161
4.5.2	Design and Fabrication of Extruded Aluminum Door, Entrance, and Store Front Framing	162
4.5.3	Design and Fabrication of Packaged Entrances	165
4.6	**Fabrication of Louvers**	**165**
4.7	**Finish Hardware for Swinging Doors**	**167**
4.7.1	Hardware Types	168
4.7.2	Hardware Materials and Finishes	168
4.7.3	Preparation of Doors and Frames for Finish Hardware	169
4.8	**Frame Supports and Anchors**	**169**
4.9	**Installation of Metal Doors and Frames in New Construction**	**170**
4.9.1	Preparation for Installation	171

4.9.2	Installation of Formed Frames and Doors	171
4.9.3	Installation of Extruded Aluminum Swinging Doors, Frames, and Store Fronts	173
4.9.4	Installation of Glazing and Louvers	173
4.9.5	Adjustments, Cleaning, and Protection	174
4.10	**Design, Fabrication, and Installation of Revolving Doors**	**174**
4.11	**Why Metal Doors and Frames Fail**	**175**
4.11.1	Improper Design	175
4.11.2	Inappropriate Door or Frame Type or Design Selection	177
4.11.3	Inappropriate Hardware Selection	179
4.11.4	Improper Fabrication	179
4.11.5	Improper Installation	180
4.11.6	Unforeseen Trauma	182
4.11.7	Natural Aging	182
4.12	**Repairing Metal Doors and Frames**	**182**
4.12.1	General Requirements	184
4.12.2	Repairing Specific Types of Damage	185
4.13	**Metal Doors and Frames in Existing Construction**	**186**
4.13.1	Door, Frame, and Hardware Materials, Finishes, Types, and Fabrication	187
4.13.2	Standards, Controls, Delivery, and Handling	187
4.13.3	Preparation for Installation	188
4.13.4	Installation of Doors and Frames	189
4.13.5	Installation of Glazing and Louvers	190
4.14	**Where to Get More Information**	**190**

CHAPTER 5 Wood Doors and Frames 192

5.1	**Fabrication and Operation of Wood Doors and Frames**	**193**
5.1.1	Standards	193
5.1.2	Controls	194
5.1.3	Materials and Finishes	197
5.1.4	Fabrication of Doors, Flush Panels, and Louvers	197
5.1.5	Wood Store Fronts	203
5.1.6	Fabrication of Wood Frames	204
5.1.7	Finish Hardware	206
5.1.8	Warranties	208
5.2	**Installation of Wood Doors and Frames in New Construction**	**208**
5.2.1	Delivery and Handling	208
5.2.2	Preparation for Installation	209
5.2.3	Installation of Frames	209
5.2.4	Installation of Doors	210
5.2.5	Glazing and Installation of Louvers	211

5.2.6	Adjustments, Protection, and Cleaning	211
5.3	**Why Wood Doors and Frames Fail**	**211**
5.3.1	Improper Design	212
5.3.2	Inappropriate Door or Frame Type or Design Selection	212
5.3.3	Inappropriate Hardware Selection	213
5.3.4	Improper Fabrication	214
5.3.5	Improper Installation	214
5.3.6	Natural Aging	215
5.4	**Repairing Wood Doors and Frames**	**215**
5.4.1	General Requirements	217
5.4.2	Repairing Specific Types of Damage	219
5.5	**Wood Doors and Frames in Existing Construction**	**221**
5.5.1	Door and Frame Materials, Finishes, Types, and Fabrication	221
5.5.2	Standards, Controls, Delivery, and Handling	221
5.5.3	Preparation for Installation	222
5.5.4	Installation of Doors and Frames	222
5.5.5	Installation of Glazing and Louvers	223
5.5.6	Patching Adjacent Materials and Surfaces	223
5.6	**Where to Get More Information**	**223**

CHAPTER 6 Windows and Sliding Glass Doors 225

6.1	**General Requirements for Windows and Sliding Glass Doors**	**226**
6.1.1	Controls	226
6.1.2	Glazing	228
6.2	**Specific Requirements for Steel Windows**	**228**
6.2.1	Standards for Steel Windows	228
6.2.2	Materials, Products, and Finishes for Steel Windows	229
6.2.3	Steel Window Grades	230
6.2.4	Steel Window Types	230
6.2.5	Steel Window Hardware	232
6.3	**Specific Requirements for Aluminum Windows and Sliding Glass Doors**	**233**
6.3.1	Standards for Aluminum Windows and Sliding Glass Doors	233
6.3.2	Materials, Products, and Finishes for Aluminum Windows and Sliding Glass Doors	233
6.3.3	Aluminum Window and Sliding Glass Door Grades and Performance Classes	234
6.3.4	Aluminum Window Types	235
6.3.5	Aluminum Window and Sliding Glass Door Hardware	236
6.4	**Specific Requirements for Wood Windows and Sliding Glass Doors**	**237**
6.4.1	Standards for Wood Windows and Sliding Glass Doors	237

6.4.2	Materials, Products, and Finishes	238
6.4.3	Performance Requirements for Wood Windows and Sliding Glass Doors	238
6.4.4	Wood Window Types	239
6.4.5	Hardware for Wood Windows and Sliding Glass Doors	240
6.5	**Miscellaneous Window and Sliding Glass Door Components and Accessories**	**241**
6.5.1	Thermal Barriers	241
6.5.2	Weather Stripping	242
6.5.3	Frames and Trim	243
6.5.4	Window Sills and Stools	243
6.5.5	Sliding Window and Door Tracks	243
6.5.6	Window Mullions	243
6.5.7	Muntins	244
6.5.8	Storm Windows	244
6.5.9	Venetian Blinds	244
6.5.10	Insect Screens	244
6.5.11	Accessories	246
6.6	**Fabrication of Windows and Sliding Glass Doors**	**246**
6.7	**Installation of Windows and Sliding Glass Doors in New Construction**	**248**
6.7.1	Delivery and Handling	248
6.7.2	Preparation for Installation	248
6.7.3	Installation	249
6.7.4	Adjustments, Cleaning, and Protection	249
6.8	**Why Windows and Sliding Glass Doors Fail**	**250**
6.8.1	Improper Design	251
6.8.2	Inappropriate Window or Sliding Glass Door Grade or Window Type Selection	252
6.8.3	Inappropriate Hardware Selection	252
6.8.4	Improper Fabrication	253
6.8.5	Improper Installation	254
6.8.6	Natural Aging	255
6.8.7	Improper Modifications	255
6.9	**Repairing Windows and Sliding Glass Doors**	**256**
6.9.1	General Requirements	258
6.9.2	Repairing Specific Types of Damage	261
6.10	**Windows and Sliding Glass Doors in Existing Construction**	**264**
6.10.1	Standards, Controls, Delivery, and Handling	265
6.10.2	Windows and Sliding Glass Doors and Associated Hardware, Glazing, and Miscellaneous Components	265
6.10.3	Preparation for Installation	266
6.10.4	Installation of Windows and Sliding Glass Doors, and Their Glazing	267
6.11	**Where to Get More Information**	**268**

CHAPTER 7 Glazed Curtain Walls　　　　　　　　　　270

7.1	**General Requirements for Glazed Curtain Walls**	271
7.1.1	Systems	271
7.1.2	Standards	272
7.1.3	Controls	273
7.1.4	Warranties	276
7.1.5	Performance Requirements	276
7.1.6	Materials	279
7.1.7	Finishes	280
7.1.8	Design and Fabrication	281
7.1.9	Glazing in Curtain Walls	283
7.1.10	Miscellaneous Components and Accessories	283
7.2	**Installation of Glazed Curtain Walls in New Construction**	285
7.2.1	Preparation for Installation	285
7.2.2	Installation	286
7.2.3	Adjustments, Cleaning, and Protection	287
7.3	**Why Glazed Curtain Walls Fail**	287
7.3.1	Improper Design	288
7.3.2	Improper Fabrication	291
7.3.3	Improper Installation	292
7.3.4	Improper Maintenance	293
7.3.5	Natural Aging	294
7.4	**Repairing Glazed Curtain Walls**	294
7.4.1	General Requirements	295
7.4.2	Repairing Specific Types of Damage	296
7.5	**Glazed Curtain Walls in Existing Construction**	300
7.5.1	Standards, Controls, Delivery, and Handling	301
7.5.2	Glazed Curtain Wall Materials, Finishes, Types, and Fabrication	301
7.5.3	Preparation for Installation	301
7.5.4	Installation of Glazed Curtain Walls	302
7.6	**Where to Get More Information**	303

CHAPTER 8 Glazing　　　　　　　　　　305

8.1	**General Requirements for Glazing**	305
8.1.1	Standards	306
8.1.2	Controls	306
8.1.3	Warranties	309
8.1.4	Performance Requirements	310
8.2	**Materials**	311
8.2.1	Glass Types and Quality	311
8.2.2	Plastic Glazing Materials	319
8.2.3	Miscellaneous Glazing Materials	320

8.3	**Glazing in New Construction**	**321**
8.3.1	Delivery, Storage, and Handling	322
8.3.2	Preparation for Glazing	322
8.3.3	General Glazing Requirements	323
8.3.4	Specific Glazing Requirements	326
8.3.5	Cleaning and Protection	329
8.4	**Why Glazing Fails**	**330**
8.4.1	Bad Materials	330
8.4.2	Improper Design	331
8.4.3	Inappropriate Type Selection	334
8.4.4	Improper Fabrication	335
8.4.5	Improper Installation	335
8.4.6	Failure to Protect and Properly Maintain Glazing	337
8.4.7	Improper Modifications	338
8.5	**Glazing in Existing Construction**	**339**
8.5.1	Standards, Controls, Warranties, Performance Requirements, Delivery, and Handling	340
8.5.2	Glazing Materials and Accessories	340
8.5.3	Preparation for Glazing	341
8.5.4	Glazing	343
8.6	**Where to Get More Information**	**344**

Appendix Data Sources	345
Glossary	353
Bibliography	364
Index	398

Series Foreword

To spite a national trend toward renovation, restoration, and remodeling, construction products producers and their associations are not universally eager to publish recommendations for repairing or extending existing materials. There are two major reasons. First, there are several possible applications of most building materials; and there is an even larger number of different problems that can occur after products are installed in a building. Thus, it is difficult to produce recommendations that cover every eventuality.

Second, it is not always in a building construction product producer's best interest to publish data that will help building owners repair their product. Producers, whose income derives from selling new products, do not necessarily applaud when their associations spend their money telling architects and building owners how to avoid buying their products.

Finally, in the *Building Renovation and Restoration Series* we have a reference that recognizes that problems frequently occur with materials used in building projects. In this book and in the other books in this series,

Simmons goes beyond the promotional hyperbole found in most product literature and explains how to identify common problems. He then offers informed "inside" recommendations on how to deal with each of the problems. Each chapter covers certain materials, or family of materials, in a way that can be understood by building owners and managers, as well as construction and design professionals.

Most people involved in designing, financing, constructing, owning, managing, and maintaining today's "high tech" buildings have limited knowledge on how all of the many materials go together to form the building and how they should look and perform. Everyone relies on specialists, who may have varying degrees of expertise, for building and installing the many individual components that make up completed buildings. Problems frequently arise when components and materials are not installed properly and often occur when substrate or supporting materials are not installed correctly.

When problems occur, even the specialists may not know why they are happening. Or they may not be willing to admit responsibility for problems. Such problems can stem from improper designer selection, defective or substandard installation, lack of understanding on use, or incorrect maintenance procedures. Armed with necessary "inside" information, one can identify causes of problems and make assessments of their extent. Only after the causes are identified can one determine how to correct the problem.

Up until now that "inside" information generally was not available to those faced with these problems. In this book, materials are described according to types and uses and how they are supposed to be installed or applied. Materials and installation or application failures and problems are identified and listed, then described in straightforward, understandable language supplemented by charts, graphs, photographs, and line drawings. Solutions ranging from proper cleaning and other maintenance and remedial repair to complete removal and replacement are recommended with cross-references to given problems.

Of further value are sections on where to get more information from such sources as manufacturers, standards setting bodies, government agencies, periodicals, and books. There are also national and regional trade and professional associations representing almost every building and finish material, most of which make available reliable, unbiased information on proper use and installation of their respective materials and products. Some associations even offer information on recognizing and solving problems for their products and materials. Names, addresses, and telephone numbers are included, along with each association's major publications. In addition, knowledgeable, independent consultants who specialize in resolving problems relating to certain materials are recognized. Where names were not available for publication, most associations can furnish names of qualified persons who can assist in resolving problems related to their products.

It is wished that one would never be faced with any problems with new buildings and even older ones. However, reality being what it is, this book, as do the others in the series, offers a guide so you can identify problems and find solutions. And it provides references for sources of more information when problems go beyond the scope of the book.

Jess McIlvain, AIA, CCS, CSI
Consulting Architect
Bethesda, Maryland

Preface

Architects working on projects where existing construction plays a part spend countless hours eliciting data from materials producers and installers relating to cleaning, repairing, and extending existing building materials and products and for installing new materials and products over existing materials. The producers and installers know much of the needed information and generally give it up readily when asked, but they often do not include such information in their standard literature packages. As a result, there has been a long-standing need for source documents that include the industry's recommendations for repairing, maintaining, and extending existing materials and for installing new materials over existing ones. This book is one of a series called the *Building Renovation and Restoration Series* that was conceived to answer that need.

In the thirty-plus years I have worked as an architect, and especially since 1975, when I began my practice as a specifications consultant, I have often wondered why there are so few comprehensive sources of data to help architects, engineers, general contractors, and building owners deal with the many types of existing building materials. It is often necessary to consult several sources to resolve even apparently simple problems, partly because authoritative sources do not agree on many subjects. The time it takes to do all the necessary research is enormous.

I have done much of that kind of research myself over the years. This book includes the fruits of my earlier research, augmented by many additional hours of recent searching to make it as broad as possible. I have included as many industry recommendations about working with existing doors, windows, store fronts, curtain walls, and related glazing as I could fit in. The data here, as is true for that in the other books in this series, come from published recommendations of producers and their associations, applicable codes and standards, federal agency guides and requirements, private-sector commercial guide specifications, contractors who actually do such work, the experiences of other architects and their consultants, and from the author's own experiences. Of course, no single book could possibly contain all known data about the subjects addressed here or discuss every potential problem that could occur. Where data are too voluminous to include in the text, references are given to help the reader find additional information from knowledgeable sources. Some sources of data about historic preservation are also listed.

This book, as do the others in the series, explains in practical, understandable narrative, supported by line drawings and photographs, the industry's recommendations on how to extend, clean, repair, refinish, restore, and protect the kinds of existing items that are the subject of this book and how to install in existing buildings items similar to those discussed.

All the books in this series are written for building owners; architects; federal and local government agencies; building contractors; university, professional, and public libraries; members of groups and associations interested in preservation; and everyone who is responsible for maintaining, cleaning, or repairing existing building construction materials. The books in this series are not how-to books meant to compete with publications such as *The Old-House Journal* or the books and tapes generated by the producers of the television series "This Old House."

I hope that, if this book doesn't directly solve your current problem, it will lead you to a source that will.

What This Book Contains

This book discusses most of the more commonly used types of doors, windows, store fronts, curtain walls, and related glazing usually found in buildings. It includes common methods of fabricating, finishing, and installing those items, and many of the industry's recommendations about repairing them and installing them in existing buildings.

How to Use This Book

This book is divided into eight chapters that discuss the subject areas suggested by their titles. Chapter 1 is a general introduction to the subject that offers suggestions about how a building owner, architect, engineer, or general building contractor might approach solving problems associated with existing building materials. It offers advice about seeking expert assistance when necessary and suggests the types of people and organizations that might be able to help. Chapter 1 also includes some general recommendations about demolition and removal of parts of existing buildings to permit installation of doors, windows, and other items discussed in this book and some general requirements regarding alterations.

Each of the other seven chapters include:

- A statement of the nature and purpose of the chapter.
- A discussion of materials commonly used to produce the items discussed in that chapter, and the finishes generally used on them.
- A description of the methods generally recommended to be used to fabricate and install the items that are the subject of that chapter.
- A discussion of how and why the items that are the subject of that chapter, or their finishes, might fail.
- A discussion of methods for repairing the items covered in that chapter, including their finishes.
- An analysis of methods recommended for installing the items of concern in that chapter in or on an existing building.
- An indication of sources for additional information about the items discussed in that chapter.

This book has an Appendix that contains a list of sources of additional data. These sources include manufacturers, trade and professional associations, standards-setting bodies, government agencies, periodicals, book publishers, and others having knowledge of methods for restoring building materials. The list includes names, addresses, and telephone numbers. Sources from which data related to historic preservation may be obtained are identified with a boldface **HP.** There is also a Glossary, which includes many of the more uncommon words and usages in it.

The items listed in the Bibliography are annotated to show the book chapters to which they apply. Entries that are related to historic preservation are identified with a boldface **HP**. Many of the publications and publishers of entries in the Bibliography are listed in the Appendix.

Building owners, engineers, architects, and general building contractors will in most cases each use this book in a somewhat different way. The following suggestions give some indication of what some of those differences might be.

Owners

A building owner who is consulting this book is probably doing so because his or her building has experienced failure of one of the items discussed here. If the problem is an emergency, the owner should immediately read sections 1.3 and "Emergencies" under section 1.4.1.

After the failure is temporarily under control, a more systematic approach is suggested. An owner may tend to want to turn directly to the chapter containing information about the item that seems to have failed. One who has experience in dealing with such problems may be able to approach the problem in this manner. For others it is better first to read and become familiar with the contents of Chapter 1, including those parts that do not at first seem applicable to the immediate problem.

After reading Chapter 1 the owner may want to consider seeking professional help but should first read the chapter covering the item that has failed. If at some point that chapter refers to another chapter, it is important also to read the cross-referenced material. An owner not experienced in dealing with the type of failure at hand should not ask for professional help before reading the chapter covering the failed item, and applicable cross-referenced material. It is always better to know as much as possible about a problem before asking for help.

Architects and Engineers

An architect's or engineer's approach will depend somewhat on his or her professional relationship with the owner. For example, an architect who has been consulted about a failure may approach the problem differently depending on whether he or she was the existing building's architect of record, especially if the failure occurred within the normal expected life of the failed item. Then there may be legal as well as technical considerations. This book, though, is limited to a discussion of technical problems.

An architect's or engineer's first impulse may be to rush to the site to determine the exact nature of the problem. That approach may be reasonable for one who has extensive experience with failures of the kind being experienced. Someone with little such experience, however, should do some homework before submitting to queries by a client or potential client.

A start might consist of reading the chapters here that deal with the problem at hand and examining the sources of additional information recommended there. Then, if the problem is still beyond the architect's or engineer's expertise, he or she should read Chapter 1 and decide whether outside professional help is needed. An architect or engineer with some related experience might delay that decision until after studying the problem in the field. One who knows little about the subject, however, will probably want to have a knowledgeable consultant present on the first site visit. Chapter 1 offers suggestions about how to go about making that decision.

An architect's or engineer's approach will be slightly different when he or she has been commissioned to renovate an existing building. Then, extensive examination of existing construction documents and field conditions is called for. When failures of the kind addressed in this book have contributed significantly to the reasons for renovating and the architect or engineer is not thoroughly versed in dealing with such conditions, it is reasonable to consider seeking professional assistance throughout the design process. In this event the architect or engineer should read Chapter 1 first, then refer to other chapters as needed during the design and document-production process. Even when there is a consultant on the team, the architect or engineer should have enough knowledge to understand what the consultant is advising and to know what to expect of the consultant.

Building Contractors

How a building contractor uses this book will depend on which hat the contractor is wearing at the time and the contractor's expertise in dealing with existing items of the type discussed here. For the contractor's own buildings the suggestions given above for owners apply, except that the contractor will probably have more experience with such problems than many owners do.

A contractor's approach might be similar to that described above for architects and engineers when he or she is asked by a building owner to repair a failed item of the type discussed in this book.

When repairs are part of a project for which a contractor is the general contractor of record, the problem is one of supervising the subcontractor who will actually repair the failed item. Sometimes even a knowledgeable contractor will find it helpful to compare the methods and materials a subcontractor proposes against the recommendations of an authoritative source, such as those listed in this book. It is also sometimes useful to verify a misgiving the contractor might have about specified materials or methods. In each of these cases, when a question arises with which the contractor is not thoroughly familiar he or she should read the chapter covering the subject at hand and check the other resources listed. Before

xxiv Preface

selecting a subcontractor for repair work, a contractor might want to review Chapter 1.

Even a contractor with extensive experience in repairing failed items of the types covered in this book will frequently encounter unusual conditions. Then the contractor should turn to the list of sources of additional information at the end of the appropriate chapter and examine the Appendix and Bibliography to discover who to ask for advice.

Disclaimer

The information in this book was derived from data published by trade associations, standards-setting organizations, manufacturers, and government organizations, and statements made to the author by their representatives; from interviews with consultants, architects, and building contractors; and from related books and periodicals. The author and publisher have exercised their best judgment in selecting data to be presented, have reported the recommendations of the sources consulted in good faith, and have made every reasonable effort to make the data presented accurate and authoritative. But neither the author nor the publisher warrant the accuracy or completeness of the data nor assume liability for its fitness for a particular purpose. Users are expected to apply their own professional knowledge and experience or consult with someone who has such knowledge and experience when using the data contained in this book. The user is also expected to consult the original sources of the data, obtain additional information as needed, and seek expert advice when appropriate.

A few manufacturers and their products are mentioned in this book. Such mention is intended solely to indicate the existence of such products and manufacturers, not to imply endorsement of the mentioned manufacturer or product. Neither is endorsement implied of other products or statements made by a manufacturer.

Similarly, handbooks and other literature produced by various manufacturers and associations are mentioned. Such mention does not imply that the item referred to is the only available one or even the best of its kind. The author and publisher expect the reader to seek out other manufacturers and appropriate associations to ascertain whether they have similar literature or will make similar data available.

Acknowledgments

A book of this kind requires the help of many people to make it valid and complete. I would like to acknowledge the manufacturers, producers' associations, standards-setting bodies, and other organizations and individuals whose product literature, recommendations, studies, reports, and advice helped make this book more complete and accurate than it would otherwise have been.

At the risk of offending the many others who helped, I would like to single out the following people who were particularly helpful:

Henry N. Carrera, CSI, Benjamin Moore and Company, Colonial Heights, Virginia.

John W. Harn, Harn Construction Co., Laurel, Florida.

R. A. McDermott, CSI, Benjamin Moore and Company, Montvale, New Jersey.

Vincent R. Sandusky, Painting and Decorating Contractors of America, Falls Church, Virginia.

Sally Sims, Librarian, National Trust for Historic Preservation Library and University of Maryland Architectural Library, College Park, Maryland.

Everett G. Spurling, Jr., FAIA, FCSI, Bethesda, Maryland.

All photographs and drawings are by the author.

CHAPTER 1

Introduction

Doors, windows, store fronts, glazed curtain walls, and related glazing are affected by many factors, including the construction of the building; the humidity and temperature during and after their installation; how the walls and partitions into which they are installed are prepared to receive them; the items themselves, including the materials from which they are made, their finishes, design, fabrication, and installation; and how well the items are protected after they have been installed. For interior doors, windows, and store fronts to remain in good condition the building's shell and elements such as exterior doors, windows, store fronts, and glazed curtain walls used to close openings in the shell must turn away wind and water and protect interior spaces from excessive temperature and humidity levels and fluctuations. For doors, windows, store fronts, and glazed curtain walls to resist failure, the materials used in them and their finishes must be selected and the items must be designed and installed to resist the effects of weather and exclude water penetration.

This chapter includes a brief listing of the items included in this book and the general types of failures they usually suffer, and outlines steps a building owner or manager can take to solve problems associated with each type of failure. It also covers the relationship of architects, engineers, general building contractors, specialty contractors, manufacturers, damage (sometimes called forensic) consultants, and hardware consultants to the owner or manager and to each other when they are working on projects where doors, windows, store fronts, glazed curtain walls, or associated glazing have failed. This chapter then outlines orderly ways in which those professionals can approach the problem-solving process.

This chapter includes an approach to determining the nature and extent of door, window, store front, glazed curtain wall, and associated glazing failures and suggests the type of assistance a building owner, building manager, architect, engineer, or contractor might seek to help solve a problem. It also presents general guidelines for demolition and alterations relating to the failure of doors, windows, store fronts, and glazed curtain walls, and to installing those items in or on an existing building.

Chapters 2 through 8 contain detailed remedies for the failures discussed and sources of additional data.

1.1 What This Book Covers

Chapter 2 identifies problems with a building's structure, walls, and partitions that might cause failures in its doors, store fronts, windows, curtain walls, or related glazing.

Chapter 3 covers the materials used in doors and frames, store fronts, windows, and glazed curtain walls, and the finishes normally used on them.

Chapter 4 discusses metal doors, frames, and store fronts. Metals included are steel, stainless steel, aluminum, and copper alloys.

Chapter 5 covers wood doors and frames. Door types include swinging, sliding, and revolving ones.

Chapter 6 includes wood, steel, and aluminum windows of all types. Both unclad and vinyl- and aluminum-clad wood windows are discussed.

All types of aluminum glazed curtain walls are discussed in Chapter 7.

Chapter 8 covers both glass and plastic glazing. Glass-installation methods discussed include ones for glazing into wood, steel, stainless steel, aluminum, and copper alloys, and of structural glass curtain walls made using silicone adhesive sealants.

Chapters 4 through 8 include discussions about the materials used in the items covered in each particular chapter and the finishes normally used on those items. Included also is information about the design, fabrication, and installation of the items covered in that chapter. In addition, Chapters 3 through 8 cover the kinds of problems that can occur with the materials, finishes, and items addressed in them, and indicate the industry's recommendations for dealing with many of those failures. The principles discussed apply to most similar items found today in buildings.

1.2 Failure Types and Conditions

Failure in a door, window, store front, glazed curtain wall, or associated glazing is sometimes a symptom of an underlying problem rather than a failure of the item itself. Recognizing that a failure has occurred often requires no special expertise (Fig. 1-1). A sticking door will be hard to open, for example. A leak in a window or glazed curtain wall will result in visible water on the inside. Discovering the cause of the failure and determining the proper remedy, however, often require detailed knowledge

Figure 1-1 The double-hung window shown in this photograph has obvious problems.

about the behavior of the materials used to manufacture the failed item and the materials' finishes, the fasteners used to anchor the item in place, the design of the item itself, flashing in the vicinity of the failed item, and sometimes even of the building's structural system.

When a door, window, store front, glazed curtain wall, or associated glazing fails, reviewing the appropriate chapter of this book or the other books in this series where the failed item is discussed might help an owner, building manager, architect, engineer, or building contractor identify the cause of the failure and solve the problem. If, after reading the material presented in this book and examining the referenced additional data, the reader still does not feel competent to proceed without doing so, seeking professional help is in order.

1.3 What to Do in an Emergency

In an emergency it is often necessary to act first and analyze later. When action must be taken immediately to stop damage that is already occurring or is imminent, whatever is necessary should be done. Emergency action, for example, might consist of taping cardboard or nailing plywood over a broken window.

A word of caution is in order, however. Unless doing so is absolutely unavoidable, no irreversible remedial action should be taken. Small repairs that cannot easily be removed can become major cost items when permanent repairs are attempted (Fig. 1-2). Screwing plywood to an aluminum window, for instance, can result in hard-to-fill holes.

1.4 Professional Help

If, after reading the chapters of this book or the other books in this series that address the identified problem area, the next step is still unclear, or it is impossible to be sure of the nature of the problem, consult with someone who has experience in dealing with the kind of failure being experienced. Who that should be will vary with the knowledge and experience of the person seeking the help.

Sometimes obtaining more than one outside opinion will be desirable. Certainly this is the case when doubt exists about the opinion obtained. A second opinion is almost always desirable for someone who has little experience in dealing with failed doors, windows, store fronts, glazed curtain walls, or their glazing.

Figure 1-2 The boards nailed over this window kept the rain out but stayed there for years, because the nailing heavily damaged the leaking window.

1.4.1 Help for Building Owners and Managers

The owners of some types of buildings routinely employ building managers, sometimes called just managers, whose responsibilities may vary considerably. Some only obtain tenants and collect rent. Others may also be responsible for maintenance, repairing failed building components, and many other duties. Some managers may be responsible only for the cleaning and repair of glazed curtain walls or some other single part of a building. Depending on his or her responsibilities, a building manager may be called something else entirely. Some titles commonly used are construction manager, maintenance contractor, and service contractor. The suggestions here are applicable regardless of the title used.

Although building managers may speak or act on behalf of owners, they are seldom granted the right to sign contracts in an owner's name.

Even when that arrangement is satisfactory with the owner, other members of the building industry are generally reluctant to participate in such an arrangement. Therefore, legal agreements for repairs, remodeling, and the like are usually made between the owner and the other parties engaged in design, maintenance, repair, or construction work. When building managers, or just managers, are mentioned in this book, it has been assumed that the actions they take are subject to approval by the owner, even though a statement to that effect is not necessarily included in each such reference. When this text speaks of owners or building managers it means either one or the other or both, as the particular agreement between the owner and the manager dictates. There has been no general attempt to separate their responsibilities, although some separation may occasionally be indicated.

Building owners and managers can turn to several types of expert help when problems occur with doors, windows, store fronts, glazed curtain walls, and related glazing.

Emergencies. In emergencies, a building owner or manager should seek help from the most available professional person or organization that can prevent the damage from continuing. But, even in an emergency, it is better to seek help from a known organization than simply to thumb through the telephone book.

The first step is to call someone who can remedy the immediate problem. Worry about protocol later. An exception should be made, however, if the building is still under a construction warranty. Then the owner or manager should go to the person responsible for the warranty. This is usually a general contractor but may be a specialty contractor.

When the building is not under warranty, if the owner or manager has a relationship with a building construction professional who is able to solve the immediate problem, that is the place to start. An owner or manager who has relationships with several building professionals should start with the most appropriate one. For example, if the failed item was recently installed by a specialty contractor under direct contract with the owner—that is, there was no general contractor involved—the owner, or manager with the owner's concurrence, should call that specialty contractor.

Unless an architect or engineer is the only building professional the owner or manager knows, or the owner or manager is unable to find someone to stop the damage from continuing, calling one of them in an emergency is probably not appropriate. The best most of them will be able to do is recommend an organization that might stop the damage from continuing. Going through such professionals could waste valuable time.

In some parts of the country, a call to a local organization representing building industry professionals will net a list of acceptable organizations specializing in repairing the kind of damage that is occurring. Such local organizations might include a chapter of the Associated General Contractors of America, an association representing the producers or installers of the failed item, or the American Institute of Architects. Local chapters can be found by calling the affiliated national organization. Most of the applicable ones are mentioned in the appropriate chapters of this book. The names, addresses, and telephone numbers of appropriate national organizations are listed in the Appendix. Sometimes the national organization can furnish the names of local organizations that can help.

It might be possible to find a local damage consultant who has some knowledge of the kind of item that has been damaged, but this person may be able only to recommend someone to repair the damage.

In each case, the methods that would be used under more routine circumstances for contacting the various entities would apply in an emergency. There might not, however, be time to do the ordinary checking of references.

Architects and Engineers. Even in non-emergency situations, building owners or managers with easily identifiable problems related to doors, windows, store fronts, glazed curtain walls, and their glazing often do not need to consult an architect or engineer. An exception occurs when repairing a failed door, window, store front, glazed curtain wall, or associated glazing is part of a general remodeling, renovation, or restoration project. Even when damage is evidenced solely by a failure of a door, window, store front, glazed curtain wall, or associated glazing, repairs requiring manipulation of structural systems or that might cause harm to the public may require a building permit. Repair of damage to the doors, windows, store fronts, glazed curtain walls, or associated glazing themselves may also require a permit. In many jurisdictions, building officials will consider permit applications only when they are accompanied by drawings and specifications signed and sealed by an architect or engineer. Sometimes the scope of a problem may warrant hiring an architect or engineer to prepare drawings and specifications even when authorities having jurisdiction do not require such documents. Occasionally, design considerations will make hiring an architect desirable whether or not doing so is required by law.

When the owner, or manager with the owner's concurrence, hires an architect or engineer, the standards one would usually follow when hiring such professionals apply. In addition to the usual professional qualifications applicable to any project, however, architects and engineers commissioned to perform professional services related to an existing building should have experience in the type of work needed. It is seldom a good idea to hire an architectural firm with experience solely in single-family housing to renovate an existing high-rise office building.

While the expected life span for many door, window, store front, glazed curtain wall, or associated glazing products may be the same as that of the building they are a part of, nothing remains pristine forever. When one such item that has not reached its expected life span fails in a building that was designed by an architect, the owner, or manager with the owner's concurrence, should consider whether to inform the architect and request assistance in solving the problem. Sometimes, the architect will have a legal or contractual obligation to help, but most architects will be interested enough to become actively involved even when they are not required to do so by law or contract provision. Discussion of the architect's potential legal or contractual obligations in such cases is, however, beyond the scope of this book. The architect may help determine the cause of the failure and suggest solutions, or help the owner or manager determine which other professionals or manufacturers to contact, and then critique their recommendations. Reimbursement for the architect's services may or may not be appropriate. Unless the owner has a long-term relationship with, or is currently a client of, the architectural firm, a frivolous complaint is likely to result in a bill for services rendered. For example, original paint peeling from a fifteen-year-old steel door is probably not the architect's responsibility under the original owner-architect agreement. On the other hand, an architect should not demand payment for work related to legitimate complaints about failures caused by the architect's negligence, ignorance, or incompetence.

General Building Contractors. Door, window, store front, glazed curtain wall, and associated glazing-failure problems that do not involve a building's structure or design are often best handled directly by a building contractor. Depending on the scope, the contractor may be either a general contractor or a specialty contractor. When the problem is isolated and involves one discipline, a single specialty contractor may be all that is needed. When multiple disciplines are involved, a general contractor may be needed to coordinate the activities of several specialty contractors. Owners or managers with experience in administering contracts may be able to coordinate the work without a general contractor.

When selecting a general contractor, it is best to select one the owner or manager knows. If there is no such contractor, the owner or manager should seek advice in finding a competent firm. Such advice might come from satisfied building owners, architects, engineers, damage consultants, or other professional consultants the owner knows, or from a professional organization such as the American Institute of Architects. As a last resort an owner or manager might ask for a recommendation from a contractors' organization. The owner or manager should bear in mind, however, that asking for a recommendation from an organization that is supported by and which represents the firm being sought will at best net a list of firms in the area. This is not a good way to get a recommendation. Saying so is not an indictment of contractors or their associations. The same advice applies to finding doctors, lawyers, and architects. You find the best ones by word of mouth.

If the project is large enough to warrant doing so, an owner or manager might consider seeking competitive bids from a list of contractors. Few door, window, store front, glazed curtain wall, or associated glazing repair projects are that large or complicated, however. Negotiation is better, if reputable firms can be found with which to negotiate. If all the parties are unknown, competitive bidding may be the only way to get a reasonable price.

When a door, window, store front, glazed curtain wall, or associated glazing fails in a building that is still covered by a construction warranty, or if the failed item is still under a special warranty, the owner or manager should contact the building contractor responsible for the warranty. The contractor's legal and contractual obligations under warranties may be complex and difficult to determine in specific cases. At any rate they are beyond the scope of this book. Under ordinary circumstances, the owner or manager should not initially have to contact the contractor's subcontractor or supplier unless the general contractor cannot, or will not, do so. The owner or manager should consult with the architect and the owner's attorney for guidance about warranties and legal matters.

Even after all warranties have expired, if a door, window, store front, glazed curtain wall, or related glazing that has not been installed for a long (long is subjective) time fails, the owner or manager should probably contact the general contractor who built the building and request assistance in solving the problem. If dissatisfied with the original contractor for some reason, the owner or manager might consider calling another contractor, but doing so may be costly. A new contractor will not be familiar with the building's peculiarities.

Manufacturers. Although some are more helpful than others, the manufacturer of an item that has failed or an association representing the manufacturers of such products are often knowledgeable sources of recommendations for dealing with door, window, store front, glazed curtain wall, or associated glazing failures. They may be the best-qualified agency to determine the cause of a failure.

In very simple cases, asking the manufacturer for advice may not be necessary. In more complex cases, however, contractors and architects will often bring in the manufacturer of the failed door, window, store front, glazed curtain wall, or associated glazing material to help determine the cause of and solution for a failure. When they do not bring in the manufacturer, the owner or manager should probably do so. When a failure is extensive or its repair will be expensive or highly disruptive to the owner's activities, the owner or manager may want to bring in a second manufacturer who makes a product similar to the failed one, to verify the advice given by the first manufacturer. Knowledgeable, reputable people often do not agree about the causes of failures or the methods needed to repair them and prevent future failures. In some cases an item itself may have failed, or the manufacturer's installation instructions may have been faulty. Then, the owner or manager might be justified in not agreeing to throw good money after bad by dealing further with that manufacturer's products or advice. A new product or method may be needed to solve the problem.

A few manufacturers install the doors, windows, store fronts, glazed curtain walls, or associated glazing they make, but most do not. When a manufacturer does not install its products, refer to the discussion under the following section, "Specialty Contractors."

When a needed repair to a door, window, store front, glazed curtain wall, or associated glazing is small and simple and there is no general contractor or related specialty contractor involved in the project, it is often best for a building owner or manager first to contact the manufacturer of the damaged item. Even those manufacturers who do not or cannot install or repair a damaged item themselves can often help an owner or manager find someone to repair it. An owner, or manager with the owner's concurrence, who does not wish to deal with the original manufacturer or does not know the name of the original manufacturer, can use the methods suggested earlier for finding a general contractor to find a reputable manufacturer of similar products. In addition, general contractors that the owner or manager knows to be reputable are a source of recommendations for manufacturers, if there is no symbiotic relationship between the general contractor and the manufacturer. Specialty contractors who install items similar to those to be repaired are also excellent sources for the names of manufacturers.

Product manufacturers may or may not agree to participate in a competitive bidding procedure. Most will not be interested in making small quantities of a product that they do not usually produce. Thus, for small failures it may be necessary to deal with the original manufacturer, regardless of the owner's or manager's wishes not to do so.

When a failed door, window, store front, glazed curtain wall, or associated glazing is under warranty, the manufacturer of the installed product will probably be required to make needed repairs or provide a replacement item, but the owner or manager should not contact the manufacturer directly if a general contractor was involved in the original project. The general contractor is usually responsible for the warranty, and may choose to have repairs made by, or a replacement item provided by, an entirely different subcontractor. There is normally no direct contractual relationship between the owner and the manufacturer in such a case.

When there is no general contractor or specialty contractor involved, the owner or manager should contact the manufacturer, who is usually directly responsible for honoring any warranty that may exist.

When a door, window, store front, glazed curtain wall, or associated glazing has failed because it was improperly manufactured or installed by its manufacturer but is no longer under warranty, the owner, or manager with the owner's concurrence, may want to consider using a different manufacturer to effect repairs or fabricate or install a replacement item.

When the replacement of a door, window, store front, glazed curtain wall, or associated glazing is either part of or constitutes a remodeling, renovation, or restoration project, consultation with an architect is in order. In such cases the architect's advice regarding manufacturers of products necessary to perform the work should be sought. Where an architect is not engaged, the services of a general or specialty contractor will probably be necessary.

Specialty Contractors. As mentioned earlier, only a few manufacturers of doors, windows, store fronts, glazed curtain walls, or associated glazing materials install the items they make. When the manufacturer does not install its product the general contractor will sometimes do so, using his own forces, or sometimes will employ a specialty contractor to install it.

When a needed repair to a door, window, store front, glazed curtain wall, or associated glazing is small and simple, there is no general contractor involved in the project, and the manufacturer of the item does not install or repair its products, it may be best for a building owner or manager to deal directly with a specialty contractor. The methods suggested earlier for finding a general contractor can also be used to find a reputable specialty contractor. In addition, general contractors that the owner or manager knows to be reputable are a source of recommendations for specialty contractors, so long as there is no symbiotic relationship between the general contractor and the specialty contractor. Manufacturers of items to be repaired are excellent sources for the names of specialty contractors, and vice versa.

Competitive bidding is an acceptable procedure to use when dealing with specialty contractors on small projects.

When a failed door, window, store front, glazed curtain wall, or associated glazing material is under warranty, the specialty contractor who installed it will probably be required to make the repairs, but the owner or manager should not contact the specialty contractor directly when there was a general contractor involved in the original project. The general contractor is usually responsible for the warranty, and may choose to have repairs made by an entirely different subcontractor. There is normally no direct contractual relationship between the owner and the specialty contractor in such a case.

When there is no general contractor, the owner or manager should contact the specialty contractor, who is usually directly responsible for honoring any warranty that may exist.

When the door, window, store front, glazed curtain wall, or associated glazing has failed because it was improperly installed and is no longer under warranty, the owner, or manager with the owner's concurrence, may want to use a different specialty contractor to make the repairs or install a replacement item.

Specialty Consultants. There are several types of specialty consultants available for work related to doors, windows, store fronts, glazed curtain walls, and associated glazing. They include damage consultants (sometimes called forensic consultants), curtain wall consultants, window consultants, door consultants, and hardware consultants.

The last four types of consultants listed are often associated with new construction projects. All five types may be involved with renovation and remodeling projects where major work replacement or the remodeling of doors, windows, store fronts, glazed curtain walls, or associated glazing is to be done. In such a case, whether these consultants are needed will depend somewhat on whether the owner decides to hire an architect or must hire one because of legal requirements. If an architect is to be employed it is usually, though not always, the architect who will engage the specialty consultants.

There are four sets of circumstances in which an owner or building manager may want to look for a specialty consultant knowledgeable about doors, door hardware, windows, store fronts, glazed curtain walls, or associated glazing.

First, a specialty consultant may be called for when a failure has occurred and either a knowledgeable architect or contractor or a reputable, trustworthy manufacturer's representative is not available for consultation. Contractors and architects who have great knowledge about relatively recent materials and their failure may know little about the types of materials and systems used in older buildings. Or there may not be a manufacturer who has knowledge of a product that is no longer manufactured. A requirement for historic preservation may be sufficient cause to look for a specialty consultant.

The second case is when the owner's contractor and architect and the manufacturer of a failed item cannot agree on the cause of the failure, the means appropriate to make repairs, or on which party, if any, should be held responsible for the failure. For some failures there may be no one to blame. No one can prevent door, window, store front, glazed curtain wall, or associated glazing from being damaged by an earthquake, for example. For other failures, though, blame can often be placed. Battles about responsibility can be long and difficult to resolve. Too frequently, they end up in court. Sometimes an outside "expert" can help resolve conflicts and prevent the parties involved from having to resort to litigation.

Third, a specialty consultant may be needed when the people who are already involved cannot determine with certainty the cause of the damage or the proper method for making repairs.

Fourth, when the people involved lack sufficient expertise in the specialty areas involved it may be appropriate to hire a specialty consultant, whether the project consists of all new construction or is a remodeling, renovation, or repair job. A particular architect, for example, may be highly proficient in most of the skills needed to carry out the project at hand, but have little knowledge of the kinds of door hardware required. Many architects are not sufficiently versed in finish hardware requirements to preclude their hiring a hardware consultant.

Consultants should be able to determine the nature of a problem and identify its true cause, find a solution to the problem, select the proper products to use in making repairs, write specifications and produce drawings related to the solution, and oversee the repairs.

It may not be easy to find a damage consultant who specializes in door, window, store front, glazed curtain wall, or associated glazing damage. It may then be necessary to engage another specialty consultant in addition to a general damage consultant. Unfortunately, all specialty consultants were not created equal. Many who present themselves as consultants are actually building product manufacturers' representatives trying to increase their sales, or specialty contractors seeking to enlarge their businesses. While most of them are reputable, and some are competent to give advice, few are sufficiently knowledgeable to identify or advise an owner or manager about solving underlying substrate or structure problems. Following the recommendations of an incompetent consultant can cause problems that will linger for years.

Selecting a consultant can be filled with uncertainty and potential for harm. Fortunately, there is help to be had in some fields. There are no licensing requirements, and no nationally recognized associations representing general damage consultants or consultants who specialize only in window problems. As a result, a building owner or manager who needs to hire a general damage consultant or window consultant must qualify that consultant with little help. One way to do so is to hire an architect or engineer and let them select and qualify the consultant, subject to the owner's approval, of course.

Asking a specialty contractor to recommend a consultant may not be a good idea. Although some hire consultants themselves, many specialty contractors feel that consultants are a necessary evil at best. Their opinion probably stems from the tendency of some manufacturers' representatives to call themselves consultants, then oversell their abilities and knowledge.

Fortunately, both hardware and door consultants can be found who have been prequalified by an industry association. The Door and Hardware Institute (DHI) certifies Architectural Hardware Consultants (AHC) and Certified Door Consultants (CDC). The prequalification is a lengthy apprenticeship period. Certification requires satisfactory completion of a two-day written examination. No one is permitted to use the AHC or CDC titles until they have been qualified to do so and certified by DHI. DHI will bring legal action against anyone who misuses those titles.

Regardless of who makes the recommendation or does the hiring, each consultant should have a demonstrated expertise in dealing with the problems at hand. AHC or CDC certification, for example, demonstrates such expertise. When a consultant is not certified by a recognized organization, obtaining references from the consultant's satisfied clients is an appropriate prequalification tool. A licensed architect or engineer who has extensive experience in dealing with existing construction might be acceptable.

However, even a consultant who is highly qualified to deal with door, window, store front, glazed curtain wall, or associated glazing failures in relatively new construction may know little about the kinds of materials that might be encountered in very old buildings. This lack of knowledge is especially a problem when historic preservation is involved. A consultant for that type work needs to demonstrate knowledge in dealing with old materials and systems and of the special requirements associated with historic preservation. It may not be possible to find a consultant expert in dealing with old materials who is also knowledgeable in general construction principles or about newer materials or products. In such a case it may be necessary to find two or more consultants with complementary knowledge.

Many door, window, store front, glazed curtain wall, and associated glazing failures are the result of several problems. For example, many failures are caused by structure movement. The owner's consultant, therefore, must be able to determine whether the movement responsible for the problem is normal movement that has not been accounted for in the door, window, store front, glazed curtain wall, or associated glazing design or installation, or is a structure failure that might require repairing the structure itself. The consultant need not, however, be a structural engineer or know how to repair the structure. Usually, the ability to make the diagnosis is sufficient.

Thus, a specialty consultant should have extensive knowledge of the product he or she has been engaged to advise on, in addition to an understanding of general construction principles and structure physiology. The consultant should know enough about buildings as a whole to be able to identify all underlying problems, not just the obvious ones or those directly associated with the failed item itself (Fig. 1-3). The owner or building manager must determine each consultant's limits and, when overall knowledge is lacking, engage such other consultants as may be necessary to ensure that the consultants can collectively determine all factors involved in the failure and offer solutions to all existing problems.

Finally, no consultant should have a financial stake in the outcome of an investigation. The owner or manager needs to be sure that each consultant is an independent third party who is selling a professional service, not the installation or repair of a product. Neither product manufacturers' representatives nor contractors who want to make the repairs meet this qualification.

Figure 1-3 The reason for the leak caused by the pipe shown in this photo is obvious, but stopping it requires knowledge of more than one discipline. At first glance, the solution seems to be to move the pipe, but further investigation reveals that the building's heating system layout will not permit it.

1.4.2 Help for Architects and Engineers

Architects and engineers usually get involved in making minor repairs to doors, windows, store fronts, glazed curtain walls, and related glazing only when the failure has occurred in a prestigious building or in one they designed. However, such professionals are often engaged when the repairs required are extensive or part of a larger renovation, restoration, or remodeling project.

Architects who do not have extensive experience in dealing with door, window, store front, glazed curtain wall, and related glazing problems or working with existing such items should seek outside help from someone with that kind of experience. The nature of this help, and the person selected to consult, will depend on the type and complexity of the problem.

Other Architects and Engineers. One source of professional consultation for architects and engineers is other architects or engineers who have experience with the type of door, window, store front, glazed curtain wall, or related glazing problem at hand. The qualifications needed are similar to those outlined for specialty consultants in section 1.4.1. The other architect or engineer need not be currently participating in an architectural or engineering practice. Specifications consultants and qualified architects and engineers employed by government, institutional, or private corporate organizations should not be overlooked.

Product Manufacturers and Industry Standards. Manufacturers and the associations that represent them are often the most knowledgeable sources of data about failure in their products and related corrective measures. They can often provide sufficient information to let a knowledgeable architect or engineer deal with door, window, store front, glazed curtain wall, or related glazing failures.

Just as in any other situation in which an architect or engineer consults producers or their associations for advice, the architect or engineer must compare a manufacturer's statements with those of other manufacturers and with industry standards, then exercise good judgment in deciding which claims to believe. The architect or engineer must study every claim carefully, especially ones that seem extravagant, and double-check everything. Claims by the fabricator or manufacturer of a failed product that something other than their product itself is responsible for a failure should be examined especially carefully. Securing a second opinion may be helpful.

The discussion under "Manufacturers" in section 1.4.1 contains additional requirements.

Specialty Consultants. When dealing with door, window, store front, glazed curtain wall, or related glazing failures or installation of such new products in existing buildings, architects often employ independent consultants of the types discussed under "Specialty Consultants" in section 1.4.1. The discussion there applies no matter who is hiring the consultant.

1.4.3 Help for General Building Contractors

Whether a general building contractor will need to consult an outside expert will depend on the complexity of the problem, the contractor's own experience, and whether the owner has engaged an architect or specialty consultant. A duplication of effort is unnecessary unless the contractor intends to challenge the views of the owner's consultants or the content of the drawings or specifications.

A general building contractor acting alone on a project where the owner has not engaged an architect, engineer, or consultant must base the need for hiring consultants on such factors as the contractor's experience and expertise with the types of problems that will be encountered, the specialty subcontractor's experience and expertise in dealing with the types of problems involved, and the complexity of the problems.

Hiring a specialty consultant may complicate a general contractor's relationship with subcontractors. Such duplication of effort is seldom justified and is often a bad idea. An experienced, qualified specialty subcontractor is not likely to appreciate having a specialty consultant hired to tell the subcontractor how to do the job. The general contractor would be better off finding a qualified subcontractor and relying on that subcontractor's advice. If no such subcontractor is available and the general contractor is not experienced with the problem at hand, hiring a consultant may be necessary, regardless of the feelings of the subcontractor. Muddling along to salve feelings is wholly incompetent and unprofessional. Instead, the contractor might want to recommend that the owner employ a qualified architect or engineer to specify the repairs and let the owner and the design professional hire necessary specialty consultants.

Should a contractor want to employ a specialty consultant, the related recommendations in sections 1.4.1 and 1.4.2 apply.

When working with an owner who has not hired an engineer or architect, it is often necessary for a general contractor to consult with the manufacturer of a failed door, window, store front, glazed curtain wall, or related glazing material. The manufacturer's representative can often identify the causes of failure and recommend repair methods. Even when the contractor has engaged a specialty subcontractor, the manufacturer is often a source for a second opinion about such matters. The same precautions mentioned in sections 1.4.1 and 1.4.2 for owners and architects dealing with manufacturers apply when a contractor deals directly with one of them.

1.5 Prework On-Site Examination

On-site examinations before work begins are important tools in helping to determine the type and extent of failure in a door, window, store front, glazed curtain wall, or related glazing and damage to underlying construction it might portend. Who should be present during an on-site examination depends on the stage at which the examination will take place.

1.5.1 The Owner

The first examination should be by the owner, the owner's personnel or building manager, or all of them, to determine the general extent of the problem. This examination should help the owner decide what the next step should be and the types of consultants the owner or building manager may need to contact.

After selecting a consultant, or consultants, the owner or his representative or building manager should visit the site with the consultant, or consultants. Together they should define the work to be done. An owner's consultants may include an architect or engineer, a general contractor, one or more specialty contractors or consultants, or a manufacturer. This second site examination should be attended by a representative of each expert that the owner has engaged to help with the problem. Specialty consultants who may have been engaged by another of the owner's consultants should also be present, as should the general contractor's specialty subcontractors, if they have been selected. During the second site visit the parties should familiarize themselves with conditions at the site and offer suggestions about how to solve the problem.

1.5.2 Architects and Engineers

An architect or engineer hired to oversee the repair of failed doors, windows, store fronts, glazed curtain walls, or related glazing or to design the installation of such items in or on an existing building should—before visiting the site, if possible—determine the components used in the existing item and the type of underlying and supporting construction, and examine available shop and setting drawings for the item. The architect or engineer should then visit the site with the owner, the owner's building manager, or another designated representative of the owner to determine the extent of the work needed and begin to decide how to solve the problem. From discussions with the owner about the nature of the problem the architect should have decided whether to engage professional help or contact the failed product's manufacturer. If one or more consultants are to be used, they should visit the site with the owner, or his building manager or other representative, and the architect. If the manufacturer is to be consulted, its representative should also participate in this site visit.

If, as a result of the first site visit, the architect determines that one or more specialty consultants or manufacturers' representatives that were previously considered unnecessary are instead needed, the architect should arrange for another site visit with those consultants or manufacturers' representatives present.

During the progress of the work the architect, the architect's consultants, and the owner's building manager or other designated representative should visit the site as often as necessary to fully determine the nature and extent of the problem and arrive at a total solution. These site visits should extend the observation beyond the immediate problem to ascertain whether previously unseen damage might be present.

1.5.3 Building Contractors

Non-Bid Projects. On non-bid projects a building contractor may wear at least two hats.

The easiest situation to deal with is a negotiated price based on professionally prepared construction documents. Then the contractor should conduct an extensive site examination to verify the conditions shown and the extent and type of work called for in the construction documents. Offering a proposal based on unverified construction documents is a bad business practice that can cost much more than proper investigation would have, if the documents are later found to be erroneous.

When the owner has not hired an architect or consultant to ascertain and document the type and extent of the work, he or she must act as both designer and contractor. Then the contractor should visit the site with the owner as soon as possible and revisit it as often as necessary to determine the nature of the problem and the extent of the work to be done. The contractor may choose to hire one or more specialty consultants or specialty subcontractors or both, or to consult with a representative of the manufacturer of the failed item. If so, those individuals should accompany the contractor on the site visits and participate in formulating the contractor's recommendations. Submitting a carefully drawn proposal is an absolute must so that the owner does not expect more than the contractor proposes to do.

Even when the owner hires an architect or other consultants, the contractor should visit the site with the owner or his designated representative and the owner's consultants as soon as possible. The purpose of this visit is to ascertain the extent and type of work to be done and to recommend repair methods. The contractor should also invite any specialty subcontractors to visit the site with the owner or the owner's representative, the owner's consultants, the manufacturer's representative, and the contractor. The more input the contractor has in the design process, the better the result is likely to be.

Bid Projects. Even when the contractor is invited to bid on a project for which construction documents have been prepared, making a prebid site visit is imperative. No contractor should bid on work related to existing construction without extensively examining the existing building. Some construction contracts demand it. Some contracts even try to make the contractor responsible if he or she fails to discover a problem. Even if the courts throw out that type of clause, who can afford the time and costs of a lawsuit? A contractor should establish exactly what work is to be done before bidding. Having insufficient data may be cause for choosing not to bid a project.

1.6 Demolition and Removal of Existing Construction

When existing doors, windows, store fronts, glazed curtain walls, and related glazing have either failed or are to be installed in or on an existing building, some selective demolition, dismantling, cutting, and removal of materials is usually necessary. While many requirements related to such work differ depending on what is to be removed, cut, or demolished, there are some general steps and precautions that should be followed in every case.

1.6.1 Controls

The first, and probably the most important, control an owner has over the work to be done is to have a signed legal agreement with each consultant and contractor who will perform services for the owner. All such agreements should be reviewed by the owner's attorney and insurance representative.

A knowledgeable architect or engineer can provide excellent control over projects where existing construction is involved. But some additional steps and submittals are still required to help maintain control, even when the owner engages an architect or engineer. For example, the contractor involved is often required to prepare and submit for approval detailed drawings of the portions of the existing building where work under the contract will take place. Such drawings are often required to show the size and location of openings to be drilled or cut in existing construction. These drawings should be reviewed by a consultant, usually a structural engineer, representing the owner who is qualified to make such a review. When an architect has been engaged the structural engineer will usually, but not always, be a consultant to the architect.

When an architect or engineer has been engaged, he or she should be required to submit for approval by the owner complete design drawings and specifications showing the entire work to be done. Such drawings do not, however, preclude the need for the contractor-furnished drawings mentioned in the preceding paragraph. Whenever the term *contractor* is used in this section it means any organization that will do the work, whether it is a general contractor, specialty contractor, or manufacturer.

The contractor is also usually required to prepare and submit a proposed schedule for the work to be done. Such a schedule should be broken down to show the stages of construction just as would be done for a new building. In addition, such schedules should show the period of demolition and any phasing of activities required to permit the owner's operations continue to the extent required by the agreement between the owner and the contractor.

Before electrical, water, sewer, telephone, or other services are interrupted or disconnected, the contractor should be required to notify the owner and authorities owning or controlling wires, conduits, pipes, and other services affected.

1.6.2 Protection of Persons and Property to Remain or Be Reused

The contractor should be made solely responsible for the safety of persons and property affected by the contractor's operations. To this end he or she should provide temporary barricades, fences, shoring, warning lights, dust barriers and partitions, casing of openings, rubbish chutes, temporary closures, and other means of protection.

The contractor should ensure the safe passage of people around the area of demolition and prevent injury to people and adjacent facilities. To do so it may be necessary to erect temporary covered passageways when required by authorities having jurisdiction, and provide shoring, bracing, or supports to prevent movement, settlement, or collapse of the existing structures and adjacent facilities.

When removal of doors, windows, or other portions of an existing building is necessary to gain access to the work space or perform the work, the contractor should provide adequate temporary waterproof protection at the resulting opening. This protection should be left in place at all times, except when the openings are actually being used, until the existing construction has been permanently repaired and provides a watertight seal. Parts of the existing building's shell should not be removed without the owner's specific written agreement.

The contractor should also be required to protect and support conduits, drains, pipes, and wires remaining in use that are subject to damage by construction activities.

In addition, he or she should provide airtight covers for return-air grilles in the areas in which work will occur, to prevent dirt and dust from entering the building's air conditioning or ventilating systems.

Floor coverings and other interior finishes should be protected from marring or other damage. Such protection should be maintained and left in place until the surface being protected is no longer subject to damage by construction operations. Heavy fire-retardant building paper may be used to protect general floor areas, but the paths of wheeled traffic, including access ways, should be protected by heavy boards or plywood.

Where portions of walls, partitions, or other vertical construction, or back-up for exterior walls are to be removed in building portions that are not to be totally demolished, the floor on both sides of the construction being demolished should be covered and protected out to a distance of at least four feet or to the adjacent wall, whichever is the shorter distance. Where ceilings are to be removed the entire floor surface should be covered.

Adjacent facilities that are damaged during demolition operations should be repaired promptly.

1.6.3 Performing Demolition Work

Selective demolition of portions of an existing building should not be started until those portions have been vacated by the owner, their use has been discontinued, the owner has performed necessary preparatory work in those spaces, and the owner's approval to proceed has been granted in writing.

Before the contractor enters the building site to begin work, the owner and the contractor should agree in writing on a list of those items located where the contractor will work that will remain the owner's property, those that will become the contractor's property to dispose of, and those that will be reused in the work.

Demolition and removal of debris should be conducted with minimum interference with roads, streets, alleys, walks, and other adjacent occupied or used facilities. Such facilities should not be closed or obstructed without permission from the authorities having jurisdiction over them. Alternate routes should be provided around closed or obstructed traffic ways, if so required by governing regulations.

Demolition work should not be performed until the construction schedule mentioned earlier has been approved by the owner. Similarly, cutting and drilling of existing construction should not begin until the cutting and drilling location drawings discussed earlier have been reviewed by the owner's competent consultant and the locations have been approved by the owner.

Spaces to remain unaltered that are adjacent to the areas where demolition will take place should be completely secured and rendered dustproof before demolition work is begun.

Demolition and removal of existing construction should be carried out carefully, and with minimum interference with the use of the building or its site, inconvenience to occupants or the public, danger to persons, or damage to existing materials that will remain in place.

Demolition and removal should be limited to the minimum quantity of materials and items necessary to permit the repairs or installation of new items in or on the existing building. Particular care should be taken to prevent damage to materials and items that will remain in place or be removed and later reinstalled. The methods used should comply with governing codes and regulations. Once demolition work has begun the process should be systematic, and the work should be done as quietly as practicable and with deliberate speed.

Structural elements, walls, and partitions should be removed in small sections to prevent their collapse and possible injury to workers or property. Removal methods and storage of removed materials should not load any part of the building in excess of the loads it was designed to support.

Temporary struts, bracing, or shoring should be installed where necessary to avoid the collapse of any part of the existing building. This temporary protection should be left in place until new construction has provided adequate bracing and support.

Damage to existing materials and items that are to remain, or be removed and reinstalled later, should be repaired promptly and properly, or new equal products should be installed. Items to be reinstalled should be removed carefully, stored in a safe location, and protected from damage until reinstalled. Items that are to be removed and turned over to the owner, but not reinstalled, should be similarly treated.

Cutting and drilling of existing construction is often necessary when doors, windows, store fronts, and glazed curtain walls are to be repaired or installed in or on an existing building. Cutting should be done using hand tools or small power tools wherever possible. Holes and slots should be cut neatly, to the size required, with minimal disturbance of adjacent materials and surfaces. Round holes in concrete slabs, floors, and walls should be cut using core drills of the required sizes and types. Square and rectangular holes in concrete should first be cut by line drilling, then chipping hammers should be used to remove the material between the drill holes. The use of large air hammers is not normally permitted, because the possibility of damage to the building from their use outweighs the time or work they might save. For similar reasons, air compressors are not usually permitted inside an existing building, especially when the building is to remain occupied during the work.

The drilling or cutting of columns, beams, joists, girders, ties, or other structural supporting elements should not be permitted, unless approved by the owner in each case or indicated on structural drawings furnished by the owner's consultant. The cutting of concrete-slab reinforcement should be avoided.

Openings through the shell of a building to the exterior should be closed temporarily when not in use and patched as soon as possible.

Should products containing asbestos or other hazardous materials be encountered in the course of the contractor's work, workers, building occupants, and the public should be protected from hazards associated with such products. Compliance with governmental laws and regulations is essential. Hazardous products should be properly and legally removed and the area decontaminated to comply with applicable laws, ordinances, and regulations.

1.6.4 Disposition of Removed Materials

Before the contractor is permitted access to the area where work is to take place, it is often a good idea for the owner to remove items that are not to be used in the completed installation but which the owner wishes to keep. Such items include locks, locksets, latch sets, door closers, and other items the owner desires to save. It is also usual for the owner to remove furnishings and portable equipment from the area where the contractor will work.

Before beginning demolition, the contractor should be required to offer for the owner's use each door; window; frame; accessory; item of hardware, equipment, and furnishings; mechanical, plumbing, and electrical devices and equipment; and every other material and product to be removed and not reinstalled. The owner should furnish the contractor with a list of materials and items that are to remain the property of the owner before the contractor begins demolition or removal of any item or material. Even after this list has been given to the contractor, the owner should still retain the right to claim as his property any material or item that is removed, whether it is on the list or not.

The contractor should carefully remove and protect existing fixtures, materials, equipment, doors, frames, and other items not to be reused but selected by the owner to remain the owner's property. The contractor should transport and store such items in a location selected by the owner, usually one within the building. The contractor should be paid an additional amount if such items and materials must be transported away from the building.

Materials and items suitable for reuse that must be removed to repair damaged doors, windows, store fronts, glazed curtain walls, and associated glazing, or to install new items of this sort in or on an existing building are often salvaged and reinstalled in the building. Refer to section 1.7 for general requirements related to reuse of removed materials, and to other chapters for specific requirements.

Rubbish and debris created by demolition and alteration work and other demolished or removed materials and equipment that are not to be reused in the building become the contractor's property, unless they are selected by the owner to remain the owner's property. Such items should be removed from the site promptly, whether the contractor intends to salvage or dispose of them.

Debris should also be removed promptly, at frequent intervals, and disposed of legally. Rubbish should not be allowed to accumulate or create a fire hazard.

Sometimes owners permit legal burning, burying, selling, or other disposal of demolished materials or items on their property, but such disposal is not usually permitted.

1.7 General Requirements for Alterations

After selective demolition, including dismantling, cutting, and removing of materials, has been done, alterations to existing construction are often necessary when existing doors, windows, store fronts, glazed curtain walls, and related glazing have failed or are to be installed in or on an existing building. While many requirements related to such work are specific and differ depending on the alterations required, there are some general steps and precautions that should be followed in every case.

1.7.1 Controls

Just as controls are needed during selective demolition, some controls are necessary during alterations. When an architect or engineer has been employed, the contract drawings and specifications should clearly spell out the requirements.

Whether or not an architect or engineer is employed, the contractor should be required to prepare and submit for the owner's approval narrative descriptions of the various methods to be used in making repairs or installing new items in or on an existing building. The descriptions should be accompanied by the manufacturers' catalog data fully describing each material and product proposed for use and the industry standards they meet.

The contractor should also submit shop drawings showing the details of each condition to be encountered. They should include, but not necessarily be limited to, installation and anchoring details and the relationship to other materials and items of each material and item requiring installation or reinstallation, at each condition to be encountered.

The contractor should also be required to submit samples of materials and items proposed for use in making repairs and in renovation or remodeling.

1.7.2 Materials

Materials and items so indicated by the owner should be reused to the extent possible. Such indications usually appear in an architect's or engineer's drawings or specifications. Such instructions are sometimes simply appended to the owner–contractor agreement.

Materials and items that the drawings, specifications, or owner's instructions show to be removed and reinstalled, or that the contractor elects to remove in order to make a way to perform the required repair or installation work should be reinstalled in the same location from which they were removed, unless the owner approves some other arrangement. Materials and items that might be salvaged and reused include, but are not necessarily limited to, doors, windows, and related operating hardware and accessories; store front components; glazed curtain wall components; glazing materials; flashings; mechanical and electrical equipment and devices; decorative metalwork; and wood and metal trim, stools, sills, and casings.

Unless a removed material or item is specifically indicated on the architect's or engineer's drawings or specifications or in the owner's written instructions to be reused, or such documents permit reuse at the contractor's option, materials and items removed from an existing building to make a way for the contractor to perform the work should not be reused in the same project. Exceptions are sometimes made, but should occur only when the owner approves of them in writing.

Salvaged items to be reinstalled should first be cleaned and put in proper working order. Reused materials should be in good condition, without objectionable chips, cracks, splits, checks, dents, scratches, or other defects. Operating items should be made to work properly.

Unless they are deliberately selected to be different from the existing ones, materials and items used in patches and repairs should be of the same types, sizes, qualities, and colors as the existing adjacent materials. New matching materials should be used to close openings and repairs (1) where the architect's or engineer's drawings or specifications so require; (2) where the owner's written instructions so require; (3) where suitable salvaged materials do not exist; (4) where insufficient quantities of salvaged materials exist to complete the repair work required; (5) where reuse of removed materials is not permitted; and (6) where the contractor does not consider reinstalling removed materials to be practicable.

Where similar materials do not exist, required new materials should comply with the requirements of the architect's or engineer's drawings and specifications and the owner's written instruction.

1.7.3 Making Alterations, Patches, and Repairs

Before disconnecting electrical, plumbing, telephone, or other services, the contractor should notify the owner and the authorities owning or controlling the associated wires, conduits, pipes, and other services affected.

Existing utilities should be carefully located using methods that will not damage them. They should then be protected from harm due to the contractor's operations. The contractor should cooperate with the owner and utility companies to maintain services. Damages to existing utilities should be repaired as required by the affected utility company.

Governing regulations pertaining to environmental protection should be complied with.

The discussion in section 1.6.2 applies to repair and alteration work as well. When working close to finished surfaces, casework, cabinetwork, equipment, accessories, and devices that remain in place while the work is being done, care should be taken to protect those items from harm. Items and surfaces that are not to be discarded but must be removed to perform the work should also be protected. Damaged items that cannot be satisfactorily repaired should be removed and new, matching items installed.

Materials and items to be reused should be stored in a safe location and protected until they have been reinstalled. Such materials and items that have become damaged during the contractor's operations should be repaired or new, equal products provided. Missing parts necessary to complete each installation should also be provided.

Patching that involves various trades should be coordinated carefully so that the work of one trade does not damage the work of another.

Floor, wall, and ceiling finishes that were damaged or defaced during cutting, patching, demolition, alteration, or repair work should be restored to a condition equal to that existing before the damage occurred.

Damaged or unfinished surfaces or materials exposed during alteration, repair, or removals should be repaired and finished or refinished. When doing so is not possible they should be removed and new, or acceptable salvaged materials, should be provided, to make continuous areas and surfaces uniform.

New work, and restoration and refinishing of existing surfaces, should be done in compliance with the applicable requirements of the architect's or engineer's drawings and specifications and the owner's written instructions. Where they are not specific, the following rules should apply:

Materials used to repair existing surfaces should conform to the highest standards of the trade involved and be in accordance with approved industry standards as required to match the existing surface.

Workmanship for repair of existing materials should conform to similar workmanship in or adjacent to the space where the alterations are to be made.

Reinstallation of salvaged items where no other similar items exist should be done in accordance with the highest standards of the trade involved and in accordance with approved shop drawings.

Holes and openings in existing floor, wall, and ceiling surfaces resulting from alteration work, and those shown to be filled in the architect's or engineer's drawings or the owner's written instructions, should be properly closed and patched to match adjacent, undisturbed surfaces

CHAPTER

2

Support Systems

For convenience, in this chapter the types of doors and frames, windows, store fronts, glazed curtain walls, and associated frames, hardware, and glazing addressed in this book are all called doors and windows.

Doors and windows are usually supported either by the building's structural system, by wood- or steel-framed partitions, or by solid walls or partitions of concrete, stone, or unit masonry. When problems in the supporting structure or other building elements cause damage to doors or windows or their finishes it is necessary first to correct the underlying problem before repairing the doors or windows. This chapter addresses the potential underlying problems that may cause such damage. They include excess structural movement, failed steel or concrete structure, failed wood structure or wood wall or partition framing, failed metal wall or partition framing, failed concrete, stone, or masonry wall or partition, and other building element problems.

The failure causes discussed in this chapter may not be the most likely reasons for door or window failure, but they will probably be more costly and difficult to repair than the types of causes addressed in the other chapters. Therefore, they should either be ruled out as a failure cause or repaired if found to be at fault before repairs to doors or windows are attempted.

2.1 Excess Structure Movement

Damage to doors and windows can occur if the structure moves, especially if the movement is greater than expected. Structure movement should be suspected when doors or windows fail, especially if the evidence of failure is broken or opened joints, a deformed or out-of-square frame, or a frame that has separated from the adjoining construction. To prevent failure the design, fabrication, and installation of doors and windows must take structure movement into account.

Undue structure movement may be a symptom of structural failure (see sections 2.2 and 2.3). Some structure movement, however, is normal and unavoidable. Expected structural movement due to wind, thermal expansion and contraction, and deflection under loads may be great both in wood structures and in many modern concrete- or metal-framed buildings, most of which are purposely designed to have light, flexible structural frames that are less rigid than the structural systems in most older buildings. Exterior column movement is a particular problem. Movement may be especially great in high-rise structures where both flexibility and wind loads are high. While these light modern designs are usually safe structurally, they may contribute to failures in doors and windows or their frames unless the designer is aware of the problems they impose and takes steps to head them off. Door and window frames must be designed to accommodate expected movement. Normal structure movement may result from one of the following causes:

Variable wind pressure, particularly on high-rise structures, can cause considerable structural movement.

Structure settlement will cause some movement. It is almost impossible to eliminate all building settlement, though proper design will keep such movement to a minimum.

All materials expand and contract when their temperature changes. Unfortunately, some materials change much more than others. Differential thermal expansion and contraction is a major cause of failure in building components.

Deflection of structural members and slabs will cause some movement in buildings. Usually, when the loads are removed, the structural elements return to their original shape. In concrete structures, however, the dead-load deflection known as creep is usually not reversible.

Permanent deflection can also occur in any material when the applied loads are large.

Another cause of movement in buildings is structure vibration, which is often transferred from operating equipment in the building. Vibration can loosen fasteners.

Earthquakes, hurricanes, tornadoes, and other traumatic events will cause buildings to move even when they are designed to resist failure from such causes. These kinds of movement are unavoidable and will usually result in some damage.

2.2 Failed Steel or Concrete Structures

Doors and windows are occasionally supported directly by a building's structural framing system. More often, though, they are installed in walls or partitions that are themselves supported by the structural system. When structural elements fail, supported walls and partitions fail, which causes the doors and windows to fail. A sticking door, for instance, is more likely to be caused by the frame moving than by a failure in the door itself.

As defined here, steel structural systems include steel columns, beams, girders, trusses, bar joists, floor and roof decks, and the foundations that support the structural steel. Also included within this category are preengineered and prefabricated steel buildings.

Concrete structural framing systems, as defined here, include both cast-in-place and precast concrete footings, columns, beams, girders, slabs, and such related components as stairs.

Structural systems also include composite construction made using both steel and concrete. Composites are used most often for beams and slabs, but may also be used for columns and other structural components.

Structural systems fail either because they were improperly designed or because they experience unanticipated conditions exceeding their design limitations. Structure failure may range from small to large in scope, from slight damage to complete building collapse, but most structure failure is limited in magnitude.

Since even minor structural failure may damage supported doors and windows, it is necessary before repairing such items to determine whether structural failure was responsible for the failure. Where structural failure is to blame, it is usually necessary to correct this failure before repairing the doors or windows, to prevent the repaired items from failing again. In severe cases, shoring up beams, adding columns, replacing structural members, or other reconstruction may be necessary. However, when the structural damage is self-limiting and not dangerous to people or the building, it is sometimes possible to modify the existing support system or provide a new one for the doors and windows without making major repairs to the failed structure.

Repair of a failed concrete or steel structural system is beyond the scope of this book, which has been written with the assumption that failures in such systems have been diagnosed and necessary repairs made.

Structural system failure will probably be due to one of the following causes:

The structure or its foundations have not been properly designed to withstand the loads to be applied, without excess deflection, vibration, settlement (especially differential settlement), expansion, or contraction.

An unforeseen traumatic event occurs that applies forces to the building that it was not designed to resist. Earthquakes, hurricanes, and tornadoes are examples of such unpredictable events.

2.3 Failed Wood Structure or Wood Wall or Partition Framing

Both wood and metal door frames and windows are used in buildings with wood structural systems and in wood-framed walls and partitions. In every case, if the wood framing fails, the attached doors and windows are likely to fail. Sticking doors or windows or out-of-line frames are often the first sign of wood-framing failure.

The type of wood framing elements to which doors and windows are fastened or that support such items includes joists, trusses, beams, columns, posts, bearing studs, and other portions of the structure, as well as partition framing. This section is not intended to supply enough information to build wood framing but rather to highlight those aspects of wood framing which, if not properly done, might cause a supported door or window to fail. The book in this series entitled *Repairing and Extending Finishes, Part I* contains a more detailed discussion of such construction. In addition, the appropriate sources mentioned in section 2.7 contain recommendations relating to building wood framing.

Except in houses, most wood structural framing systems in modern buildings are supported on either concrete or masonry piers or walls that are in turn supported on concrete footings. Such buildings often have a concrete slab as their first floor. Older buildings are more likely to have wood-floor construction over a basement or crawl space, and may be supported on masonry piers or walls set on concrete footings. Sometimes the footings are stone or masonry; the number of possibilities is vast. Failure

2.3 Failed Wood Structure or Wood Wall or Partition Framing

in the supporting walls, piers, or foundations can drastically affect a building, including applied or supported doors and windows.

Wood framing systems may also contain some structural steel components. Such steel structural elements as beams and columns are frequently used where long spans are desired in wood-framed buildings. The discussion in section 2.2 applies here also.

Wood structural systems may fail because of problems with foundations or other supports or from problems in the wood framing itself. Excess movement in wood framing can cause damage to doors and windows. Refer to section 2.1 for a discussion of excess movement problems in structural framing.

Wood framing may fail because it has been improperly designed or because it experiences unanticipated conditions that exceed its design limitations. Design limitations are dictated by material characteristics, legal requirements, and economic factors. It is not economically feasible to design every structure to handle all conditions that might occur.

While wood structure failure may range from slight damage to complete building collapse, most such failure is limited in magnitude. A single cause may generate failure at any level. For example, an undersized footing may lead to building collapse or simply cause more settlement than normal. Wood structural-system failure will probably be due to one of the following:

- The structure has not been properly designed to withstand all loads that will be applied without excess deflection, vibration, settlement (especially differential settlement), expansion, or contraction.
- The building's foundations have not been properly designed or installed, resulting in excessive or uneven settlement.
- An unforeseen traumatic event such as an earthquake, hurricane, or tornado has occurred that has applied forces to the building that it has not been designed to resist.

Wood framing that is supported on a concrete or steel structural framing system may fail due to excess movement in or failure of the concrete or steel structure. Refer to sections 2.1 and 2.2 for discussions. The reasons for failure given there apply also to the concrete and steel portions of a wood-framed building.

Problems with other building elements can also cause wood framing to fail. Refer to section 2.6 for a discussion.

Wood framing may also fail because of problems inherent in the framing itself. To help prevent such failure, wood framing should comply with the applicable building code and the standards and minimum requirements of generally recognized industry standards, such as the following:

- The American Institute of Timber Construction (AITC)'s *Timber Construction Standards* and *Timber Construction Manual.*
- The U.S. Department of Commerce (DOC)'s *PS 1—Construction and Industrial Plywood* and *PS 20—American Softwood Lumber Standard.*
- The National Forest Products Association (NFPA)'s *National Design Specifications for Wood Construction, Span Tables for Joists and Rafters,* and *Manual for House Framing.*
- The Southern Pine Inspection Bureau (SPIB)'s *Standard Grading Rules for Southern Pine Lumber.*
- The Western Wood Products Association (WWPA)'s *Grading Rules for Western Lumber, Grade Stamp Manual, A-2, Lumber Specifications Information, Western Woods Use Book,* and *Wood Frame Design.*
- The West Coast Lumber Inspection Bureau (WCLIB)'s *Standard Grading Rules for West Coast Lumber.*
- Applicable American Wood-Preservers' Association (AWPA) standards.
- Applicable federal specifications and ASTM standards.
- Applicable rules of the respective grading and inspecting agencies for species and products indicated.

Softwood materials in framing systems should comply with the U.S. Department of Commerce's *PS 20* and the National Forest Products Association's *National Design Specifications for Wood Construction.*

One cause of problems with doors and windows fastened to or supported by wood framing is the use of improperly seasoned (cured) wood. Wood for framing should be seasoned lumber with 19 percent maximum moisture content at the time of dressing. Lumber with a moisture content in excess of 19 percent can be expected to decrease in size by 1 percent for each 4 percent decrease in moisture content as it dries out. As the wood changes in size it will usually also warp and twist, especially when held in place at the ends, as are framing members.

Building a stud wall or partition across a building expansion or control joint can result in movement in the stud assembly, which can in turn affect a wood door or window placed in the wall or partition.

Interior stud partitions are usually framed with 2 by 4 lumber, but sometimes 2 by 3's are used. Exterior walls and some interior bearing walls are often framed with 2 by 6's. Partitions containing pipes are also sometimes framed with 2 by 6's. Interior wall and partition studs are ordinarily placed 16 inches on center. Exterior wall studs are usually placed at either 16 or 24 inches on center, depending on the wall's construction and the stud sizes. When an installation has lumber sizes or spacings that vary from those recommended by the industry it does not automatically mean that failure will occur. When failure has occurred, however, deviation from industry recommendations should be examined as a potential cause.

Every opening should be framed with at least two studs at the jambs. Each opening should also have a wood lintel consisting of as many members as necessary to ensure that the lintel supports the applied loads and finishes flush with the studs on each side of the wall. Three or more studs may be needed at the jambs of wide openings, especially in load-bearing walls. Bridging should be provided in stud walls and partitions where it is suggested by applicable industry standards, to ensure that the wall or partition will be stable. At least three studs should be used at corners and intersections to stabilize them and provide appropriate surfaces for fastening finishing materials. A failure to provide proper opening, corner, and intersection framing and bridging will result in a wall or partition that is not stable, which will in turn often result in door or window failure.

Loads imposed by deflecting structural elements should not be allowed to pass down into a stud wall or partition, unless the wall or partition has been designed to carry such loads. One way to prevent stress from being applied to wall or partition framing is to provide a space at the top of the framing and fill it with a compressible material. Of course, lateral stability of the wall or partition must be maintained.

Misaligned, twisted, or protruding wood framing can force supported doors or windows out of alignment or position or can otherwise damage them. Such defects may be caused by changes in lumber size. Even relatively dry wood will shrink, and thus warp or twist. Dry lumber may also expand if permitted to absorb free water, condensation, or water vapor in high-humidity conditions.

Wood framing members must be properly supported and securely fastened in place. When a portion of the framing loses stability or separates from its supports or the adjacent construction, the failure of an attached door or window frame may follow.

Repairing failed wood structural framing is beyond the scope of this series. Suggestions for repairing wood wall and partition framing are included in the book in this series entitled *Repairing and Extending Finishes, Part I*.

2.4 Failed Metal Wall or Partition Framing

Doors and windows are often fastened to and supported by metal wall or partition framing. Failure of such wall or partition framing usually results in problems with associated doors and windows.

The metal framing referred to in this section is the type of non-bearing wall and partition framing into which hollow steel or wood door frames are usually set. It includes both non-load bearing C-shaped or truss studs and metal stud-type framing designed to bear loads. The latter, commonly called cold-formed metal framing, is constructed of heavier gage components than is the non-load-bearing type and often requires heavier attachment devices.

Metal wall or partition framing may fail because of the types of concrete or steel structure failure or movement discussed in sections 2.1 and 2.2 or the other building element problems discussed in section 2.6. In addition, such framing may fail because of problems inherent in the framing itself. If, for example, the framing should be constructed improperly, lose its stability, or separate from its substrates, regardless of the cause, the failure of an associated door or window is likely. These types of failures might be caused by the framing being installed with some components out of alignment with the rest, by the placing of components too far apart, or by using too few members. Using undersized framing members can also lead to such failures.

Doors and windows may also fail if a metal framing system supporting them is installed so that unplanned-for loads from a deflecting or otherwise moving structure are passed into the framing and subsequently into the doors or windows. One way to prevent stress from being applied to wall or partition framing is to provide a space at the top of the framing and fill it with a compressible material. The lateral stability of the wall or partition must of course be maintained. Partition and wall furring should also be held away from adjacent solid substrates and structural elements, to prevent load transference and damage due to differential expansion and contraction between the framing and the substrates or structural members.

Building a stud assembly across a building expansion or control joint can result in movement in the stud assembly, which may affect a door or window mounted in it.

Failure to frame the jambs and heads of partitions and walls where doors or windows will be installed in accordance with applicable industry standards can lead to failure. Additional studs and runners should be provided where recommended, for example. All elements should be fastened together and to adjacent floor and ceiling runners with screws. The space above opening headers should be framed with short studs. Metal door frames should be spot grouted at jamb anchor clips, using joint compound, plaster, or mortar, depending on the finish.

Welded stud systems with prefabricated panels that are either too large or installed too rigidly or with insufficient provision for movement may cause attached doors or windows to fail.

Repairing failed metal wall and partition framing is beyond the scope of this book. Suggestions for such repairs are included in the book in this series entitled *Repairing and Extending Nonstructural Metals*.

2.5 Failed Concrete, Stone, or Masonry Walls or Partitions

Doors and windows are often fastened to and supported by solid walls or partitions. Failures in such walls and partitions can cause failures in associated doors and windows.

Solid walls and partitions may be part of the building's structural system or be just fillers, as are non-bearing walls and partitions. Most of the materials used in solid walls and partitions into which doors and frames are installed are concrete, stone, concrete unit masonry, or brick. Gypsum block, clay tile, and other weak masonry units are not appropriate for use where they must support loads. Some, including clay tiles and gypsum blocks, are not even adequate for use under a lintel to support the weight of masonry over the opening. In such materials opening jambs must be reinforced or other means provided to carry loads. When a door or window must be placed in such materials the jambs should be removed and solid brick, concrete masonry units, or other strong masonry units substituted.

Failures in solid walls and partitions with doors or windows installed in them will often, though not always, damage the door or window. Unless cracks in a concrete wall are severe, for example, they will probably not damage doors or windows located in the wall. The possibility that solid wall or partition failure may have caused a failure in a door or window should be investigated and ruled out, however, before repairs to the door or window are attempted. When a damaged solid wall or partition is responsible for failure of a door or window, it is often necessary to repair the solid wall or partition before repairing the door or window. When the solid wall or partition damage is self-limiting and not dangerous to the building or to people, it is sometimes possible to repair an existing door or window or install a new one without repairing the solid wall or partition.

Repairing solid walls and partitions is beyond the scope of this book, which assumes that if damaged solid walls or partitions have been discovered the necessary repairs have been made.

This book does not discuss solid wall or partition materials or construction methods. It is important to recognize, however, that the design and construction of such elements can affect doors or windows installed in them. For example, a solid wall or partition may exude materials that will affect the door or window or cause its attachments to fail. Some harmful substances that extrude from solid substrates are not, however, foreign to the substrate material. For example, it is perfectly natural for a concrete wall to evaporate water for a long time. Anchoring a door frame or window using a material that can rust can lead to failure of the anchor. Wet alkaline materials such as those found in wet mortar will harm some metals. Problems of this sort are not actually ones of substrate failure—they are really design problems, because the wrong material was selected.

The existence of some foreign substances that can cause failures in door and window materials or their finishes is the result of bad workmanship in installing the wall or partition and has nothing to do with the door or window designer or its installer's workmanship.

Associated doors and windows may be damaged if a solid wall or partition cracks severely or breaks up, its joints crack, or its surfaces spall due to bad materials, incorrect material selection for the location and application, or bad workmanship.

Damage may also occur if the solid wall or partition material is weaker than required by the standard in accordance with which it was supposed to be manufactured. This condition can result from poor manufacturing techniques or controls. Weak mortar or concrete masonry units may not hold anchors.

An associated door or window may be damaged if its solid wall or partition moves excessively because of improper design or installation. Types of excess movement that can cause problems include deflection, vibration, settlement (especially differential), expansion, and contraction.

Similar damage may occur if normal movement in solid walls or partitions is not accounted for in the design and installation of doors and windows. As is true for concrete and steel structural systems, some movement in solid walls and partitions is normal and unavoidable. This movement must be accounted for in the design, manufacture, and installation of associated doors and windows. Normal movement includes that caused by settlement, thermal expansion and contraction, creep, and vibration.

2.6 Failed Other Building Elements

Other building elements that are poorly designed or that fail can cause doors and windows to fail. Those other building elements include site grading adjacent to the building (Fig. 2-1), roofing, flashing, waterproofing, insulation, louvers and other devices that close openings, calking and sealants, and mechanical and electrical systems.

Other building elements, whether they were poorly designed or simply failed, that permit water to reach portions of doors and windows not designed to resist or shed water may thereby cause damage to the doors or windows or their finishes. Some possible sources of water intrusion include roof and plumbing leaks; failed sealants; leaks through doors, windows, louvers, and other opening closers; leaking or missing rain gutters and

Figure 2-1 The level of the grade immediately adjacent to the window shown in this photo makes leaks inevitable.

downspouts; and adjacent exterior grades that let surface water reach doors or windows (Fig. 2-2).

Similar problems will occur if other building elements are designed so that condensation is permitted to form on concealed or unprotected portions of doors or windows. Condensation can result from selecting the wrong materials, from the installation methods used for insulation and vapor retarders, or from improperly locating those elements. Refer to the book in this series entitled *Repairing and Extending Weather Barriers* for a detailed discussion of condensation and how to prevent it.

2.7 Where to Get More Information

The following AIA Service Corporation's *Masterspec*, Basic Version, sections contain helpful information about the subjects addressed in this chapter. The author assumes that *Masterspec*'s later editions will contain similar data. Unfortunately, the sections listed here contain little that will help with troubleshooting such installations.

Figure 2-2 The asphalt paving shown in this photo is actually higher than the bottom of the door. This condition occurred when a topping course was applied over existing pavement that was already too high.

The May 1987 edition of Section 03310, "Concrete Work."

The August 1987 edition of Section 03410, "Structural Precast Concrete."

The May 1986 edition of Section 03450, "Architectural Precast Concrete."

The May 1987 edition of Section 03470, "Tilt-Up Concrete Construction."

The May 1985 edition of Section 04200, "Unit Masonry."

The May 1985 edition of Section 04230, "Reinforced Unit Masonry."

The November 1989 edition of Section 04405, "Dimension Stone."

The August 1986 edition of Section 05120, "Structural Steel."

The August 1986 edition of Section 05210, "Steel Joists and Joist Girders."

The May 1989 edition of Section 05400, "Cold-Formed Metal Framing."

The August 1986 edition of Section 06100, "Rough Carpentry."

The May 1988 edition of Section 06130, "Heavy Timber Construction."

The May 1988 edition of Section 06170, "Structural Glue Laminated Units."

The August 1986 edition of Section 06192, "Prefabricated Wood Trusses."

The February 1985 edition of Section 09200, "Lath and Plaster."

The August 1987 edition of Section 09250, "Gypsum Drywall."

Every designer should have the full complement of applicable ASTM Standards available for reference. Some of those applicable to the subjects addressed in this chapter are marked with a [2] in the Bibliography.

The Commerce Publishing Corporation's *The Woodbook* is a wood-products reference book published annually. It contains specifications, application recommendations, span tables, and other data about a number of wood products. Most of it consists of fliers and product data published by wood-products producers and their associations. Unfortunately, purchasers of *The Woodbook* are expected to pay for this manufacturers' literature, much of which is obtainable at no cost from other sources.

The Forest Products Laboratory's *Handbook No. 72—Wood Handbook* contains a detailed discussion on wood shrinkage.

The Gypsum Association has published several documents that include requirements for framing and furring systems that support gypsum board products. Those applicable to the framing and furring discussed in this chapter are marked with a [2] in the Bibliography. Of particular interest is the 1986 publication *Recommended Specifications: Recommendations for Installation of Steel Fire Door Frames in Steel Stud–Gypsum Board Fire-Rated Partitions* (GA-219-86).

The Metal Lath/Steel Framing Association's publications *Lightweight Steel Framing Systems Manual* and *Specifications For Metal Lathing and Furring* should be available to everyone responsible for metal framing or furring systems.

The National Forest Products Association's *Manual for House Framing* contains a comprehensive nailing schedule and other significant data about wood framing. It is a useful tool for anyone who must deal with wood construction.

Ramsey/Sleeper's *Architectural Graphic Standards* contains data about nailing arrangements and nail sizes for many framing situations.

The United States Gypsum Company's publications *Gypsum Construction Handbook* and *Red Book: Lathing and Plastering Handbook* are excellent data sources for information about the framing and furring that should underlie gypsum board or plaster finishes.

The Western Lath, Plaster, and Drywall Contractors Association (formerly the California Lathing and Plastering Contractors Association) is responsible for a 1981 publication called *Plaster/Metal Framing System/ Lath Manual*, which is an excellent source of information about metal partition framing. A new edition was published in late 1988 by McGraw-Hill.

See also the other entries in the Bibliography that are marked with a [2].

CHAPTER

3

Wood and Metal Materials and Finishes

This chapter discusses the various wood and metal materials used to make the items addressed in this book and outlines some reasons those materials and finishes might fail. It also offers suggestions for repairing those materials and finishes when they do fail and for refinishing existing metal and wood. For convenience, the doors and frames, store fronts, windows, glazed curtain walls, and associated hardware discussed in Chapters 4 through 8 will all be called doors and windows in this chapter.

3.1 Wood

The standard for wood materials and their finishes that are used in swinging, interior sliding, and folding doors and their frames is the Architectural Woodwork Institute's (AWI) *Architectural Woodwork Quality Standards, Guide Specifications and Quality Certification Program,* fifth edition. Future editions will likely also be the standard of their day. The information in this section merely summarizes the AWI standards. Those desiring more information should obtain a copy of the above-mentioned AWI publication.

There is no definitive standard for the wood materials used in windows and sliding glass doors, except that they be capable of producing windows and doors that will perform according to the standards set in the National Wood Window and Door Association's I.S. 2, "Industry Standard for Wood Window Units" and I.S. 3, "Industry Standard for Sliding Patio Doors."

3.1.1 Wood Door and Panel Materials and Their Quality

AWI has established three grade levels for doors: Premium, Custom, and Economy. It has also established two face-quality grades for wood used in doors, I and II. (There is also a grade III, but it is not applicable to doors.) In the AWI grading system the quality of the facing material is directly tied to the door grade. Thus, a Premium grade architectural flush door requires a Grade I veneer. Either a factory or a field finish may be used.

Facings for Transparent Finishes. There are many species of wood used as veneer facings for transparent finish doors. Among the more common are ash, birch, butternut, cherry, elm, lauan, maple, African mahogany, Honduras mahogany, walnut, teak, red oak, and white oak.

The way that veneer is cut from a log has an effect on its appearance. Types of cuts include rotary, plain (flat) sliced, quarter-sliced, half-round sliced, and rift cut. Refer to AWI Quality Standard 200, "Panel Products," for a description of the various cuts and their effects. Not all cuts are permitted in every door-quality grade.

Veneer matching patterns also affect the appearance of panels. Matching patterns often used for doors include book match, slip match, random match, and end match. Again, refer to AWI Standard 200, "Panel Products," for more information.

It is worth noting here that not all matching patterns are possible in all cuts.

Facings for Opaque Finishes. In paint-grade doors, any closed-grain hardwood species may be used for door veneer. Medium-density overlay plywood, which has an impregnated paper surface, and particleboard, fiberboard, or hardwood may also be used.

Plastic Facings. Plastic facings on doors and panels include low-pressure laminates such as melamine, vinyl, and polyester, and high-pressure laminates.

3.1.2 Wood Window and Sliding-Glass-Door Materials

Older wood windows and sliding glass doors may have been made from any convenient softwood or hardwood. The majority of wood windows and sliding glass doors made today are manufactured from kiln-dried clear ponderosa pine or another similarly fine-grained northern white pine. The wood should have a moisture content of 6 to 12 percent when the window is fabricated.

When a natural finish is desired or a transparent finish will be applied, windows and sliding glass doors may be made from ash, cypress, mahogany, or teak. The latter three are sometimes used without a finish.

Most modern wood windows and sliding glass doors are clad with sheet aluminum on surfaces that will be exposed to the exterior. Sheet vinyl is also used for cladding, but not as frequently as is aluminum, especially in buildings other than houses. Both have the advantage of low maintenance.

Aluminum cladding material is sheet aluminum mechanically bonded to the wood. The finish is usually baked-on enamel. Vinyl cladding is a rigid polyvinyl chloride sheath that complies with the requirements of ASTM Standard D 1784.

The required exterior trim for clad windows and sliding glass doors is usually furnished by the window or door manufacturer to match the windows or doors. The trim for aluminum-clad windows and sliding glass doors may be aluminum-clad wood, aluminum extrusions, or formed aluminum. The trim for vinyl-clad windows and sliding glass doors is either vinyl-clad wood or hollow vinyl extrusions.

The few modern wood windows and doors that are not clad are usually prime painted in the factory for later field painting.

3.1.3 Wood Frame and Trim Materials

Either softwood or hardwood can be used for door frames and trim and for interior and exterior window trim that is not a part of the window itself. Hardwood is usual, however, only when the finish is a transparent one. Softwood is also sometimes given a transparent finish, but it is more often painted.

The moisture content of the wood in door frames and door and window trim is critical. Material that is too wet when fabricated and installed will warp, sometimes excessively. Wood that is too dry will absorb moisture and swell. In general, the moisture content of exterior door frames should not exceed 15 percent. For interior door frames and both exterior and interior door and window trim, the moisture content should be between 5 and 12 percent.

Many species of trees are used for softwood door frames and for framing and blocking associated with doors or windows. Some woods are regional ones used only in their native localities. Often, such wood is marked at the mill with its type and grade, the mill's name, and the grading agency. The markings, however, should be, and almost always are, concealed in the installation. Because such marks are often omitted from wood that is to receive a transparent finish, it may be difficult to determine the exact species of a softwood. Fortunately, the specific identification of a painted softwood is usually not necessary. When matching the color or grain of a transparent-finished wood is desired, the wood is visible and can usually be identified by someone knowledgeable about such things.

In softwood door frames and wood trim and casings, and for window trim that is not furnished with the windows, the materials and their grades vary from region to region. The following list gives some examples. Keep in mind, though, that these are only examples and that many other species and grades may be found.

Concealed trim and blocking: Douglas Fir C Select or an equivalent Western Wood Products Association (WWPA) softwood.
Concealed framing: Douglas Fir-Larch or Hem-Fir (WWPA Construction or Standard).
Painted frames, casings, and trim: ponderosa pine (WWPA C Select, Clear, or Better).
Stained interior frames and trim: ponderosa pine (WWPA C Select, Clear, or Better).

Any hardwood material used as a door facing may also be used to make door frames and trim. The more commonly used ones are birch, cherry, elm, maple, walnut, teak, red oak, and white oak.

Any hardwood from which the windows were made may also be used for associated trim.

3.1.4 Preservative Treatment

Painted or clad wood to be used in exterior swinging-door frames and trim, windows and associated trim, and sliding glass doors and their frames and trim should be treated with a water-repellent preservative after it has been milled. This treatment should comply with the requirements of the National Wood Window and Door Association's NWWDA I.S.4, "Industry Standard for Water-repellent Preservative Non-pressure Treatment for Millwork."

Unfortunately, it may not be possible to treat natural-finished or transparent-finished wood with some preservatives, because they will stain or discolor the wood.

3.2 Carbon Steel and Stainless Steel

There are four basic types of steel: carbon, alloy, tool, and stainless. Alloy and tool steels are seldom found in doors and windows, or in their frames or hardware, or in other related items.

Carbon-steel hollow-metal doors and frames are widely used in the construction industry.

Carbon-steel windows are found in many commercial and industrial buildings and in some housing units.

Stainless steel is used quite frequently to manufacture doors and frames, cased-opening frames, windows, and frames for glazing that will be placed in harsh environments, and sometimes even in those for use in mild environments when the appearance of stainless steel is desirable. Stainless steel finish hardware is common.

The carbon steel and stainless steel in doors, frames, and hardware are first formed at the mill into a useful shape. The four basic methods used to produce these shapes are casting, rolling, forging, and drawing, which includes extruding. The different shaping methods produce items with their own individual characteristics.

The shape of a stainless steel item, and thus the method used to form it, often dictate its finish. The standard shape designations assigned to stainless steel by the American Iron and Steel Institute (AISI) are plates, sheet, strip, bars, wire, pipe and tubing, and extrusions. The method of production (rolled, forged, cold finished, hot finished, and so on) and the sizes of each shape are also a part of the designation. The National Association of Architectural Metal Manufacturers (NAAMM) *Metal Finishes Manual for Architectural and Metal Products* contains a table showing the classifications and their characteristics.

Castings and extrusions are used directly for making finish hardware items. Finish hardware is also produced by further working sheets, strips, and plates using hammer forging, power pressing, or machining.

Stainless steel extrusions are sometimes used as door and frame components, but most carbon-steel and stainless steel doors and frames are made from formed shapes. Sometimes sheets, plates, angles, and bars are used directly to fabricate door and frame components. But, most carbon-steel and stainless steel door and frame components are made by forming sheets or strips into the desired shapes (Fig. 3-1). Forming methods include both hot- and cold-rolling in strip mills, hammer forging, power pressing, machining, brake forming, and shear forming. Probably the most common method is cold brake forming.

The book in this series entitled *Repairing and Extending Nonstructural Metals* contains more discussion about the production of carbon steel and stainless steel and the forming of them into useful shapes.

Figure 3-1 Formed shapes.

3.2.1 Heat Treatment

Carbon steel and stainless steel are both heat treated after their initial cooling, to change their physical properties. Normal heat treatments include quenching and tempering.

Working carbon steel or stainless steel cold, as in a rolling mill, makes them hard and brittle. The more working, the harder and more brittle they become. When hardness and brittleness are not desirable, the material may be annealed, to relieve the stresses imposed during the rolling process, making it softer and easier to machine.

Carbon steel can be made especially hard by casehardening or nitriding, stainless steel by casehardening.

3.2.2 Carbon-Steel and Stainless Steel Products

Carbon-steel and stainless steel sheets and shapes used to make doors and frames should be capable of having items produced from them in which the exposed surfaces are flat, smooth, and free from surface blemishes. When finished, such surfaces should not exhibit pitting, seam marks, roller marks, undesirable roughness, oil-canning, stains, discolorations, or other imperfections. After such metals have been formed into door or frame components they should be capable of withstanding the degrees of stress of every kind that will be imparted to them without their becoming deformed or damaged in any way.

Many carbon-steel and stainless steel products are used in producing doors, frames, finish hardware, and anchors and fasteners. The following is a partial list of such products. After each item is the industry standard which that product is usually required to meet.

- Carbon-steel castings: ASTM Standard A 27.
- Steel plates, shapes, and bars: ASTM Standards A 6 and A 36.
- Cold-finished steel bars: ASTM Standard A 108.
- Stainless and heat-resisting steel plate, sheet, and strip: ASTM Standard A 167.

- Stainless and heat-resisting steel bars and shapes: ASTM Standard A 276.
- Mild-steel plates, shapes, and bars: ASTM Standard A 283.
- Carbon-steel bolts: ASTM Standard A 307.
- Stainless steel pipe: ASTM Standard A 312.
- Cold-rolled carbon sheet steel: ASTM Standard A 366.
- Welded stainless steel tubing: ASTM Standard A 554.
- Cold-rolled carbon and high-strength, low-alloy steel sheets: ASTM Standard A 568.
- Hot-rolled, commercial quality steel sheet and strip: ASTM Standard A 569.
- Hot-rolled structural steel sheet and strip: ASTM Standard A 570.
- High-strength, low-alloy steel: ASTM Standard A 572.
- Hot-rolled carbon-steel bars and bar-size shapes: ASTM Standard A 575.
- High-strength, low-alloy steel sheet and strip: ASTM Standard A 607.
- Cold-rolled structural steel sheet: ASTM Standard A 611.
- Carbon-steel bars: ASTM Standard A 663.
- Hot-wrought carbon steel bars: ASTM Standard A 675.
- Stainless steel castings: ASTM Standard A 743.

In addition to the ASTM standards, stainless steel is often referred to by its American Iron and Steel Institute (AISI) designation, AISI identifies stainless steel as either martensitic, ferritic, or austenitic. Martensitic and ferritic stainless steels both fall into the AISI 400 series. Type 430 is often used in sheet-metal applications. All stainless steels contain between 12 and 27 percent chromium. Austenitic stainless steels, which include the AISI 200 and 300 series, also contain nickel, and sometimes manganese or molybdenum. Stainless steel doors and frames are usually made from AISI Type 302, 303, or 304 material. Type 302 is often called 18/8 stainless steel, because it contains about 18 percent chromium and 8 percent nickel. Type 304 is considered interchangeable with Type 302. The 200-series stainless steels contain manganese in addition to chromium and nickel. Types 201 and 202, which are similar to Types 301 and 302, are often used interchangeably with them.

3.3 Aluminum

As it comes from the pots, aluminum is about 99.5 percent pure, but even that level can be improved by further refinement, to 99.99 percent purity. Unfortunately, pure aluminum has a tensile strength of only about 7,000 psi. In addition, it is ductile and soft. To make it more usable than these characteristics permit, aluminum is alloyed with other metals.

Alloys can be divided into two basic categories: casting and wrought. Wrought alloys can in turn be subdivided into heat-treatable and non-heat-treatable alloys.

3.3.1 Alloy Designations

The Aluminum Association's method of designating an aluminum alloy uses a four-digit number. The first number is for the major alloying element and establishes the series. The second one designates modifications to the alloy. The third and fourth are arbitrary numbers that identify alloys within a series.

The 1000 series alloy is 99 percent or higher pure aluminum. When some alloys in the 2000 series are heat treated, they have properties that are equal or superior to those of some grades of steel. Structural-aluminum shapes and aluminum fasteners are often made from alloys in this series. Many common nonstructural-aluminum items, including some door and frame components, are made from Alloy 3003. Series 4000 alloys are used to produce some dark anodic finishes. Series 5000 alloys are often used to produce weather stripping. Series 6000 alloys, which are all heat treatable, are probably the most used to produce structural shapes, nails, bars, hardware, door frames, curtain walls, and store front elements.

3.3.2 Temper Designations

Temper designations consist of a letter and one or more numbers. The letter designates whether the alloy is as fabricated (F), annealed (O), strain hardened (H), solution heat treated (W), or thermally treated (T). Alloys with a T designation are heat treatable, those with an H are not.

The first number denotes the type and degree of treatment. Thus, T6 means that the metal has been solution heat treated, then age hardened.

The second number, which is used only in the H designation, indicates the degree of strain hardening. Thus, H24 indicates a strain-hardened, partially annealed alloy hardened to the 1/2 hard state.

The temper designations follow the alloy designations. Thus, a store front section might be said to have been made from Alloy 6063-T6.

3.3.3 Heat Treatment

Aluminum is heat treated to improve its workability and make it stronger. The technique of heat treatment can be divided into two categories. The first, called age hardening, is only effective when used on heat-treatable alloys. The second, annealing, is effective on both heat-treatable and non-heat-treatable aluminums.

3.3.4 Aluminum Products

Aluminum is usable in producing door and window components only after it has been milled into a useful shape. There are four basic methods used to produce such shapes: casting, rolling, extruding, and drawing.

Many aluminum products are used in producing doors, frames, and finish hardware. The following is a partial list of such products. After each item is the industry standard that the product is usually required to meet.

- Castings: ASTM Standard B 26 or B 108.
- Sheet: ASTM Standard B 209.
- Plate: ASTM Standard B 209.
- Drawn tubes: ASTM Standard B 210 or B 483.
- Bars, rods, and wire: ASTM Standard B 211.
- Extrusions: ASTM Standard B 221 or B 429.
- Forgings: ASTM Standard B 247.

Some cast, rolled, extruded, and drawn products are used exactly as they come from the mill to fabricate door and window components (Fig. 3-2). Others require only a degree of additional finishing. But many components are made by further fabricating milled products into the desired shapes. The methods used include additional roller forming, forging, bending, stretch forming, and machining. Shapes made from aluminum sheets are often produced by additional roller forming after they come from the mill. Forging is used to produce many aluminum-finish hardware items.

Some aluminum sheet metal and plate components used in doors and windows are shaped by bending. Some alloys can be bent to a sharp 90-degree angle or more; others resist such severe bending. Bending is often used to create closers and corners in store fronts and curtain walls. A well-

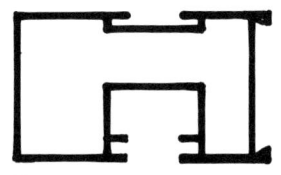

Figure 3-2 Extrusions similar to the one shown in this drawing are often used without further fabrication.

Figure 3-3 It would have been highly impracticable to use an extrusion where the coped aluminum trim shown in this photograph was installed.

made bent section can simulate extruded shapes and be much cheaper than a special extruded shape made for a non-repetitive member (Fig. 3-3).

Stretch forming is sometimes used to make irregularly shaped, curved aluminum-finish hardware items.

Many door, frame, and hardware components are first produced by one of the other methods, then worked to their final form by machining.

3.4 Copper Alloys

Of the approximately 1,100 recognized alloys of copper, more than half contain either zinc or tin. Those with zinc have been traditionally known as brass, those with tin as bronze. Most copper alloys used today to make doors, frames, and related items are called bronze, even though many of them are alloys of copper and zinc and are thus really brass.

3.4.1 Alloy Designations

The Copper Development Association (CDA) is the administrator of the Unified Numbering System (UNS) for copper alloys developed jointly by ASTM and the Society of Automotive Engineers (SAE). The CDA also has its own numbering system, which consists of three digits. The UNS adds a letter *C* prefix to the three digits and attaches a *00* suffix. For example, Muntz metal is alloy 280 in the CDA system, C28000 in the UNS.

The CDA places the copper alloys used most commonly in the construction industry into three groups. The A group contains copper, including Alloy 110, which is 99.9 percent copper and Alloy 122, which is 0.02 percent phosphorus.

The B group contains the common brasses and extruded architectural bronze. It includes alloys 220 (commercial brass), 230 (red brass), 260 (cartridge brass), 280 (Muntz metal), and 385 (architectural bronze).

The C group contains the so-called white bronzes. It includes alloys 651 (low-silicon bronze), 655 (high-silicon bronze), 745 (nickel silver), and 796 (leaded nickel silver).

Casting alloys are special alloys made specifically for that purpose. They fall in the CDA 800 and 900 series of alloy designations.

Other alloys are also used, of course, especially when color matching is required. The CDA publication *Copper, Brass, Bronze Design Handbook, Architectural Applications* contains a chart showing which alloys match in color. When a door, frame, or associated item contains several different shapes or types of components, such as sheets, extrusions, fasteners, or castings, the different parts will probably be made from different alloys, even though the colors match. For example, possible matches for the color of Alloy 230 (red brass) sheets include Alloy 385 (architectural bronze) extrusions, Alloy 836 (a casting alloy) castings, Alloy 280 (Muntz metal) fasteners, and Alloy 655 (high-silicon bronze) filler metals.

3.4.2 Copper Alloy Products

Some rolled, extruded, and drawn products are used exactly as they come from the mill to fabricate doors and frames. Others require only additional finishing. But many components are made by fabricating milled products into shapes, such as the one shown in Figure 3-4. The methods used to fabricate milled products into the kinds of shapes shown include additional roll forming, forging, bending, brake forming, explosive forming, and spinning. Not all methods are applicable to all alloys. Refer to the CDA pub-

Figure 3-4 A typical frame shape.

lication *Copper, Brass, Bronze Design Handbook, Architectural Applications* for an indication of the alloys that can be formed by each method.

Some shapes are best produced by additional roll forming. Others should be produced using one of four forging methods: hammering, pressing, hydroforming, or stamping. Many copper-alloy sheet metal and plate components used in doors and frames are shaped by bending them in machines made for the purpose. The most frequently used method is brake forming.

Explosive forming and spinning (see the Glossary) are sometimes used in making copper-alloy finish hardware items.

3.5 Other Metals

Other metals are used to produce and coat finish hardware items for use with wood and metal doors and windows and to coat other steel, aluminum, and copper-alloy items. Anchors and fasteners are also manufactured using other metals.

The very hard and highly corrosion-resistant metal chromium is used as a coating on steel and other metals. Chromium is so hard that it resists polishing. Therefore, to produce the familiar highly polished chrome-plated surfaces chromium must be deposited over a highly polished steel surface, usually over an intermediate nickel layer.

Nickel, a silver-white metal highly resistant to corrosion, is sometimes used as a finish coating, but it is more often applied over steel as an undercoating for chromium.

The corrosion-resistant metal cadmium is highly toxic. It is used to coat steel items that will be concealed in the final work.

Zinc is used extensively as a coating for steel and as an alloying metal with copper. It is also used alone to produce some kinds of hardware.

3.6 General Requirements for Finishes on Metals

The finishes used on carbon steel, stainless steel, aluminum, and copper alloys are divided into three broad categories: mechanical finishes, chemical finishes, and coatings. Coatings are then further subdivided into inorganic and organic types.

Except for the category called as fabricated, finishes on aluminum and copper alloys are called process finishes, with one possible exception. Aluminum sheets that have been given an organic or laminated coating in the mill (coil coated) fall into a gray area. They are called process finishes by some people, mill finishes by others.

Standards for finishes for carbon steel and stainless steel have been established by the American Iron and Steel Institute (AISI), for aluminum by the Aluminum Association (AA), and for copper alloys by the National Association of Architectural Metal Manufacturers (NAAMM). Their designations, summarized here, are discussed in complete detail in the NAAMM publication *Metal Finishes Manual for Architectural and Metal Products.* Anyone who must deal with new or existing metals or their finishes should obtain a copy. Determining the specific category into which an existing metal finish falls may be necessary if matching it is required, but identifying specific finishes within the types included in the standard designations is a job for professionals.

3.7 Mechanical and Chemical Cleaning of and Finishes for Steel

Mechanical finishes include those left by the manufacturing process (mill finishes) and ones created by cold rolling with polished rollers, polishing, and buffing.

Mechanical cleaning includes processes used to remove mill scale and oils and to prepare the metal to receive applied finishes.

3.7.1 Mechanical Finishes for and Cleaning of Carbon Steel

The mill finish produced by hot rolling carbon steel, which is called a black, as-rolled finish, is usually overlaid by scale and rust. Carbon steel produced by the cold-rolled process is almost always coated with grease and oil and is often so smooth that it will not hold paint well. Therefore, mill-finished steel must be cleaned and treated before it can be given an organic coating.

There are two basic classes of mechanical cleaning methods used on carbon steel. The methods in the first class effectively remove mill scale and rust, but not grease and oil. This class includes the following methods recommended by the Steel Structures Painting Council (SSPC):

Hand-tool cleaning, in accordance with SSPC-SP-2. This highly labor-intensive method is best suited for use in spot cleaning.

Power-tool cleaning, in accordance with SSPC-SP-3. The grinders, sanders, brushes, and abrasives used in this method will often so damage thin metals that they become useless. The use of this method is therefore generally limited to thick materials.

The methods in the second class are probably best for removing mill scale and rust but will also remove oil and grease and roughen the surface enough to make paint adhere well. This class includes the following four methods recommended by the SSPC:

White metal blast cleaning, in accordance with SSPC-SP-5.
Commercial blast cleaning, in accordance with SSPC-SP-6.
Brush-off blast cleaning, in accordance with SSPC-SP-7.
Near-white blast cleaning, in accordance with SSPC-SP-10.

Flame cleaning (SSPC-SP-4) and weathering (SSPC-SP-9) are no longer recommended by the SSPC.

3.7.2 Chemical Finishes for and Cleaning of Carbon Steel

Chemicals are used on carbon steel to clean and pretreat it for the application of other finishes.

There are essentially four chemical-cleaning methods in general use: solvent cleaning (SSPC-SP-1), pickling (SSPC-SP-8), vapor degreasing, and alkaline degreasing.

Carbon steel is usually given a conversion coating to change (convert) the chemical nature of its surface so that paint and coatings will adhere more readily. Acid phosphate solutions are the most commonly used materials to produce these conversion coatings.

3.7.3 Mechanical Finishes for and Cleaning of Stainless Steel

The mill finish on stainless steel sheet and strip will vary depending on whether the steel was hot or cold rolled. A hot-rolled mill finish, called a No. 1 Sheet Finish, is comparatively rough and dull.

There are two levels of sheet or strip finish that can be achieved by the cold-rolling process. The first, called the No. 2D Sheet Finish or No. 1 Strip Finish, is produced by rolling, descaling, pickling, and a final pass through unpolished rollers.

The second, called No. 2B Sheet Finish or No. 2 Strip Finish, is produced by further rolling stainless steel with the previously mentioned No. 2D Sheet Finish (No. 1 Strip Finish). It is more likely to be used in doors and frames than is the rougher No. 2B Sheet (No. 2 Strip) Finish.

A so-called Bright Annealed Finish can be obtained by annealing stainless steel sheet and strip that has been given a No. 2B Sheet (No. 2 Strip) Finish.

The standard mill finish for stainless steel plate is dull and non-reflective. Bars have a special mill finish that is applicable only to them. Pipe and tubing may have a mill finish that resembles the No. 1 Strip Finish if they are hot rolled or forged, or a No. 2 Strip Finish if they are then further finished by cold rolling. Extrusions have a mill finish resembling a No. 1 Sheet Finish.

Stainless steel is often mechanically polished as a part of its finishing operation. Sometimes a patterned finish is applied. Non-standard finishes are also available.

The five levels of polished finish commonly used on stainless steel are No. 3, a semifinished surface that is usually more highly polished in a final product; No. 4, a general-purpose bright, directional finish that is probably the most common stainless steel finish for doors and frames; No. 6, a soft, satin finish that is also widely used; No. 7, a reflective satin, directional finish; and No. 8, a mirrorlike finish.

A pattern may be imparted to stainless steel sheets by passing them between matched rollers embossed with the desired design or by a variation of this process. Other methods of patterning stainless steel sheets include various types of cold rolling, polishing, and grinding. Cross-brushed, matte, frosted, geometrically patterned, and many other pattern variations are available.

3.7.4 Chemical Finishes for and Cleaning of Stainless Steel

Chemicals are generally used on stainless steel to prepare the metal for the application of other finishes, but some chemical processes are used decoratively.

Bare stainless steel that is to receive an organic finish should be washed with solvent and prepared according to the paint or coating material manufacturer's recommendations.

Conversion coatings used to darken (blacken) stainless steel may make it blue, dark brown, or black, depending on the coating and process used. The source of the color is an oxide that the conversion coating causes to form on the stainless steel.

Another coloring effect can be obtained by flash coating stainless steel with nickel or copper.

3.8 Mechanical and Chemical Cleaning of and Finishes for Aluminum

The Aluminum Association's designated types of mechanical finishes for use on aluminum include as fabricated, buffed, directional textured, non-directional textured, and patterned. Within those types there are many finishes. Each one in the first four categories is given a designation consisting of the letter *M* followed by a two-digit number. The first digit corresponds to the four types of finish, the second to the specific finish within that type. Thus, M21 is a smooth, specular, buffed finish, M31 a fine satin, directional, textured finish.

Chemical finishes for use on aluminum are similarly designated. Their four types are non-etched cleaned, etched, brightened, and conversion coatings. The letter used for chemical finishes is *C*. The first digit corresponds to the four types, the second to the specific finish. Thus, C21 is a fine matte, etched finish, C31 a highly specular brightened finish.

Anodic coatings are designated by the letter *A*. The first digit corresponds to one of the headings: general, protective and decorative, architectural class II, and architectural class I. Thus, A31 is a clear, architectural class II anodic coating, A41 a clear, architectural class I anodic coating.

Other coatings are similarly denoted. The letter *R* indicates resinous and other organic coatings; *V*, vitreous coatings; *E*, electroplated and metallic coatings; and *L*, laminated coatings.

When specifying a particular finish, all the designations are used together. The entire designation is preceded by AA- to identify it as an Aluminum Association designation. For example, the finish AA-M22C22A42 has a specular, buffed mechanical finish (M22), a medium-matte etched chemical finish (C22), and an integrally colored architectural class I anodic coating (A42). The NAAMM publication *Metal Finishes Manual for Architectural and Metal Products* contains the complete identification system.

3.8.1 Mechanical Cleaning of and Finishes for Aluminum

Mechanical finishes include those left by the manufacturing process and those created by grinding, polishing, sandblasting, and rolling. The NAAMM categories include as fabricated, buffed, directional textured, non-directional textured, and patterned mechanical finishes. There are several subdivisions in each category.

Aluminum with an as fabricated finish will display some imperfections, which must be taken into account when deciding whether to use it without further finishing. A mill finish is often used on concealed aluminum surfaces.

Buffed finishes are created by a process of buffing alone, or by grinding, polishing, and buffing.

Directional textured and non-directional textured mechanical finishes are often used on aluminum doors and frames, especially beneath an anodic coating.

Thin aluminum sheets can be given a patterned finish by rolling as fabricated sheet between rollers shaped to the desired design. There is no specific AA category for patterned finishes. The most common use of patterned sheets on doors and frames is as flush door facings.

After it has been washed with mineral spirits or turpentine, bare new aluminum that is to receive an organic finish should either be allowed to weather for one month or be roughened with stainless steel wool.

3.8.2 Chemical Cleaning of and Finishes for Aluminum

Chemicals are used to prepare aluminum for the application of other finishes, to act as a final finish themselves, or to be part of a total finishing process that involves other steps.

The Aluminum Association lists four types of chemical finishes for aluminum: non-etched cleaned, etched, brightened, and chemical conversion coatings.

There are a number of chemical methods commonly used to clean aluminum without otherwise altering the metal. Cleaning is essential if the aluminum is to receive an applied finish.

Etching is necessary under some finishes.

A natural oxide forms a thin film on the surface of bare aluminum, which prevents applied finishes from bonding. The application of certain chemicals changes (converts) the chemical nature of the oxide film so that it provides a good bond for paint, organic coatings, and laminates. The products of such chemical treatment are called conversion coatings or conversion films.

3.8.3 Anodic Coatings

Anodizing is an electrolytic, anodic oxidation treatment that produces a thicker coating on a metal than is natural. This coating protects the metal against further oxidation and abrasion. Anodized coatings can be dyed to produce a colored, decorative finish. Many metals can be anodized, but of those available for manufacturing doors and windows aluminum has proven to be the only practical one.

The anodizing process consists of immersing the aluminum in a tank containing an acid, then passing an electric current between the metal and the acid. The resulting oxide on the aluminum may be clear, opaque, or translucent, depending on the chemicals used and the aluminum alloy. The sizes of immersion tanks available limits the size of the aluminum items that can be anodized.

The thickness of an anodic coating on aluminum can be controlled. Thin anodic coatings, called flash coatings, are used primarily as a pretreatment for paint or other organic coatings. The anodic coatings that are usually called anodizing are much thicker.

Even the thickest of anodic coatings seldom provides a complete finish for a piece of aluminum. Usually, before the anodizing process begins a mechanical finish is applied (see section 3.8.1), followed by a chemical etched finish (see section 3.8.2). In addition, oil, grease, soil, and other contaminants must be removed completely before aluminum can be anodized effectively. Anodized aluminum that is not to receive an organic coating should be sealed, to overcome the natural porosity of its anodic coating and protect the underlying metal. Sealing is usually accomplished by boiling in either pure water or in a nickel-acetate solution.

Types of Anodic Coatings on Aluminum. There are five commonly used processes for anodizing aluminum: sulfuric acid process, chromatic acid process, oxalic acid process, phosphoric acid process, and boric acid process. The first is the only one that is widely used on aluminum for doors and frames.

Colors. Anodizing by the sulfuric acid process can be easily treated to make color-anodized aluminum. The coloring process may be done in several ways. A few gold colors are produced by impregnating the anodizing with dyes or pigments. Other colors may be produced in the same way, but they are often not colorfast.

Some colors are produced by electrolytically depositing pigments in the anodizing.

Probably the most colorfast and widely used colors are the integral ones, produced by carefully selecting the alloy, chemicals, and methods to be used. Most integral colors are proprietary and must be requested by name.

Standard Designations for Anodic Coatings. The Aluminum Association classifies anodic coatings as general, protective and decorative, architectural class II, and architectural class I. The first two categories apply to general industrial work and do not apply to aluminum products of the types addressed in this book.

Architectural class II coatings must be not less than 0.4 mils thick and weigh 17 milligrams per square inch or more. They range from these lower limits up to the thinnest and lightest permitted for architectural class I coatings.

Coatings in the architectural class I category must weigh at least 27 milligrams per square inch and be at least 0.7 mils thick, but many are much heavier and thicker. Some so-called hardcoat anodic coatings are as much as 3 mils thick. Hardcoat coatings are produced using proprietary alloys and processes. They are also harder, denser, and heavier than those produced by the other anodizing processes.

3.9 Mechanical and Chemical Cleaning of and Finishes for Copper Alloys

The NAAMM designations for the types of mechanical finishes on copper alloy include as fabricated, buffed, directional textured, non-directional textured, and patterned. Within these types there are many finishes. Each finish in the first four categories is given a designation consisting of the letter *M* and a two-digit number. The first digit corresponds to one of the four types of finish. The second number designates the specific finish within that type. Thus, M21 is a smooth, specular buffed finish, M31 a fine, satin directional textured finish.

Chemical finishes for use on copper alloys are similarly designated, but there are only two types: non-etched cleaned and conversion coatings. The letter used for chemical finishes is *C*. The first digit corresponds to the types, the second to the specific finish. Thus, C11 is a degreased cleaning finish, C51 a cuprous chloride–hydrochloride acid patina.

Coatings are similarly denoted. The letter *O* is used for clear organic coatings, *L* for laminated coatings.

All the appropriate designations are used together when describing a particular finish. For example, the finish M31-M34-07x has a fine satin (M31), hand rubbed (M34), directional textured mechanical finish, and a clear thermoset organic coating (07x). To complete the designation, the organic coating, represented by the *x*, must be described completely, giving its characteristics and manufacturer.

3.9.1 Mechanical Finishes for and Cleaning of Copper Alloys

Mechanical finishes include both those left by the manufacturing process and those created by grinding, polishing, sandblasting, and rolling.

Most copper alloys with an as fabricated finish will display some imperfections, which must be taken into account when deciding whether to use them without further finishing. They are usually installed only in concealed locations.

Copper-alloy finishes created by a process of grinding, polishing, and buffing are called buffed finishes.

Probably the most common mechanical finish on copper alloy products is the smooth, satiny finish called directional texturing, which consists of many very fine parallel lines.

Non-directional texturing, which is produced by abrasive blasting, is seldom used on doors or frames.

Thin copper-alloy sheets can be given a patterned finish by rolling as fabricated sheets between rollers shaped to the desired design. There is no specific standard category for patterned finishes.

After being washed with mineral spirits, bare copper and copper-alloy materials that are to receive an organic finish should have stains, mill scale, and other foreign materials removed by sanding.

3.9.2 Chemical Cleaning of and Finishes for Copper Alloys

Chemicals are used on copper alloys for two primary purposes: to clean them of foreign matter and to change the metal's color and provide a final finish. There are other chemical treatments used on copper alloys, but they are not extensively used in the kinds of items discussed here. The two chemical finishes we are concerned with are non-etched cleaned and conversion coatings.

Non-Etched Cleaned Copper Alloys. There are a number of chemical methods commonly used to clean copper alloys without altering the metal. Some such cleaning is necessary if the metal is to receive another finish.

Conversion Coatings for Copper Alloys. Conversion coatings are used to change the color of a copper alloy and provide a final finish for the metal. A copper alloy oxidizes naturally as it ages, changing its color and producing a coating that alters the appearance of the material and protects it from further oxidation. Conversion coatings are an attempt to duplicate the normal aged appearance and protective coating of a copper alloy by accelerated chemical means. The coatings produced are oxides or sulfides of the metal.

There are two basic types of conversion coatings commonly used: those that produce a patina (verde antique) finish and those that produce the oxidized finish known as statuary bronze.

There are seven types of materials used to produce the more common conversion coatings. Refer to the NAAMM publication *Metal Finishes Manual for Architectural and Metal Products* for a complete list of these coatings and their uses.

Patinas are more difficult to control than statuary bronze finishes. They often have variations in color, especially over large surface areas, and sometimes fail to adhere to the metal. They are also likely to stain adjacent materials.

Oxidized (statuary) finishes are somewhat more stable than patinas, and their color is easier to control. They come in three tones: light, medium, and dark. A range of color should be expected, however, even within a single tone group.

3.10 Inorganic Coatings on Metal

The inorganic coatings commonly used on metals include metallic, vitreous, and laminated coatings.

3.10.1 Metallic Coatings

Some metals used in doors and windows and their hardware are coated or plated with another metal. The base metal may be either ferrous or nonferrous, but the coating metal is usually nonferrous. In most cases the two metals are bonded together but remain as separate metals; that is, they do not form an alloy.

Some metallic coatings, such as galvanizing and aluminizing on steel, are intended primarily only to protect the metal, although they may be left as the final finish in inconspicuous locations. When a decorative effect is desired, these kinds of metallic coatings are usually given a finish coat of another material. Chromium plating and some other metallic finishes both decorate and protect the underlying metal.

There are six basic methods of coating a metal with other metals: hot-dipping, electroplating, spraying (metallizing), cladding, alloying (cementation), and fusion welding.

In the hot-dip process the underlying metal is coated with a second metal by immersing it in a molten bath of the second metal. Zinc and aluminum are both applied to steel in this way.

Zinc, cadmium, aluminum, and nickel are often deposited on steel by electroplating. Chromium and copper are similarly deposited over nickel plating on steel or aluminum. Chromium is also deposited over nickel applied to copper over steel.

Most metal coatings may be applied to metals by spraying, but probably the most frequent use of this method is to apply zinc and aluminum to steel. This is the only method, other than touch-up repair of galvanizing by brush, that can be effectively used in field applications.

The other available methods are not generally used on the metals in doors and windows.

The most common ways of coating materials used in steel doors and windows and their hardware with other metals are the hot-dip and electroplating processes. Galvanizing (zinc coating) and aluminizing (aluminum coating) are the most common coatings used on steel.

Nonferrous metals in door and window hardware may be coated with other metals by either hot-dipping or electroplating.

Galvanized Steel. Steel coated with zinc, whether electrolytically or by the hot-dip process, is called galvanized steel. Most galvanized steel used on doors and frames, and all galvanized steel used in windows, is produced by the hot-dip process.

The general requirements for hot-dip galvanized-steel sheet products are contained in ASTM Standard A 525. Hot-dip galvanized-steel sheet is available in several qualities, including commercial, lock forming, drawing, special killed (deoxidized, and thus hole-free) drawing, and structural (physical quality).

Several types of zinc coatings are available, including regular spangle, minimized spangle, iron–zinc alloy, wiped, and differential. The most commonly used type of zinc coating for doors and windows is regular spangle, but some is further treated by being wipe coated or galvannealed. Different zinc-coating designations are used for each coating type. Coating designations for the regular spangle type begin with a *G* prefix, for example. The most commonly used coating designation for sheet material used to produce doors and frames is G 90, which requires that the coating have 0.90 ounces of zinc on each square foot of metal surface. Designation G 60, which requires 0.60 ounces per square foot, is sometimes used where the metal is in a more protected location. Materials with other coating thicknesses are also used occasionally. Zinc coatings with weights up to 2.35 ounces per square foot may be used in extreme environments. Zinc coating weights for sheet materials always refer to the combined weights of the coatings on both sides of the sheet.

The galvanized steel products most commonly used to fabricate doors and windows are shown in the following list. (Other galvanized steel products may also be used.) The standard that follows each item is the usual one for that product.

Rolled, pressed, or forged steel shapes, plates, bars, and strip 1/8 inch thick and heavier: ASTM Standard A 123.

Iron and steel hardware, including castings; rolled, pressed, and forged articles; bolts and their nuts and washers; screws, rivets, nails, and similar items: ASTM Standard A 153.

Assembled steel products: ASTM Standard A 386.

Structural steel sheet: ASTM Standard A 446.

Commercial quality carbon-steel sheet: ASTM Standard A 526.

Unprotected galvanized metal will form a coating called white rust, which may be removed by wire brushing, sanding, or blasting. Normal rust appearing on a galvanized surface indicates a failure of the galvanizing. Such rust should be removed, exposing the bare metal. The zinc coating should then be restored with a galvanizing repair paint specifically formulated for the purpose.

To help prevent white rust, fabricators coat galvanized surfaces with oils, waxes, silicons, or silicates, which must be removed before an organic finish can be applied. Some sources say that bare galvanized surfaces should be cleaned using mineral spirits or xylol. Others suggest using a solvent wash and specifically recommend against using mineral spirits. Some say to use petroleum spirits. Others say not to use petroleum-based solvents.

Galvanized steel is often bonderized, which leaves its surface ready for immediate painting.

Authorities disagree about the proper pretreatment of galvanized metal that is to receive paint. For example, some sources say that it should be permitted to weather for at least six months before it is painted. Others argue that weathering is a bad idea, because galvanized metal weathers unevenly, causing poor paint adhesion. They claim that unless the metal is directly exposed it will not weather appreciably anyway, and even more preparation will then be required after the metal has been allowed to weather, so that the weathering will make the preparation more difficult to accomplish properly.

Most sources say that after oils, white rust, and other contaminants have been removed, bare galvanized metal should be pretreated before the first coat of an organic finish is applied to it. Galvanized steel that will be factory coated is often pretreated with complex oxides, zinc phosphates, or chromates in accordance with ASTM Standard D 2092. However, even the industry sources that recommend pretreatment before field painting disagree on the proper material to use. Some recommend a weak acetic acid, others a proprietary acid-bound resinous or crystalline zinc-phosphate preparation or phosphoric acid. Still others insist that acetic acid not be applied to galvanized metal. The only available option when trying to decide on the type of pretreatment, or indeed whether any pretreatment is needed, is to ask the paint's manufacturer for a recommendation. At least then there will be someone to complain to if the paint fails.

Aluminized Steel. Aluminized steel is a steel sheet coated with aluminum. It is expected to conform with the requirements of either ASTM Standard A 792, which requires that the coating be 55 percent aluminum and 45 percent zinc; ASTM Standard A 875, which requires 95 percent zinc and only 5 percent aluminum and misch metal; or ASTM Standard A 463, which is most often used in prepainted applications and requires a minimum coating weight of aluminum of 0.65 ounces per square foot. Most aluminized steel sheet is between 0.12 and 0.046 inch thick, but both thicker and thinner materials are sometimes used.

When it will be factory coated with a finish, aluminized steel conforming with ASTM Standard A 875 is often pretreated using complex oxides, zinc phosphates, or chromates, in accordance with ASTM Standard D 2092.

The type that conforms with ASTM Standard A 792 is often pretreated with chromates.

Other Metal-Coated Metals. Many other types of metal-coated metals are used in door and window hardware, anchors, and fasteners.

Steel fasteners and hardware items are sometimes electroplated with cadmium to provide electrolytic separation between the steel and other materials to prevent galvanic corrosion. Since cadmium is toxic, cadmium-plated steel items should not be used where they will come into contact with people, especially children, or with food. Cadmium-plated steel should comply with the requirements of ASTM Standard A 165.

In the process called zincating, aluminum plate is immersed in a zincate bath to coat it with a thin film of zinc. Zincating is generally used to prepare aluminum for electroplating. The standard for zincating on aluminum is ASTM Standard B 253.

Chrome is plated onto aluminum after zincating, or plating the aluminum with copper, brass, or nickel.

Iron, carbon steel, and stainless steel are also chromium plated to produce hardware items. Applicable standards include ASTM Standards B 177, B 254, B 320, and B 650.

Copper can best be plated with chromium after a coating of nickel has been applied. Chromium-plated copper has many uses in the construction industry, including its use in finish hardware and fasteners.

Nickel plating is sometimes used over steel as an intermediate coat beneath chromium or copper plating. Nickel plating over copper is used primarily as a base for chromium plating. Nickel is seldom used alone in the building industry.

Brass-plated steel is sometimes used as finish hardware and for similar items.

Copper plating may be used on aluminum either as a base for chromium plating or as a finish itself.

Some alloys of aluminum have superior corrosion resistance. Others have high strength. To take advantage of both characteristics, sheets of the stronger alloys are sometimes clad with sheets of the more corrosion-resistant ones. The cladding material is metallurgically bonded to the core material. This product class is called Alclad. Most clad products come in sheets, but other shapes, such as tubes and rods, are also available. Alclad 3004, for example, is a commonly used sheet material. Alclad 3003, used for extruded bars, rods, wire, shapes, and tubes, is a 3003 alloy clad on both sides with a 7072 alloy.

3.10.2 Vitreous Coatings

The only type of vitreous coating on metal that is commonly used in doors and windows is porcelain enamel, which is used on metal panels, mostly in glazed curtain walls.

Porcelain enamel is a form of vitreous organic coating that displays most of the characteristics of glass. It is fused to a backing metal at a high temperature. Materials baked at temperatures of 800 degrees Fahrenheit or higher comply with the ASTM definition of porcelain enamel, but most of it is fired at temperatures between 1450 and 1550 degrees Fahrenheit. For most building-related purposes, porcelain enamel is applied over low-carbon steel or aluminized steel. For other purposes it is also applied over stainless steel and aluminum. It is seldom used over copper alloys in buildings.

Porcelain enamels are hard, abrasive resistant, and non-porous. Water and atmospheric pollutants will barely penetrate them. Their color is as permanent as that of any material used in building construction. They are available in a great number of colors, textures, and patterns.

Porcelain Enamel on Steel. The steel to which porcelain enamel is applied is either decarbonized enameling steel, conventional cold-rolled sheet, or a special material produced for the purpose.

Almost all porcelain enamel on steel is applied to sheets that are from 14 to 22 gage in thickness before the enamel is applied. Most panels are four feet by eight feet or smaller, although larger panels are possible.

Most applications require two coats of enamel. The first (ground) coat forms a permanent bond with the base metal. The second (top, or cover) coat contains the coloring elements and frits that give the porcelain enamel its color, gloss, and corrosion resistance.

Porcelain enamel on steel should comply with the recommendations of the Porcelain Enamel Institute. Their applicable standards include

Guide to Designing with Architectural Porcelain Enamel on Steel.
S-100(65), "Specification for Architectural Porcelain Enamel on Steel for Exterior Use."
"Color Guide for Architectural Porcelain Enamel."
"The Weatherability of Porcelain Enamel."
Bulletin T-2, "Test for Resistance of Porcelain Enamel to Abrasion."
Bulletin T-20, "Image Gloss Test."
Bulletin T-21, "Test for Acid Resistance of Porcelain Enamels."
Bulletin T-22, "Cupric Sulfate Test for Color Retention."

In addition, several ASTM standards are applicable to porcelain enamel on steel, including standards C 282, C 283, C 286, C 313, C 346, C 448, C 538, C 540, and E 97.

Porcelain Enamel on Aluminum. The porcelain enamel used on aluminum is slightly different from that used on steel. It is formulated to fire at lower temperatures and does not always require a ground coat. Some colors and patterns, however, do require that a ground coat be applied.

Only certain alloys and tempers of aluminum are suitable to receive porcelain enamel, because the baking process tends to heat treat and anneal aluminum.

As is true for steel, almost all porcelain enamel over aluminum is applied to sheets that must be stiff enough to prevent warping during the enameling process. Alternatively, the sheets can be given a coat of porcelain enamel on the concealed side to balance the stresses imposed by the porcelain enamel on the exposed side, or they can be reinforced to prevent warping. The sheets must also be smooth and free of die lines that will show through the enamel.

In any case, porcelain enamel on steel should comply with the recommendations of the Porcelain Enamel Institute. Their applicable standards include

ALS-105(69), "Recommended Specification for Architectural Porcelain Enamel on Aluminum for Exterior Use."
"Color Guide for Architectural Porcelain Enamel."
"The Weatherability of Porcelain Enamel."
Bulletin T-2, "Test for Resistance of Porcelain Enamel to Abrasion."
Bulletin T-20, "Image Gloss Test."
Bulletin T-21, "Test for Acid Resistance of Porcelain Enamels."
Bulletin T-22, "Cupric Sulfate Test for Color Retention."
Bulletin T-51, "Antimony Trichloride Spall Test for Porcelain Enameled Aluminum."

In addition, several ASTM standards, including standards C 282, C 283, C 286, C 313, C 346, C 448, C 538, C 540, C 703, and E 97, are applicable to porcelain enamel on aluminum.

3.10.3 Laminated Coatings

Laminated coatings include all the adhesive-bonded plastic coatings used today over sheet metals. Plastic films used in these coatings include, but are not necessarily limited to, polyvinylchloride (PVC) and polyvinyl fluoride (PVF). These materials are laminated to steel, galvanized steel, aluminum, after those metals have been given a conversion coating to prepare them to receive the coating, and other nonferrous metals. Sheets with laminated coatings are available for a wide variety of uses, including both interior and exterior applications.

3.11 Organic Coatings

As defined in the Glossary, organic coatings are used both to protect and to decorate wood and metal. The variety and number of different opaque and transparent organic coatings available today are enormous and constantly growing. This section discusses both opaque and transparent organic coatings.

Opaque coatings, which include paint, are used routinely on steel and wood doors and windows. They are also used occasionally on stainless steel and aluminum, but almost never on copper alloys on the kinds of items discussed in this book.

Transparent organic coatings are used frequently on wood doors and windows, often on copper alloys, and occasionally on stainless steel—but rarely on aluminum, except as temporary strippable coatings or a lacquer on clear anodized aluminum.

It might be possible to guess the type of an existing organic coating material by looking at it, but positive identification usually requires laboratory examination of samples.

According to the National Association of Architectural Metal Manufacturers (NAAMM), paint is an organic coating. Even though paint (coating) manufacturers and contractors do not all agree with the NAAMM, we will accept their definition for this book. General references in this book to organic coatings are intended to include paint as well as all other types of organic coatings.

3.11.1 Standards

The current standard for high-performance coatings such as the various fluorocarbon polymers is the American Architectural Manufacturers Association's (AAMA) publication AAMA 605.2, "Voluntary Specifications for High Performance Organic Coatings on Architectural Extrusions and Panels." Fluorocarbon polymer coatings and any material sold in competition with them should be required to comply with the latest version of AAMA 605.2.

The AAMA also publishes a performance standard for organic coatings on aluminum: AAMA 603.8, "Voluntary Performance Requirements and Test Procedures for Pigmented Organic Coatings on Extruded Aluminum."

3.11.2 Manufacturers and Products

Organic coatings are manufactured and distributed by national, regional, and local manufacturers. Specific manufacturers are not mentioned in this book. Refer to section 3.16 for suggestions about finding manufacturers.

Coatings may be applied on metal and wood doors and windows either in the factory or the field. Many metal door and window manufacturers will only furnish doors with their standard finishes, but others supply custom finishes.

Most wood-door manufacturers furnish doors only unfinished or with their standard finish applied. Wood window and sliding-glass-door manufacturers furnish their products with aluminum or vinyl cladding (see section 3.1.2) or primed for field painting. A particular wood-door manufacturer's standard finish may not comply with the requirements in AWI Standard 1500, "Factory Finishing," which defines thirteen factory-finishing systems. Systems 1 through 8 in that system are transparent, systems 9 through 13 opaque. Unfortunately, architects, engineers, and building owners often accept a door manufacturer's standard organic coating without question, and the results are not always good. Unless the particular door manufacturer's standard finish is known to be acceptable, it is generally better to require that factory-applied finishes conform with the requirements for a specific selected AWI finish system even if such a requirement means that someone other than the manufacturer must finish the doors. AWI Standard 1500 lists the applicable industry performance standards for each AWI finishing system.

There are also available other factory-applied wood-door finishes that are not included in AWI Standard 1500 and not standard with a particular manufacturer. When one of them is desired, the system must be identified completely so that the finisher knows exactly what is expected.

With so many organic coating types available, it is difficult for anyone who is not a coatings professional to determine which is right for a particular situation. In addition, the only widely recognized standards for many coatings are those contained in federal specifications, which are often out of date, inapplicable to a particular situation, and sometimes require products that no manufacturer offers.

There are hundreds of ASTM standards related to organic coatings, but many of them cover only component materials. Since a single coating may contain a large number of different chemicals, using ASTM standards to control them is generally impractical. For wood doors it is possible to use the same ASTM standards the AWI uses to specify performance for field-applied coatings, but doing so may be impracticable because of the many factors involved in making field applications, including the level of the applicator's experience. Also, many manufactured organic coating systems for field application on wood have not been specifically tested to ensure compliance with the AWI's requirements for factory-applied coatings.

In addition, environmental control laws and other legal restrictions can drastically affect organic coating selection in areas where such laws are in force. In general, products that will comply with applicable laws and regulations are available, but their selection requires a thorough understanding of the regulations and knowledge of available organic coating products.

Therefore, most of us must rely on reputable manufacturers for help in selecting organic coatings. Those responsible for selecting or accepting organic coatings should obtain, review, and understand the manufacturer's technical literature for each product to be used, including the product's contents and application instructions.

Organic coatings that will be applied to metal in the field must be formulated to dry properly in normal air. Organic coatings that will be applied to metal at the factory can be formulated to be either an air drying or baking finish.

Organic coatings for finishing wood doors and windows, whether to be factory or field applied, are usually formulated to dry properly in normal air.

Coatings used on an exterior surface may be either chalking or nonchalking, but in either case they should be mildew resistant.

The quality and expected service life of organic coatings are directly related to their color and gloss retention; adhesion; corrosion, mildew, and graffiti resistance; hiding power (opacity); and the percentage of solids that will remain on the finished surface after the material has dried. The more solids there are, the thicker the dried film will be, and the greater its opacity. Conversely, the more thinner an organic coating contains, the lower its quality is likely to be. The composition of a coating's solids also has an effect on its quality.

Codes, regulations, and laws often control the acceptable lead content of an organic coating used in residential applications and other locations where it may be accessible to children.

The NAAMM says that the optimal organic coating would have good flowing and leveling characteristics. It should result in a film thickness of not less than 1.0 mils but preferably 2.5 mils or thicker, with a high solids content. It should also be fast drying and have a permeable primer but have high resistance to moisture and gas penetration through its finish coat. Further, it should exhibit good adhesion characteristics, be flexible, be hard and resistant to abrasion, and be durable. Finding all these characteristics in one material is not likely, but some may be more important in a particular application than are others. The finish should be selected that has the greatest number of desired characteristics.

3.11.3 Opaque Organic Coating Composition

An old enamel may be nothing more than a pigmented varnish. Modern enamels, however, are practically indistinguishable from paints. One difference between paint and other opaque organic coatings is that the other coatings may have materials added, or their basic composition altered, so that forced drying or baking is required for proper curing. With this exception, modern paints and other opaque organic coatings are composed of four different groups of components: vehicle; volatile or thinner, also called the solvent; pigment; and additives. Each group is made up of several different ingredients and serves a different function. Transparent coatings of course lack pigments.

Opaque organic coating composition is discussed at length in the book in this series entitled *Repairing and Extending Finishes, Part II*.

Most opaque organic coatings are identified by their binder type. The more common ones in use today in opaque organic coatings on metal doors and windows are acrylic, alkyd, elastomers, epoxy, latex, oil, oil–alkyd combinations, epoxy emulsion, phenolic, polyester, rubber, urethane, vinyl chloride copolymers, and combinations of those.

The more common binder types used today in opaque organic coatings on wood doors and windows are acrylic, alkyd, epoxy, latex, oil–alkyd combinations, polyester, urethane, and combinations of the others. Those included in AWI Standard 1500 as being factory finishes are catalyzed and standard lacquer (System 9), conversion varnish (System 10), catalyzed polyurethane (System 11), pigmented polyester (System 12), and pigmented polyester and urethane (System 13).

3.11.4 Transparent Organic Coating Composition

Transparent Organic Coatings for Use on Metal. The most common transparent organic coating binder types used on metal are acrylic, alkyd, cellulose, epoxy, nitrocellulose, silicone, and urethane.

Temporary clear (strippable) coatings are used on some metals, particularly aluminum, to protect the finish during handling and installation. These coatings must usually be removed quite soon after the metal has been installed, especially in exterior applications. Heat and sunlight will make some of them very difficult to remove if they are left exposed.

Transparent Organic Coatings for Use on Wood. The transparent coatings discussed in this section include oiled, stained, and natural finishes for wood. The wood's color, grain, or both are visible through transparent coatings. The wood may appear in its natural color or its color may be darkened by a finish, lightened by a bleach, or altered by a stain. The grain may be enhanced, diminished, or concealed by the finish system. The components of these finishes include bleaches, transparent stains, undercoats, and finish coats. The undercoats and finish coats are often lacquer, varnish, shellac, or polyurethane but may also be oil.

Transparent coatings on wood doors and frames are achieved using the following materials, either alone or in combination:

Wood Stain. Stains used to color wood in interior applications are made from color pigments suspended in either linseed oil or another drying oil. Stains may also be thinned, to make them easy to apply. Stained wood may be finished with linseed oil or one of the clear finishes discussed in this section.

There are also several types of wood stains designed for exterior use. Semi-transparent stains are oil-based materials that dye the wood fibers as they penetrate. They cannot be used over previously painted or sealed surfaces and they must be renewed frequently, perhaps as often as once a year. Pigmented semi-transparent exterior wood stains—especially those in dark colors—tend to hide the natural bleeding that occurs when they begin to fail, and they may therefore need renewal somewhat less frequently than other stains.

The other types of exterior wood stains include semi-solid stains, oil-based solid-color stains, acrylic solid-color stains, weathering stains, bleaching stains, deck stains, and others. Most are weather resistant; many contain water repellents and wood preservatives. They usually require recoating less frequently than the semi-transparent stains, but they tend to fade fairly rapidly. Oil-based stains may require recoating within a year or two. Wood stains containing 100 percent acrylic latex tend to last longer and fade slower than oil-based stains, however, and may go virtually unattended for three years or longer.

No exterior wood stain will last as long as or resist fading as well as paint will.

Linseed Oil. Boiled linseed oil constitutes the classic oil finish. It is inexpensive and easy to apply, maintain, and repair. Normal applications require reducing the oil by mixing it with turpentine in ratios ranging from equal parts of each to twice as much oil as turpentine.

Varnish. Varnish is a homogeneous mixture of a resin, drying oil, drier, and solvent. When varnish dries it forms a transparent or translucent film that may have a flat, satin, or high-gloss finish. It is available in several qualities and in colors ranging from nearly clear to dark brown. The higher-quality products expand and contract without cracking. Varnish is classified as either short-oil, medium-oil, or long-oil, depending on the number of gallons of oil it contains for each hundred pounds of resin.

Other materials are added to some varnishes to impart qualities other than those natural to the material. For example, synthetic silica and other pigments are added to change the natural high-gloss sheen of varnish to a low-gloss finish. Varnish with additives included to give the material a resistance to salt water is called spar varnish.

Varnish is used as a clear finish and, when reduced by solvents, as a sealer for wood and plywood. Varnishes for interior use usually include alkyd resins, but epoxy-ester varnishes are also available.

Many varnishes formulated for exterior use are made from tung oil and phenolic resins, though other formulations are also used. Sometimes varnishes intended for interior use are applied on the exterior, where they may function satisfactorily if well protected from the weather.

Shellac. Shellac produces a finish similar to that produced by varnish. It is available in either white or orange. White shellac can be tinted with alcohol-soluble aniline dyes, which makes it especially valuable for blending in repairs with old work. Orange shellac produces a finish with a deeper tone than does an undyed white material. Its relatively short shelf life can make shellac unusable within four to six months after manufacture.

Lacquer. Lacquer, which is available in formulations that will produce either a flat or a glossy finish, is seldom used for field-applied finishes on building surfaces. It is more likely to be used as a factory finish. Lacquers have the distinct disadvantage of not being usable over existing finishes. They can actually be used as paint removers.

Polyurethane. Some references and some manufacturers classify transparent polyurethane finishes as varnishes. Others call them lacquers. But, although they have some characteristics of each, they are actually neither. Urethane finishes are available as oil-modified or moisture-curing types.

Oil-modified polyurethanes are clear materials only. They are sometimes used in exterior applications but are better suited to interior locations. Urethanes are water resistant and highly durable. Clear urethanes are available in formulations that will produce either gloss or matte finishes. They tend to not hold a gloss, however, when used on the exterior. Some sources say that they do not bond well to existing finishes, others say they may be so used. In any case, they cannot be used over shellac, paste wood fillers, and some other finishes. The individual manufacturer's recommendations should be followed when determining where a urethane finish can be used.

Moisture-curing polyurethanes may be clear or pigmented to produce a colored opaque finish. Since they are dependent on water to cure, they do not work well in very dry climates.

Wood Preservatives and Sealers. Materials called wood preservatives, which come as clear, semi-transparent, or opaque, and penetrating wood sealers, which are also actually wood preservatives and not transparent finishes, are both beyond the scope of this book. Some of these may resemble one or another transparent finish, however, and their failure may be similar to that of a transparent finish. The expected life span of many such materials is one year or less. When they are not renewed within that interval, the wood being protected may show evidence of water and weather damage such as bleeding, mildew growth, and general blackening.

3.11.5 Organic Coating Systems

A single coat of an organic coating, used alone, will almost never be adequate, so a system of materials must be used in combination to form a satisfactory finish application.

In general, all coats in a system must be compatible to have a successful installation. For example, the success of most organic coatings requires that an incompatible primer be removed completely and a new one applied. There are a few exceptions, however. Some paint top coats can be separated from an incompatible primer by barrier coats that are compatible with both the primer and the top coats.

Most manufacturers of organic coatings routinely furnish specific recommendations for the different components in coating systems that use their products. Each proposed organic coating system should be reviewed by the manufacturers of the products to be used to ensure compatibility of the various coats of the system with each other and with the materials they will contact, and the suitability of the system for prevailing conditions. When the system's manufacturer suggests that different products or combinations of products be used than those selected, its suggestions should be followed.

Most opaque organic coating systems consist of at least a primer and a top coat. Often, one or more intermediate coats is also required.

For transparent finishes over wood, the number of coats applied and the composition of them depends on the desired effect. Generally, at least three coats of the finish material are needed to provide acceptable durability and appearance, although some of those coats may be cut with the proper thinning agent. Where a color different from the natural one of the material being covered is desired, adding a stain coat will be necessary. On open-grain woods, a wood filler is usually used to provide a smooth surface.

Unless there is a compelling reason to do otherwise, all the products used in a coating system should be produced by the same manufacturer, or at least be supplied by the manufacturer of the finish coats. Thinners should be approved by the manufacturer of the material being thinned. Undercoats and primers, whether shop or field applied, should be approved by the finish coat manufacturer.

When the manufacturer's standard factory-applied coating is acceptable it should be applied in the door or window manufacturer's standard way. If more control over the factory finishing of wood doors is desired, an AWI finish system may be used as the standard. The AWI transparent finish systems are Finish 1, standard lacquer; Finish 2, catalyzed lacquer; Finish 3, conversion varnish; Finish 4, catalyzed vinyl lacquer; System 5, catalyzed polyurethanes; Finish 6, synthetic penetrating oil; Finish 7, water-reducible acrylic lacquer; and Finish 8, clear polyester.

Primers must be formulated to adhere readily to both the substrate and the succeeding coat, and be compatible with each. Oil-based and alkyd primers should not be used on galvanized metal surfaces, for example, because their presence can create a chemical reaction that will leave a soap on the metal's surface, preventing the paint from adhering. Some primers, such as rust-inhibitive ones on metals, also serve the major secondary function of protecting the substrate. Some primers serve to isolate essentially incompatible paints and substrates. Primers on galvanized metal surfaces, for example, separate the galvanizing from the top coats, many of which contain chemicals that can react with the zinc coating on the metal.

Ferrous metals that will be painted in the field are usually shop primed, unless they are galvanized or aluminized. Even galvanized or aluminized steel is sometimes shop primed. Galvanized surfaces to which a shop coat of paint will be applied should first be given a phosphate or equivalent chemical treatment, to provide a bond for the paint.

Several types of paint used extensively as primer paint on ferrous metals are red lead mixed pigment, alkyd varnish, linseed-oil paint, conforming with Federal Specification TT-P-86G, Type II; red lead iron oxide, raw linseed oil, alkyd paint, conforming with the Steel Structures Painting Council's SSPC-Paint 2-64; and basic lead silico-chromate-base iron oxide, linseed oil, alkyd paint, conforming with Federal Specification TT-P-615, Type II. These materials are usually applied at the rate necessary to form a final primer thickness of about 2.0 mils. Two coats are usually applied to surfaces that will be in contact with exterior masonry or concrete. Because they contain lead, some of the above listed paints may not be acceptable in some jurisdictions for some uses.

Primers for galvanized metals are usually metallic zinc paints or zinc dust–zinc oxide paints, conforming with Federal Specification TT-P-641G.

3.11.6 Miscellaneous Materials

Many related materials are necessary to produce effective organic-coating finishes. They include crack and seam fillers, wood filler, plastic wood, turpentine, linseed oil, mineral spirits, denatured alcohol, lacquer thinner, and others. For standards applicable to these and other materials needed for paint and transparent systems, refer to section 3.16.

3.11.7 Special Paints

Paints specifically formulated to handle unusual conditions are available. There are, for example, paints intended for use on cleaned metal, which are generally called direct-to-metal paints. There are also direct-to-rust paints that are epoxy mastic materials formulated for direct application over rust, and rust-conversion coatings that convert ferric oxide (rust) to a stable organic iron compound that then becomes part of the coating. Other special paints include aluminum, bituminous, emulsified asphalt, and chlorinated rubber paints. These materials are used on metal expected to be exposed to particularly harsh conditions.

Paints that are specifically formulated to handle unusual conditions are routinely used on metals where they will be in contact with concrete, masonry mortar, plaster, pressure-preservative-treated wood, or other corrosive materials. Special paints include bituminous paints, which are usually required to comply with the requirements of the Steel Structures Painting Council's SSPC-Paint 12 for cold-applied asphalt mastic. Bituminous paints are not attractive and are therefore usually used only when they will be concealed in the completed work.

Zinc chromate is another special paint used to protect concealed aluminum and copper alloys from corrosion.

3.11.8 Powder Coatings

Powder coatings are applied by covering a metal surface with a fine powder, then fusing the powder into a continuous film by heating it to between 300 and 400 degrees Fahrenheit. Most powder coatings used today are based on epoxy, polyester, or acrylic resins. Because powder coatings are relatively new in this country, they will probably not be found on buildings constructed before 1982.

3.11.9 Film Laminate

Both interior and exterior finishes on copper alloys are sometimes protected by a laminated film called Incracoat. This film, developed by the Copper Research Association, consists of a 1-mil-thick layer of polyvinyl fluoride bonded to the copper alloy with an adhesive.

3.11.10 Preparation of Surfaces to Receive Organic Coatings

Surface preparation will often determine whether an organic coating application will be successful. Some sources indicate that as many as 80 percent of paint failures are caused by improper surface preparation.

Regardless of the type of organic coating to be applied, the substrate must first be properly prepared to receive it, and the conditions under which it is applied and cured must be in accordance with its manufacturer's recommendations.

Woods and metals that are to receive an organic coating, and the conditions under which the coating will be applied, should be inspected carefully. Unsatisfactory conditions should be corrected before work begins.

Preparation of Wood Surfaces. Surfaces should be cleaned before the first coat of an organic coating is applied, and if necessary between coats, to remove oil, grease, dirt, dust, and other contaminants. In general, the cleaning of wood should be done using soap and water, solvents, scrapers, and sandpaper, as required, in accordance with the recommendations of the coating manufacturer.

In addition to cleaning, wood should be sanded smooth and even, before the first coat is applied and between coats as well. Dust should be removed after sanding. Residue should be removed from knots, pitch streaks, cracks, open joints, and sappy spots. On wood surfaces to be painted, a thin coat of white shellac should be applied to pitch and resinous sapwood before the prime coat is applied. Small, dry, seasoned knots should be scraped, cleaned, and covered with a thin coat of white shellac or another recommended knot sealer before the prime coat is applied. Holes and imperfections in finished surfaces should be filled with putty or plastic wood filler after the primer has been applied. Such surfaces should be sandpapered smooth after they have dried.

New wood that is to receive a field-applied organic coating should be primed or sealed immediately upon delivery and stained, if appropriate, before it is installed. The edges, ends, faces, and backsides of casings, trim, and other wood should be primed. The primer should be an enamel undercoat, a penetrating sealer, or a varnish of the same material that is to be used for the first coat of the exposed finish. The priming of wood in or on a building should never be done during masonry erection.

In wood to be coated, nails should be set and screws countersunk. After the prime coat has dried, nail and screw holes, cracks, open joints, and other defects should be filled with putty or wood filler. Where a transparent coating will be used, the filler should be tinted to match the color of the wood.

When surfaces cannot be put in proper condition by customary cleaning, sanding, and puttying, organic coating work should not proceed until the proper conditions have been achieved using other accepted methods.

Water-stained surfaces to which paint is to be applied should first be primed. The primer should be an oil-based paint product specifically recommended by both its manufacturer and the finish coat's manufacturer to conceal water-stain marks so that they will not show through finish-paint coats. This prime coat should be in addition to the number of coats recommended by the paint manufacturer to produce the paint system to be applied.

Before application of an organic coating is started, the area should be swept clean with brooms and excessive dust removed. An area should not be swept with brooms after coating operations have started there. Then the necessary cleaning should be done using commercial vacuum-cleaning equipment.

Surfaces to be coated should be kept clean, dry, smooth, and free from dust and foreign matter that would adversely affect the coating material's adhesion or appearance.

Each coat of a coating system should be inspected and found to be satisfactory before the next coat is applied.

Preparation of Metal Surfaces. Hot-rolled ferrous metal comes from the mill with a coating of mill scale. Carbon steel from the mill has both mill scale and rust. A few types of paint can be applied directly over scale and rust, but the vast majority of organic coatings require complete removal of those materials. Cold-rolled steel and stainless steel surfaces must be degreased before being finished and may also need roughening. Aluminum ordinarily leaves the mill with a coating of oil and grease. And the production of metal doors and windows will leave weld deposits and other contaminants on their surfaces. All such foreign materials must be removed or properly treated before an organic coating is applied.

Metal surfaces that are to receive an organic coating must be cleaned of oil, grease, dirt, loose mill scale, and other foreign substances, using solvents or mechanical cleaning methods. The recommendations of the paint or coating manufacturer and the Steel Structures Painting Council should be followed. Bare and sandblasted or pickled steel should be given a metal treatment wash before the primer is applied.

3.11.11 Factory-Applied Organic Coatings

Factory-Applied Organic Coatings on Wood. Factory-applied organic coatings should be applied strictly in accordance with the recommendations of the manufacturer of the coating product, following the finisher's standard practices.

Factory Applied Organic Coatings on Metal. Factory-applied organic coatings on metal should be applied strictly in accordance with the recommendations of the manufacturers of the coating product and the metal, the fabricator's standard practices, and applicable industry standards.

Factory-applied organic coatings are usually baking-finish epoxy or other enamel coatings applied by mechanical methods such as hot spraying, airless spraying, electrostatic spraying, dipping, flow coating, or roller coating. The latter is a coil-coating system that is often used to finish sheet steel and aluminum used in manufacturing building panels and similar items.

Baking-finish organic coatings are cured at high temperatures, either by force drying or baking.

Clear organic coatings are seldom applied to new metals in the field, because it is usually more economical and easier to apply them in the shop.

Some coatings are not appropriate in every circumstance. For example, regardless of who recommends it, methacrylate or another lacquer should never be used on color-anodized aluminum. The lacquer will tend to alter the color and make the anodizing appear more like paint than anodizing. In addition, lacquer never completely adheres to anodized aluminum. Where it does not fully adhere, it will appear white. On clear anodized aluminum, the difference is not noticeable, but on color-anodized aluminum the white spots will produce a blotchy, wholly unappealing effect. If protecting color-anodized aluminum is necessary, a strippable coating should be used. Strippable coatings, however, are expensive to apply and remove. If left on too long in the sun, they may become almost impossible to remove. Usually, covering the aluminum with a loose plastic film, paper, or even grease will offer ample protection.

Organic coatings on stainless steel, aluminum, and copper alloys are usually applied by such mechanical methods as hot spraying, airless spraying, electrostatic spraying, dipping, flow coating, or roller coating.

3.11.12 Shop Primer-Coat Application for Field-Applied Paint

A coat of paint is often applied in the factory or shop to wood doors and windows, and to ungalvanized ferrous-metal doors and windows that will be painted in the field. Galvanized and aluminized steel are not usually given a shop coat but may sometimes be.

Anchors and other devices that will be embedded in concrete, stone, or masonry are not usually given an organic coating, except for coatings used strictly to prevent metal corrosion.

Shop Primers for Wood. On doors, the surfaces that should be primed include both faces, the four edges, and surfaces machined for hardware. Wood door frames and trim are not usually given a shop coat but sometimes are. Window surfaces that are not aluminum or vinyl clad should be primed. Window trim is not usually shop primed.

Primer paint should be applied only to cleaned wood surfaces. Immediately after surface preparation, the primer should be brushed or sprayed on in accordance with its manufacturer's instructions, at a rate that will produce a uniform dry-film thickness of about 2.0 mils for each coat. The painting methods used should result in full coverage of joints, corners, edges, and exposed surfaces.

The paint used must be compatible with the finish coats.

Factory Primers for Metal. Primer paint should be applied only to cleaned, degreased metal surfaces. Immediately after surface preparation, the primer should be brushed or sprayed on, in accordance with its manufacturer's instructions, at a rate that will produce a uniform dry-film thickness of about 2.0 mils for each coat. The painting methods used should result in full coverage of joints, corners, edges, and exposed surfaces.

Only one shop coat is usually applied on fabricated metal items, except that two coats are normally applied on surfaces that will become inaccessible after assembly or erection of the metal item. It is always a good idea to change the color of the second coat so that an observer can distinguish it from the first and thereby determine if all the required coats have been completely applied.

Since new stainless steel, aluminum, and copper-alloy items are seldom given field-applied organic-coating finishes, few come from the factory with a shop primer on them.

3.11.13 Preparing the Site for Field Application of Organic Coatings

Hardware, accessories, machined surfaces, plates, fixtures, and similar items should be removed before surface preparation for field-application of organic coatings begins. Alternatively, such items can be protected by surface-applied tape or another type of protection. Even when such items are to be coated they are often removed, to make coating them and adjacent surfaces easier. After the coating application in each area has been completed, the removed items can be reinstalled.

Mildew should be removed and neutralized. Except where such chemicals are prohibited by law, this can be done by scrubbing the affected surfaces thoroughly with a solution made by mixing two ounces of a trisodium phosphate-type cleaner, eight ounces of sodium hypochlorite (Clorox), and one gallon of warm water. Where use of these chemicals is not permitted, a commercially available mildewcide can be substituted. If necessary, a scouring powder may be used to remove mildew spores, but care must be taken to prevent damage to the surface being cleaned. The cleaned surfaces should be rinsed with clear water and allowed to dry thoroughly.

The surfaces being dealt with should be cleaned before the first coat of paint is applied and if necessary between coats, to remove oil, grease, dirt, dust, and other contaminants. Activities should be scheduled so that contaminants will not fall onto wet, newly coated surfaces. Paint should not be applied over dirt, rust, scale, oil, grease, moisture, scuffed surfaces, or other conditions detrimental to the formation of a durable coating film.

Before the application of paint is started in an area, that space should be swept clean with brooms and excessive dust removed. After the application process has begun in a given area, that space should not be swept with brooms. Necessary cleaning then should be done using commercial vacuum-cleaning equipment.

Surfaces to be painted should be kept clean, dry, smooth, and free from dust and foreign matter that would adversely affect the coating's adhesion or appearance.

3.11.14 Field Touch-up of Shop Coats

The first step in preventing metal-finish failure due to corrosion is to make sure that the factory or field finish originally applied is touched up where abraded, to prevent corrosion from occurring at scratches and abrasions. Initial touch-up should be done immediately after erection and before calking and application of the first field coat. The touch-up material should be the same as the primer. Surfaces to be covered should be touched up before they are concealed. Touch-ups in paint should be sanded smooth.

Before touching up is started, surfaces should be cleaned. Mortar, damaged or unwanted paint, dirt, grease, dust, efflorescence, salts, stains, and other contaminants should be completely removed. The cleaning materials used should not harm the metal, metal finishes, or adjacent surfaces. Abrasives or caustic or acid cleaning agents should not be used. Initially, cleaning should be done using only clear water and mild soap.

3.11.15 Applying Organic Coatings in the Field

Organic coating materials should arrive at the application site ready for application, except for tinting and thinning, and in their original, unopened containers bearing the material's name or title, the manufacturer's name and label, the federal specification number or other standard to which the material complies, manufacturer's stock number, date of manufacture, the contents by volume for major pigment and vehicle constituents, thinning instructions, application instructions, color name and number, and fire-hazard data, when applicable.

Organic coating materials should be mixed, thinned, and applied in accordance with the manufacturer's latest published directions. Materials should not be thinned unless the manufacturer specifically recommends doing so. The applicators and techniques used should be those best suited for the substrate and the type of material being applied.

Sealers and undercoats should not vary from those recommended by the paint or finish manufacturer.

Unless the manufacturer recommends otherwise, organic coating materials should be stirred before application, to produce a mixture of uniform density, and be stirred as required during application. Surface films should not be stirred into the coating material. Such film should be removed and, if necessary, the material strained before it is used. Shaking to mix coating materials should be done only if specifically permitted by its manufacturer.

Workmanship should be of a high standard. Application should be done by skilled mechanics using the proper types and sizes of brushes, roller covers, and spray equipment. Equipment should be kept clean and in proper condition. The rollers used for a field-applied gloss finish and the corresponding primer should have a short nap.

The coating material should be applied evenly and uniformly, under adequate illumination. Surfaces should be completely covered, smooth, and free from runs, sags, holidays, clogging, and excessive flooding. Completed surfaces should be free of brush marks, bubbles, dust, excessive roller stipple, and other imperfections. Where spraying is either required or permitted, the coating should be free of streaking, lapping, and pileup.

The number of coats recommended by the manufacturer should be the minimum number used. When properly applied, this number of coats should produce a fully covered, workmanlike, presentable job. Each coat should be applied in heavy body, without improper thinning.

Primer coats may be omitted from previously primed surfaces, but every other coat recommended should be applied to each surface to be covered. When stains, dirt, or undercoats show through the final coat of a paint, defects should be corrected and the surface covered with additional coats until the coating film presents a uniform finish, color, appearance, and coverage. When defects show through a transparent finish, the finish should be removed, the defect corrected, and the surface recoated.

Organic coatings should be applied at such rates of coverage that the fully dried film thickness for each coat will not be less than that recommended by the manufacturer. Special attention should be given to ensuring that edges, corners, crevices, and exposed fasteners receive a dry-film thickness equivalent to that on flat surfaces.

The first coat of an organic coating system should be applied to surfaces that have been cleaned, pretreated, or otherwise prepared for finishing as soon as is practicable after preparation, before surface deterioration can begin.

Water-based coatings should be applied only when the temperature of the surfaces to be coated and of the surrounding air is between 50 and 90 degrees Fahrenheit, unless the coating manufacturer's printed instructions say otherwise.

Solvent-thinned coatings should be applied only when the temperature of the surfaces to be coated and of the surrounding air is between 45 and 95 (some sources say 50 and 120) degrees Fahrenheit, unless the coating manufacturer's printed instructions recommend otherwise.

Organic coatings should not be permitted to freeze or be applied in snow, rain, fog, or mist; when the relative humidity exceeds 85 (some sources say 90) percent; to damp or wet surfaces; or to extremely hot or cold substrates, unless the coating manufacturer's printed instructions recommend otherwise. The coating of surfaces exposed to hot sun should be avoided. Coating may be continued during inclement weather, however, if the areas and surfaces to be coated are enclosed and heated during the application and drying periods to within the temperature limits specified by the coating's manufacturer.

Once application of organic coatings has been started within a building, a temperature of 65 degrees Fahrenheit or higher should be provided in the area where the work is being done. Wide variations in temperature that might result in condensation on freshly coated surfaces should be avoided.

Concealed surfaces of wood members should be back primed before they are installed.

It is necessary to sand lightly between each succeeding enamel or varnish coat. High-gloss paint should be sanded between coats, using very-fine-grit sandpaper. Dust should be removed after each sanding, to produce a smooth, even finish.

Wood door manufacturers' warranties usually require that the top and bottom edges of, and cutouts in, job-finished wood doors be sealed with the same finish applied to their faces. Regardless of warranty considerations, the tops, bottoms, and side edges of exterior doors should be finished to match the exterior faces of the doors. Edge and cutout sealer should be applied after the doors have been fitted but before the faces are finished. Sealer should not be allowed to run onto the faces or edges of the doors.

Except with oil finishes, whenever a transparent finish is applied on the interior of a building over elm, oak, hickory, walnut, or another open-grained wood, a paste wood filler should be applied and then wiped across the grain as it begins to flatten. A circular motion should be used to secure a smooth, filled, clean surface that leaves filler in the wood's open grain.

When the wood is to be stained, the filler is often slightly tinted with the stain to avoid emphasizing the grain. When the overall surface will not be stained, the filler can be stained slightly to emphasize the grain. After the filler has dried overnight, the surface should be sanded until smooth before the next coat is applied. Steel wool should not be used. The filler should be allowed to dry for at least twenty-four hours before stain is applied.

Wood surfaces that will be stained and given a transparent finish should be covered with a uniform coat of stain, using either a brush or a clean cloth. Excess stain should then be wiped off.

When a classic oil finish is desired on wood, the boiled linseed oil to be used should be thinned with turpentine. The mixture should range from equal parts of oil and turpentine to twice as much oil as turpentine. The actual mixture will depend solely on the personal preferences of the applicator. An oil finish may be applied over a stain or directly to bare wood. The oil–turpentine mixture should be brushed liberally onto the wood until no more is absorbed. Then the excess should be wiped off with clean cloths and the surface polished to a sheen. Sufficient pressure should be exerted to melt the oil. This can be achieved with an orbital sander or by hand pressure. After a forty-eight-hour drying period the entire process should be repeated, at least five times.

Weather-stripped doors and frames should be finished before door equipment is installed.

The first coat or primer recommended by the paint manufacturer may be omitted on doors that have been shop primed if the primer is touched up in the field to repair abrasions and other damage.

Primed and sealed wood surfaces that show evidence of suction spots or unsealed areas in the first coat should be reprimed, to assure a finish coat with no burn-through or other defects due to insufficient sealing.

Copper alloy and aluminum surfaces should be protected from corrosion where they will be in contact with dissimilar metals, concrete, stone, masonry, or pressure-treated wood, by coating the contact surfaces with bituminous paint. Alternatively, copper and aluminum may be coated with a zinc-chromate primer paint.

Before a sealant is applied, protective strippable coatings and tapes on aluminum or copper alloys should be removed from the portion of the metal that will be in contact with the sealant.

Clear coatings are frequently used on copper-alloy surfaces. They should be applied strictly in accordance with their manufacturers' instructions and the recommendations of the CDA.

Sufficient time should be allowed between the application of successive coats of organic coatings to permit proper drying. A minimum of twenty-four hours is required between interior coats, forty-eight hours between exterior coats. Surfaces should not be recoated until the previous coat has dried until it feels firm, does not deform or feel sticky under moderate thumb pressure, and application of another coat does not cause lifting or loss of adhesion of the undercoat.

The edges of paint adjoining other materials or colors should be made sharp and clean, in straight lines, and without overlapping.

Organic coatings should be cured in the proper humidity and temperature conditions recommended by their manufacturers. Coatings should be kept dry until cured, unless their manufacturer specifically recommends otherwise.

At completion of coating operations, coatings should be examined and damage should be touched up and restored, and left in proper condition.

While coating work is being done, discarded coating materials, rubbish, cans, and rags should be removed from the site at the end of each work day.

When coating has been completed, glass and other coating-spattered surfaces should be cleaned. Paint should be removed using scraping or other methods that will not scratch or otherwise damage finished surfaces.

3.12 Why Wood and Metal Materials and Finishes Fail

Anyone who has to deal with a failed existing wood or metal item or a failed finish on such an item should first learn as much as is reasonable about the failed material and finish. One way to do so is to read the information in this chapter and the pertinent references listed in section 3.16, then consult the manufacturer of the material. For finish types not discussed in this book the first step should be to contact the manufacturer.

There are many reasons a wood or metal material or finish might fail, including those related to structure movement or failure, wall or partition framing failure, solid supporting wall failure, and other building element problems. Such causes are discussed in Chapter 2. When a metal material or finish fails, the types of problems discussed there should be investigated and either ruled out as causes or be repaired.

Several failure causes directly related to the materials and finishes themselves are addressed in this section.

3.12.1 Bad Materials

There are several ways in which the wood and metal materials used in producing doors and windows may have been bad, and several different associated consequences.

Bad Metal Materials. Metal materials may be improperly manufactured. Sheet thicknesses may not be consistent, for example, or the carbon content of the steel may be incorrect. Or an alloy used to manufacture a door or window may have been improperly constituted and thus not as strong as it should be. There are many such possibilities. The failure of doors and windows because the metal materials used in them were bad is not a major problem, however, when compared with the other possible failure causes.

Bad Wood Materials. Wood materials may be improperly manufactured. Veneer thicknesses may be incorrect, for example. Or the adhesive used to laminate facing sheets may not have been the correct type (Fig. 3-5), or may have been bad or improperly used. Bad wood materials are not a significant cause of door or window failure, however.

Figure 3-5 The door in this photo was supposed to have been an exterior-grade one, but the adhesives used were clearly not up to the task.

Bad Chemical or Inorganic Coating Materials. When a chemical or inorganic coating material is bad, the results usually become apparent before the finished item is installed. Metals whose finishes have failed for this reason are normally eliminated in the construction process, however, and therefore seldom pose a problem in existing buildings.

Bad Organic Finishing Materials. There are many ways in which an organic coating material may be bad and thus responsible for a failure of the finish. The coating material may have been improperly manufactured, for example. Such cases are certainly not unheard of, and this possibility should be considered when an organic coating fails. The number of incidents of bad coating materials, however, is small compared with the number of instances of bad design and workmanship. The following types of manufacturing defects might occur:

There may have been too much oil in the organic coating, which can cause alligatoring and checking.

The organic coating material may be inconsistent in color, composition, or density.

The pigments in an opaque organic coating may be incompatible with its other ingredients.

An organic coating may contain too much pigment for the binder. Such materials will chalk excessively.

An organic coating's formula may be such that it does not dry, or dries very slowly. Using old or defective materials can have the same effect, as can not having enough drier in a paint or using a poor quality solvent that vaporizes too slowly. A slow-drying oil will also cause paint to dry slowly. Poorly formulated paint may never dry. A poorly formulated baking-finish organic coating may not cure at the usual temperature and may therefore not be properly cured when delivered to the construction site. Old shellac may not only dry slowly but may never dry completely.

3.12.2 Selecting Inappropriate Materials

Selecting wood or metal materials that are inappropriate for their use can lead to failure. This problem includes such errors as selecting a soft metal alloy for use where abrasion will occur, choosing a brittle alloy when components must be bent into curves, selecting sheet metal materials for use where extrusions would be a better solution, and selecting thin sheet metals for use where they will come into contact with such devices as lawn mowers, snow blowers, and the like.

Requiring that incompatible fasteners be used will lead to failure. Steel fasteners, for example, may corrode if used in a copper-alloy frame and steel, galvanized steel, and aluminum may corrode if they contact copper-alloy fasteners. Requiring that incompatible anchors and attachments be used is a similar problem.

3.12.3 Selecting Inappropriate Finishes

Selecting finishes that are inappropriate for the conditions will probably result in failure of the finish.

Trying to match one type of finish with another is seldom successful. A color-anodized aluminum member will almost surely not match fasteners or another aluminum member that has been finished with any coating, whether inorganic or organic. Painted fasteners will not match anodized aluminum and may not match fluorocarbon coatings. Some producers claim that such matches are possible, but the author has yet to see a successful one. Mismatching, however, is not always the case when both materials are organic coatings.

Another problem caused by painting fasteners in anodized or fluorocarbon-coated aluminum is the difference in the life spans of the two finishes. Anodizing and high-quality coatings will both last longer than the paint on the fasteners, which generally leads to an unsightly condition. Owners can seldom justify the costs of repainting fasteners in an otherwise sound finish.

Inappropriate Mechanical, Chemical, or Inorganic Coating Finishes. Improper finish selection includes selecting the wrong mechanical, chemical, anodic, or inorganic or organic coating finish for the location and conditions. For example, unpainted galvanized steel, No. 2D, 2B, and bright rolled stainless steel, mill-finished stainless steel, as fabricated (mill finish) aluminum, and as fabricated copper-alloy items should not be used where appearance is important, because they are likely to contain visual flaws and discolorations.

Requiring too thin a zinc coating for the expected exposure will result in a coating that is too thin to protect the underlying steel. Corrosion will soon result.

Proprietary finishes that cannot be matched in a fabricator's shop should not be required as the final finish for metals in doors or windows. An exception might be made where appearance is unimportant, however.

Mill-induced discolorations and imperfections are often made to appear even worse by welding and bending or other metal-forming techniques.

Requiring that a mechanical finish be used on clad sheet, such as aluminum-clad aluminum, will often result in the finishing process penetrating the thin cladding, which will defeat the purpose of the cladding.

Requiring a buffed or other highly polished finish on metals that will be used on large, flat surfaces will likely produce the rippling effect known as oil-canning. Flat, unpatterned metals are much more likely to show this effect. Using non-reflective matte, textured, patterned, or etched (embossed) sheet metals will usually reduce—and often eliminate—this effect.

Requiring that thin metal sheets have an abrasive-blasted finish will often lead to thinning of the sheet metal so that it will buckle or fracture when subjected even to small stresses. Blasting may also cause thin sheets to distort.

Requiring that aluminum on a large flat surface have a chemically brightened finish will probably produce an unsatisfactory finish, because chemically brightened surfaces are difficult to make uniform over large areas.

Using matte dipped or bright dipped chemical finishes as final finishes on copper-alloy items is not usually satisfactory. These finishes are hard to control and do not usually present uniform surfaces when used alone.

Requiring that a wire-brushed finish be used on large surfaces is not a good idea. Such finishes are difficult to maintain. Normal maintenance procedures sometimes dull them or make them appear non-uniform.

An anodic coating too thin for its intended use will soon wear and produce an unsightly appearance. An architectural class II coating on main-entrance lobby doors, for example, will soon wear away from constant use.

Failing to properly match the finishes on adjacent components in a door or window assembly can produce an unsightly condition. If, for example, a chemical finish is used on aluminum assembly components made from different alloys, the components will not match in appearance. Each different alloy is likely to look different, even when identical chemicals and processes are used. Thus, the different aluminum alloys, anodic colors, and procedures used to produce the finish on the various components of an anodized aluminum assembly must be carefully selected, to ensure that they exhibit the same appearance.

Copper alloys in the same fabrication must also be carefully selected for compatibility. Different copper alloys will produce different finishes, even when the same process and chemicals are used. Refer to the CDA publication *Copper, Brass, Bronze Design Handbook, Architectural Applications* for guidance.

Selecting the wrong metal alloy to use as a base for a porcelain-enamel finish will lead to failure of the finish. Fortunately, such a failure will usually appear immediately.

Inappropriate Organic Coatings. Selecting an organic coating material with a composition inappropriate for the location and use intended will usually lead to failure. Using residential-quality paint in a commercial building is one example; selecting latex paint for use where humidity will be high is another. Very high humidity can cause the water-soluble components in latex paints to appear as brown spots in the dried paint.

Selecting incompatible paint-system products will also lead to failure. Using a water-based top coat over a solvent-based undercoat may cause alligatoring, checking, or peeling, for example. And using a hard finish over a soft undercoat can also result in alligatoring or checking. Latex paint applied over old, chalking oil paint cannot penetrate the chalk and probably will not adhere. Applying an oil paint over a latex one can cause separation, because when they age, oil paint becomes harder and less elastic than latex. Incompatible top coats may also blister.

Selecting organic coating materials that are incompatible with their substrates, including existing finishes, may lead to peeling, flaking, cracking, or scaling in the coating.

Selecting other organic coating materials of inferior quality will eventually lead to failure. An example is selecting a paint that has excessive chalking characteristics. Selecting a chalking paint to use where the runoff will cross another material, such as brick, is another.

3.12.4 Improper Preparation for Application of Finishes

Correct preparation for application of finishes on metal and wood is essential if the failure of the finishes is to be prevented. Failure to properly prepare the substrate is probably the largest single cause of finish failure. Many preparation errors can be avoided if the recommendations of the finish manufacturer and recognized authorities are followed.

Failures will occur if surfaces to be finished are not prepared properly. One example is failing to require chemical precleaning of metal. Failing to properly clean the underlying metal before applying a finish will usually cause the finish to fail. Oil, grease, and mill scale must be removed from aluminum, for example, before an anodic coating is applied. Mill scale, dirt, oil, grease, and even fingerprints must be removed from copper alloys before a conversion coating is applied to produce a patina or statuary bronze finish. Such contaminants will mar the finished surface and may even appear worse after the finishing than before.

Failing to apply the proper pretreatment beneath a metallic coating will often result in delamination of the metal coating. Failing to require a nickel plate between chromium plating and its underlying copper is an example. Failing to require zincating on aluminum that is to receive copper plating is another.

Applying a finish over an improperly applied pretreatment or underlying finish will also usually lead to failure of the finish. Non-adherence or some other failure of an underlying pretreatment or finish may cause the next coating to fail. This can be a problem regardless of the type of pretreatment or underlying finish that has failed, but it is most likely to happen when an underlying metallic coating fails. If the zincating on aluminum does not adhere, for example, an applied chromium plating will delaminate.

Failing to require the use of either mechanical or chemical finishes or both under an anodic coating on aluminum sheet material or to use an embossed sheet will lead to an unsatisfactory finish. Anodic coatings on large, flat sheet-aluminum surfaces that have not been mechanically or chemically finished to roughen them often exhibit streaks or discolored areas that would have been concealed by an underlying texture.

Failing to remove sources of moisture and water that will affect an applied coating and to ensure that the metal is completely dry may cause an organic coating to dry slowly. It will also promote mildew, moss, and other plant growth and cause the organic coating to blister and eventually peel, flake, crack, or scale. Alligatoring in clear organic coatings, especially in shellac, is often caused by exposure to damp conditions.

Failing to properly pretreat the substrate and protect a metallic coating will usually cause an applied organic coating to peel, flake, or otherwise delaminate. Failure to properly treat and protect metallic-coated steel that is not finished with an organic coating can lead to corrosion of the metal. There is a tendency, for example, to believe that galvanized steel is immune from corrosion, but this is far from the truth (Fig. 3-6). The zinc coating is itself subject to white rust, which will interfere with paint adhesion. The zinc coating on galvanized steel is also subject to galvanic corrosion from more noble metals such as copper, and to chemical corrosion from soluble sulfites like those found in cinders. Zinc will also be corroded by salt, some concrete aggregates, and the acids in cedar and oak. The zinc coating may itself oxidize or fail to protect steel that is submerged in water or subjected to heavy layers of condensation or other water deposits. Zinc coatings will be burned away during the welding process.

Failing to apply the proper pretreatment beneath a metallic coating will often result in delamination of the coating.

Applying an organic coating over an improperly applied pretreatment or underlying finish will often lead to failure of the coating. Non-adherence or some other failure of an underlying pretreatment or finish may cause the next coating to fail. This can be a problem regardless of the type of pretreatment or underlying finish that has failed, but it is most likely to happen when an underlying metallic coating fails. If the zincating on aluminum does not adhere, for example, an applied chromium plating may delaminate.

Failing to properly prepare the area where an organic coating will be applied can contribute to failures. It is necessary to completely remove

3.12 Why Wood and Metal Materials and Finishes Fail   93

Figure 3-6 The owner thought that galvanizing was sufficient to protect the door shown in this photo. The dark rust deposits show how wrong he was.

existing paint that is more than 1/16 inch thick, because thick existing paint can cause new coats to alligator or crack. It is also necessary to remove mildew, moss, ivy and other plant growth, oil, grease, dirt, dust, rust and other corrosion, loose mill scale, wax, loose existing paint, stains, and other contaminants that will either interfere with proper application, damage the coating, or telegraph through the coating. For example, a coating applied over grease or oil may dry slowly, or not at all. Since many paints permit water-vapor transmission, improperly prepared metals, including nails and flashings, that begin to corrode may continue to corrode beneath the coating. They can eventually disappear almost completely, leaving a hollow shell of coating film. Even in less severe cases, corrosion beneath a coating may stain it or create unsightly surface irregularities.

It is necessary to smooth out rough surfaces and roughen surfaces that are so smooth that a coating will not adhere to them. Surfaces must be dry and no other condition can exist that will be detrimental to the formation of a durable coating film.

Failing to remove the gloss from an existing surface or undercoat before applying a succeeding layer of an organic coating will result in failure of the organic coating to adhere. Applying new paint over existing glossy paint without roughening the surface will create this problem. Other examples are applying paint directly over a vitreous coating or over slick laminated coatings, stainless steel with a highly polished finish, or chromium plating.

Failing to remove sources of moisture and water that will affect the coating and to ensure that the metal or wood to be coated is completely dry may cause a coating to dry slowly. It will also aid mildew, moss, and other plant growth and cause the coating to blister and eventually peel, flake, crack, or scale. Alligatoring in shellac, for example, is almost always caused by exposure to damp conditions.

Failing to remove natural salts from previously painted surfaces will result in the peeling of an applied paint. Salts that normally collect on exterior surfaces are washed off by rain where they are exposed to it. They must be removed, however, from areas where rain does not touch their surfaces, by washing them off with trisodium phosphate where its use is legal or with another material recommended by the paint's manufacturer.

Failing to remove loose, peeling, or otherwise unsound organic coating materials from an existing surface before applying a new organic coating will result in the failure of the new coating.

Also, failing to remove chalking before painting over it will result in a failure of the new paint to adhere to the old.

Failing to prime a water-stained surface before painting it will let such stains show through the new paint.

Continuing corrosion can result from failing to remove rust or other corrosion and to apply a rust-inhibitive primer or one of the several available special paints formulated for application over rust. It may be best to completely remove rusted nails and the like and install new ones. New metal must of course be properly primed before the new coating is applied.

Failing to countersink nails and screws, fill the holes, and spot prime will result in finish failure.

Another example of improper surface preparation is failing to remove discolorations caused by color extractives in wood substrates and to provide a stain-blocking primer. A further example is failing to clean and apply a knot primer to knots that are either already bleeding or may bleed in the future.

Removing existing paint by using inappropriate methods will result in failure. Open flames may char wood or deform thin metals, for example. Rotary sanders, sandblasting, and water blasting may remove too much material or otherwise damage some surfaces, particularly wood. Abrasives and strong chemicals may scratch or etch glass or destroy glazing sealants or plastic gaskets.

Failing to sand shoulders at their edges or feather the edges of existing coatings where a new coating will be applied will result in the old edges telegraphing through the new paint.

Failing to prepare wood substrates properly, including not removing residue from knots, pitch streaks, cracks, open joints, and sappy spots, will set up the conditions for failure. Other problems are failing to apply a coat of white shellac to pitch and resinous sapwood before the prime coat is applied; to scrape and clean small, dry, seasoned knots and apply a thin coat of white shellac or other knot sealer to them before the prime coat is applied; and to sand surfaces before painting or finishing them.

Failing to fill holes and imperfections in the substrate will result in failure also.

Applying a porcelain-enamel finish to items with inside or outside corners that are too sharp will lead to failure of the porcelain enamel. The minimum radius of corners should be not less than 3/16 inch in steel or 1/16 inch in aluminum.

3.12.5 Improper Finish Application

The proper application of finishes is essential to prevent their failure. Probably the biggest cause of failures from bad workmanship is that of not following the design and recommendations of the manufacturer and recognized authorities. Applying fewer than the recommended number of coats, for example, can result in failure to cover the surface and protect the substrate, and the telegraphing of underlying faults through the coating. Applying finishes under temperature, humidity, or other conditions different from those recommended or otherwise improperly applying a mechanical, chemical, or other finish will often lead to failure.

Applying paint in a building before it has been completely closed in and concrete, masonry, plaster, and other wet work has dried is asking for failure. Moving a factory-finished metal or wood item into the same type of environment may also have detrimental effects on the coating. Such conditions may cause the humidity to be too high or condensation to form on surfaces that will be coated or have already been coated. Applying most organic coatings to damp or wet surfaces can cause them to adhere poorly and blister. Applying them when the humidity exceeds the level recommended by their manufacturer can cause them to dry slowly and form blisters or not adhere. Brown spots may appear in latex paints when the humidity is too high. Properly ventilating spaces where organic coatings are being applied not only helps protect workers from potentially toxic fumes but also helps control humidity and promotes proper drying of the organic coatings. Placing factory-coated metal or wood items in environments for which their finishes were not designed will almost always lead to finish failure.

Applying an organic coating when the room temperature has been less than that recommended by the coating material's manufacturer for forty-eight hours before the installation or when the material's temperature is too low can cause failures. The manufacturer's recommended minimum temperature for field-applied paint generally ranges from 45 to 65 degrees Fahrenheit, but it will vary also with the paint type. An associated problem is that of letting the room temperature fall below the temperature recommended by the manufacturer after a coating has been applied. Low temperature is a major factor in air-drying organic coating failures, because it can affect drying time and adhesion. Low temperatures shorten drying time and make coatings hard to apply. Applying a coating material that contains ice or applying it to a frozen or frost-covered surface can lead to poor adhesion and blistering. Low temperatures can cause wrinkling and may cause the coating not to adhere.

Another major failure-producing error is applying an organic coating when the temperature of the air, the surface being coated, or the coating material is too high. High temperatures can thin coating materials and make them not cover as well as they otherwise would. A coating that is too warm when applied will set too rapidly and form too thin a film.

One source of excess heat is hot air passing over the surface being coated. This can happen, for example, when a temporary heater is placed too close to the surface. Such a passage of hot air not only raises the temperature but can also cause the surface to dry too rapidly through evaporation, trapping wet coating material beneath the dried surface. Later, when the underlying material dries, it will shrink and sometimes delaminate from the top coat. Another potential result is the trapping of vapors that evaporate from the undercoat, either between the two coats or between the lower coat and the substrate. In either case, bubbles form that may rupture and form pits or puncture wounds that can extend completely through the coating to the substrate. Applying a coating in direct sunlight or when the air temperature is too high can also cause the surface of a coating to dry too quickly, entrapping solvent vapors, which will appear as blisters. Coating materials, surfaces, or air that are too hot can also cause the surface to wrinkle or not adhere to the substrate.

The manufacturer's recommended maximum temperatures for organic coating application usually range from 85 to 120 degrees Fahrenheit, but they may vary with the coating type.

Permitting drastic temperature changes to occur in a location where a coating has been applied or into which a factory-finished metal or wood item has been placed will also cause failure. When drastic temperature changes occur before a coating has set completely, alligatoring or checking may occur when the coating materials expand or contract. If the top coat is not elastic enough it will crack.

Another temperature-related error is permitting a wide temperature differential to exist between a coating and the surface to which it is applied. Large differences may cause the coating to blister and eventually peel.

Failing to remove dust, dirt, moisture, and other contaminants between coats can cause an organic coating to become loose, peel, crack, flake, or scale. Moisture left between coats can cause a coating not to dry properly and may cause blisters that will eventually lead to interlayer peeling.

Failing to sand previous coats of a coating material when so directed by the manufacturer and to remove the dust generated by it can result in lack of adhesion of the next coat.

Using organic coating materials that are the wrong type for a particular installation will usually lead to failure. The type of failure will depend on the materials used and how severely they are mismatched. A similar problem occurs when a coat is applied that is incompatible with previous coats, an existing finish, or the substrate.

A slightly different but akin problem is that of using the wrong primer for the substrate or finish coats, which can cause many types of problems. For example, some types of paint will react with the zinc on galvanized metal, creating a type of film that will prevent paint from adhering.

Applying damaged organic coating materials will usually lead to failure, regardless of whether the damage was inherent in the manufactured materials or occurred during shipment, storage, or application. Applying paint materials that are stained, for example, will result in a discolored surface.

The same caution applies to using the individual components of a coating system. Using the wrong thinner, or a defective or poor quality one, for example, may cause a coating to fail. Such solvents may evaporate too quickly, resulting in wrinkling, alligatoring, blistering, checking, peeling, flaking, cracking, or checking of the coating. Or they may dry too slowly, which will cause the coating to also dry slowly, or never dry completely.

Using old shellac can cause a problem, because it may never dry.

Using too little thinner in a coating material can result in peeling, flaking, cracking, or scaling.

Using too much thinner may result in uneven colors. This problem is more serious with alkyd- or oil-based materials than with water-based ones. Adding too much thinner to save coating material may result in premature aging or fading of a coating or make it unable to withstand normal cleaning.

Failure to agitate and mix organic coating materials properly before and during their application can cause color separation and streaking. Insufficient mixing that results in the pigments not being properly blended may cause alligatoring, checking, peeling, flaking, cracking, or scaling. An improperly mixed coating material may also not dry.

Combining incompatible components will often produce a coating material that is doomed to fail. The type of failure it exhibits will depend on which materials were mixed together. For example, using pigments that are incompatible with other ingredients can result in alligatoring, checking, peeling, flaking, cracking, or scaling.

Adding lampblack or another shading material to professional-quality paint materials to make the paint hide better with less paint can result in changes in color, or the paint's failing to adhere properly or aging prematurely.

Adding too much oil to paint can cause alligatoring or checking and may cause it to dry improperly. This condition can occur inadvertently if spraying equipment is used when there is grease or oil in the spray line.

Failing to apply organic coatings properly, including not using enough material, improperly applying the material, and not applying the correct number of coats can lead to failure. Examples include failing to cover the surface completely or applying coats that are too thin or too thick. Applying too-thick coats will cause surface film to form before the coating beneath has dried. At first the film will be smooth across the surface, but when the underlying material dries it will shrink, which will wrinkle the surface film. Coats that are too thick may also check or even alligator. Too thick a coat, especially if it is an undercoat, can also cause peeling, flaking, cracking, or scaling. Coating materials that are applied in too-thick coats may also run or sag and may not dry properly.

Applying a coat of an organic coating to a still-wet undercoat can cause the coating to dry slowly or wrinkle. It may also cause alligatoring, checking, blistering, peeling, flaking, cracking, or scaling.

Applying a hard finish coat over a relatively soft undercoat can cause a coating to wrinkle and may result in alligatoring, checking, or peeling of the top coat.

Failing to brush out a brush-applied organic coating can cause wrinkling, runs, or sags.

Improperly operating spraying equipment, and incorrectly applying a coating using it, can cause several types of problems. Too much air pressure may result in entrapping air in the applied coating, which can appear as visible bubbles beneath the surface. Water in an air line can dilute the coating material and cause it to cure improperly. Improperly adjusting a spraying device's pattern so that a sprayed coating material is applied with too much air pressure can cause a wrinkled surface. Failing to remove oil and grease from the lines can cause the problems mentioned earlier.

3.12.6 Failure of the Immediate Substrate

The failure of an underlying metal or its mechanical or chemical finish will ultimately result in the failure of an applied coating. If a substrate material should crack, for example, an applied organic coating will also crack. There are two basic types of failures experienced by metals: corrosion and fracture. Each may result from a number of different causes.

A major problem that may contribute to either corrosion or fracture, or both, is poor material selection. The failure usually happens because the wrong metal was selected for the particular item and location.

Another major reason for failure is that of not protecting susceptible metals from corrosion or fracture, either of which will always ultimately result in the failure of most finishes. Since mechanical finishes are a part of the metal, the effect on them is the same as the effect on the underlying metal. Many applied finishes, such as chromium plating, will usually be damaged immediately by fracture. The corrosion of an underlying metal that has an applied finish does not often occur unless there has been previous damage to the applied finish. When a metal cracks, most applied coatings will also crack, but there are exceptions. Some coatings, especially laminated ones, may withstand initial cracking and not show damage until the damage to the underlying metal is quite advanced.

Corrosion. Damage to metal doors and windows by corrosion is a major cause of failure. If the metal beneath a coating corrodes, the coating will fail, even if the metal does not fracture.

As generally defined, corrosion means the wearing away of a material, usually gradually. Here, though, we will broaden the definition to include the processes that erode a metal and those that change a useful metal into a useless chemical compound, thus destroying its desirable characteristics or properties.

Alloying a corrosion-resistant metal with metals that corrode readily helps the alloy resist corrosion. Thus, stainless steel is corrosion resistant, because it contains highly corrosion-resistant chromium. Most nonferrous metals resist corrosion, especially oxidation, to some extent, but no metal is truly corrosionproof.

Most of the corrosion we need to worry about when dealing with metal doors and windows affects only the surface of the metal. The types of corrosion that affect the interior structure of the metal (structural corrosion) may become a factor in some instances, however. Corrosion weakens a metal and causes pits and crevices in it. Corrosion occurring in a beginning fracture can contribute heavily to continuing fracture failure.

If left unprotected, most metals corrode rather rapidly. The types of corrosion that affect the metals used in doors and windows include oxidation, corrosion caused by weathering and exposure, chemical attack, and galvanic action, and structural corrosion.

Oxidation. The most common form of corrosion that damages steel is oxidation, which changes the steel to ferric or ferrous oxide. When either water vapor or liquid water contacts ferric or ferrous oxide, hydration occurs. The resulting formation may be one of several possible compounds, including hydrated ferric oxide, ferric oxyhydroxide, and hydroxide, all of which are commonly called rust. Rust may be yellow, bright orange, dark reddish brown, red, or black. The presence of salt, as in sea water or salt sea air, will enhance the process of oxidation, increasing its severity and the speed with which it occurs.

Nonferrous metals also oxidize, but the oxidation products that form on most of them are not destructive. The green alkaline oxide of copper that forms on bronze, for example, not only protects the metal from further corrosion but also imparts a desirable coloration.

The oxide that forms on aluminum will not necessarily protect the metal in severe environments, but under normal conditions it will prevent further oxidation. In addition, this coating is unobtrusive in appearance and only about one ten millionth of an inch thick. Anodizing, which is probably today's favorite finishing method for aluminum, amounts in principle to nothing more than increasing the thickness and effectiveness of aluminum's natural oxide film. In fact, the term *anodic oxidation* implies that the oxide film was created deliberately.

Weathering and Exposure. Wind-driven materials such as sand, dirt, and hail can erode the surface of metals. Ice crystals may also damage some surfaces. However, much of the damage that people assume to have been caused by wind-driven debris was actually caused by chemical attack.

Chemical Attack. Metals may be damaged by both liquid and airborne chemicals. They are attacked by acids, alkalies, carbonates, fluorides, and other chemicals used to erect and maintain buildings. The acids used to clean masonry are a particularly odious problem, as is acid rain. Mortar, plaster, concrete, and other wet alkaline materials are another problem. Also, the chlorides and sulfates often found in soil will severely attack metals.

Metals in industrial areas, where the air is likely to contain more pollutants than in other locations, tend to corrode more quickly than those in rural areas. They will also corrode more quickly in seacoast areas. In salt air, cast iron may corrode so badly that it collapses on itself. The effect is enhanced in southern climates where the sun is hotter.

The effect of acid rain and other pollutants is a major problem for most metals, but the damage is more pronounced on some materials than on others. Acid rain spread over a large area of copper, for example, forms a basic copper-sulfate patina that may not present the desired appearance but may actually protect the copper from further corrosion. On the other hand, copper may corrode completely if acid rain is permitted to accumulate and stand on it.

When there is doubt about whether the components of a specific material will be harmful to a particular metal, the best bet is to assume the worst and protect the metal. When corrosion has occurred and there is doubt about its cause, it is a good idea to verify whether a questionable adjacent material is at fault.

Metals can be protected from chemical attack by preventing them from contacting the offending material. One way to accomplish this is to separate the metal physically from the offending material. When separation is not practicable, the metal can be coated with a nonreactive material to protect it. Some metal surfaces that will be in contact with concrete or masonry, for example, can be covered with an alkali-resistant coating such as heavy-bodied aluminum paint. Care must be taken to ensure that the material used to separate the metal from an offending material does not itself harm the metal.

Galvanic Corrosion. Most metal corrosion results from an electrochemical process. Galvanic corrosion (electrolysis) happens when dissimilar metals are each contacted by an electrolyte such as water. The process is similar to that which takes place inside an automobile battery, where one metal acts as a cathode, the other as an anode. The presence of the electrolyte causes an electric current to flow between the two metals. The anode metal dissolves while hydrogen ions accumulate on the cathode metal.

The electrolyte can be rain water or just condensation. The effect is more severe when the dissimilar materials are close together, but even the wash-off from one material can corrode a dissimilar one. Water that has washed across copper, for example, may cause steel to corrode.

Some combinations of materials are more susceptible to galvanic corrosion than others. Galvanic tables have been developed to show which materials are likely to create galvanic corrosion. The propensity of two materials for galvanic action depends on where they fall in the galvanic table: the farther apart they are, the more severely they will corrode. For example, aluminum, which falls near one end of most tables, will corrode when it becomes wet while in contact with lead, brass, bronze, iron, or steel. When dissimilar materials are attacked by galvanic action it is the material that is highest in the galvanic table that will corrode.

Galvanic tables are available from many sources, but they do not always agree. The book in this series entitled *Repairing and Extending Nonstructural Metals* includes such a table (p. 95).

When trying to resolve a corrosion problem that persists after other possible causes have been eliminated, assume that the culprit is the adjacent material, regardless of what a published galvanic series table or any other source says.

There are two ways to prevent corrosion due to galvanic action. One is to avoid using dissimilar metals in close contact. When that is not possible, it is necessary to prevent an electrolyte from contacting both. A heavy-bodied bituminous paint is sometimes used to coat one surface. Aluminum and copper alloys may be painted with zinc-chromate paint, for instance, to prevent galvanic corrosion. Metals can also be separated by using moisture-resistant building felt or plastic sheets or other means of separation. Aluminum should not be coated with tar or creosote, however, because they contain acids that are harmful to the aluminum.

Structural Corrosion. The type of corrosion that takes place within a metal, either between its metal crystals or in pores created during the forming or fabricating process, is more insidious than the corrosion types previously discussed, because it often cannot be seen until it is too far advanced to do anything about. Failure due to this type of corrosion is usually but not always accompanied by applied stress, particularly fatigue stress.

The metals used in doors and windows are seldom subject to structural corrosion. Stress corrosion, however, which can occur in any metal, is more common in the copper alloys and some aluminum alloys. Pure metals seldom have this problem. The beginning of stress corrosion occurs during cold fabrication. In the copper alloys, annealing will often help prevent failure from stress corrosion, because it helps relieve some of the internal stresses that contribute to stress-corrosion failures. In aluminum, the previously discussed process known as age hardening will sometimes set up corrosion lenses within the metal. Correct heat treatment can remove this problem, and protecting the metal with a corrosion-resistant coating can prevent its getting worse.

Errors That Lead to Corrosion. While all forms of corrosion are not directly related to the presence of water many are, and other types of corrosion progress more rapidly or become more severe in the presence of water. Water may contact unprotected metal because of the types of problems discussed earlier in section 2.6. While such problems may not be the most probable causes of corrosion, they can be more serious and costly to fix than many other types of problems. So, the possibility that they may be responsible for such corrosion should be investigated.

The other building-element problems discussed in section 2.6 should be ruled out or, if found to be at fault be rectified, before repair of corroded metal is attempted, or a new door or window component is installed. It will do no good to repair existing or install new door or window components when a failed, uncorrected, or unaccounted-for problem of the types discussed in section 2.6 exists, because the repaired or new installation will also fail. After the problems discussed in section 2.6 have been investigated and found to be not present, or if found are repaired, the next step is to discover any additional causes for the failure and correct them.

There are several other types of errors that, if committed, will almost certainly lead to corrosion of one type or another. The most prevalent one is selecting the wrong material for the installation. Using an easily corrodible metal such as steel in a highly corrosive atmosphere, such as inside a structure that houses a swimming pool is an example. Even protected steel may corrode when corrosion products are present.

The second most-committed error that causes metal to corrode is failing to protect it. Any one of the following may be at fault:

Failing to coat or galvanize ferrous metals so that they do not rust.

Failing to protect ferrous and nonferrous metals from chemical attack.

Assuming that galvanized steel is immune from corrosion and thus failing to protect it. Permitting galvanized steel to come in contact with reactive materials will cause corrosion. The soluble sulfites in cinders, for example, will rapidly corrode zinc. The zinc coating on galvanized steel that will be in contact with people, animals, or moving equipment or machinery will wear off and the metal beneath will corrode.

Failing to separate ferrous and nonferrous metals from mortar, plaster, concrete, and other harmful materials that might cause corrosion.

Failing to separate a metal from dissimilar metals, to prevent galvanic corrosion.

Failing to renew and repair protection when damage occurs.

Fracture. A second major type of failure that occurs in metals is fracture. Eventually, even many corrosion failures show up as cracks. There are basically four types of fracture: stress-corrosion fracture, ductile fracture, brittle fracture, and fatigue fracture. Fracture is not a common cause of failure in metal doors or windows, but it may occur, especially in glazed curtain-wall construction.

While a break in the metal of a door or frame, such as one caused by a severe impact, is technically a fracture, for clarity in this book such injuries are called breaks and tears.

When a fracture occurs in metal, corrosion is often evident. It is sometimes difficult to determine if the fracture permitted the corrosion to begin or the corrosion caused a fracture to form. Usually, the two are linked in some way. Fracture may, however, result from many causes unrelated to corrosion. Fracture of a metal, regardless of its cause, will always lead to the failure of an applied organic coating.

Stress-Corrosion Fracture. Fracture failure may be directly related to corrosion, which starts the failure by enlarging existing or causing new pits and crevices in metal. Metal, especially cold-worked metal, has internal stresses created in the forming process and may have local inhomogeneities that result in anodic–cathodic adjacent areas. Those stresses tend to congregate at the locations of pits and crevices, and the combination results in cracks in the metal, beginning there. Corrosion-weakened areas and the resulting cracks can also occur at inhomogeneities.

Ductile Fracture. When a ductile metal is loaded above its elastic limit, it fails. The excessive stress may be either tensile (in tension), compressive (in compression), or shear (tangential to the section).

Brittle Fracture. When a metal with limited ductility, such as cast iron, is subjected to a sharp blow, cracks develop in the metal, beginning at the site of the initial stress.

Fatigue Fracture. Cyclic stressing over a long period of time will cause metals to fracture. Cracks occur, usually starting at the location of imperfections in the metal or at sites of corrosion.

Errors That Can Lead to Fracture. Since doors and windows are supported by adjacent construction, the cause of fracture failure in them may be traced to a failure in their supports. Potential support failures are discussed in Chapter 2. The causes of the material and finish failures discussed there may not be the most probable causes of such failures, but they are more serious and costly to fix than those discussed in this chapter. Consequently, the possibility that one of them may be responsible for a material or finish failure should be investigated.

The causes discussed in Chapter 2 should be ruled out or rectified if found to be at fault before the repair of materials or finishes is attempted or a new metal item is installed. It will do no good to repair an existing door or window or install a new one when a failed, uncorrected, or unaccounted-for problem of the types discussed in Chapter 2 exists. The new item will also fail. After the problems discussed in Chapter 2 have been investigated and found to be not present or if found have been repaired, the next step is to discover any additional causes for the failure and correct them.

The most frequently committed error leading to fracture in metals in doors and windows is selecting the wrong material for the installation. Using a ductile material where it will be loaded beyond its elastic limit is an example. A less obvious case is that of using ferrous metals, especially sheets and other thin sections, in a highly corrosive atmosphere such as that inside a swimming pool equipment room, where the air is often laden with chlorine and other chemicals that are caustic to ferrous metal and where even liquid chemicals may come into contact with it. The problem is exacerbated when the metal is loaded cyclically so that it bends back and forth. Even seemingly protected metals may corrode in such an environment. Finishes are seldom perfect—even ones that completely cover the metal may become scratched or chipped, leaving the bare metal open to attack. Corrosion along fracture lines contributes to fracture failure, and after fracture has begun, corrosion hastens its progress.

A much less frequent cause of fracture is failing to manufacture the metal properly and in accordance with recognized standards. The designer has a right to expect that a metal material will conform to the standards claimed for it and have the properties required by that standard. Improper heat treatments may leave the metal too brittle, for example. An improperly made casting may contain imperfections or impurities that cause it to fracture. A variation from the correct proportion of materials in a steel alloy might produce steel with properties quite different from those expected.

Subjecting a metal to stresses it is incapable of safely supporting and thus in effect violating the designer's intent for the item fabricated from the metal may cause the metal to buckle and fracture. Loading a metal item in ways not intended in its design can result in imposing stresses on the metal that exceed its elastic limit, for example, resulting in permanent deformation. Failing to protect a metal from imposed stresses that the material cannot safely support is a similar problem. The effect is the same, whether the excess stress is imposed deliberately or inadvertently.

Underlying Finish Failure. Delamination or other failure of an underlying mechanical or chemical finish will usually cause an applied organic coating to fail. This failure is a problem regardless of the type of the underlying finish, but it is more likely to happen when an underlying metallic coating fails. If the aluminum coating in an aluminized steel product delaminates from the steel, for example, any secondary applied finish will also fail.

3.12.7 Failure to Protect Materials and Finishes

Metal and wood materials and their finishes must be protected before, during, and after installation. Errors include failing to protect them from staining, denting, scratching, or other damage by other construction materials or procedures, and permitting an item to be abused before or during installation or after it has been installed.

Contaminants, including airborne dust and other solid particles, that are permitted to fall into chemicals during finishing or to interfere with mechanical finishing operations, or ones that fall onto wet, newly finished surfaces, will affect the final appearance of a finish and may even cause it to fail.

Failing to protect mechanical and chemical finishes and applied coatings from staining or marring during shipment or from contact with other construction materials and during subsequent construction activities may result in damage to them. Permitting contaminants, including airborne dust such as that from broom cleaning to fall onto wet, newly painted surfaces is an example.

Permitting rain to fall on an organic coating while it is still wet can pit the surface and may remove the gloss from oil- or alkyd-based paint or transparent coatings. It may even wash off the coating partially, or even completely, from the wood.

Permitting the temperature in a space where an organic coating has been applied to fall below 55 degrees Fahrenheit or rise above 95 degrees Fahrenheit will contribute to finish failure. Temperatures that are too high or too low cause expansion and contraction in the coating and substrates. Extremes of temperature may cause the organic coating to crack, especially if it is old, is naturally hard, or for some other reason is not elastic enough to withstand the stresses involved.

Permitting the deterioration of adjacent construction, such as flashing, to occur to the extent that a coated metal or wood surface becomes wet will promote failure.

Permitting abuse before or during application or after a finish has been applied may be the largest single cause of damaged finishes. Stacking galvanized steel items, for example, in such a way that moisture is allowed to accumulate on them will lead to their early corrosion. Allowing contact by vehicles is another major problem.

Permitting adjacent materials to be cleaned with cleaning materials that will harm the metal or wood should not be done. An often-occurring problem is that of cleaning adjacent masonry using acid cleaners. If this is permitted, damage to the metal from runoff or splatter is almost inevitable, even when the person cleaning the masonry attempts to take precautions to prevent contact. Acid-type cleaning materials should never be permitted in the vicinity of metal or wood finishes.

3.12.8 Failure to Properly Maintain Applied Finishes

Failure to remove grease, dirt, stains, and other contaminants regularly from mechanical or chemical finishes or coatings can result in permanent stains.

Improper cleaning of organic coatings may damage them. They may, for example, be severely affected by sandpaper, steel wool, and abrasive or caustic cleaners.

Failing to keep vines and other vegetation from growing close to or on door frames and windows (Fig. 3-7) can lead to damaged coatings and eventual damage to the substrate. Vegetation close to a building can shade it and make an environment conducive to mildew and moss growth. Vines can penetrate the surface of wood and grow into joints in both metal and wood, carrying moisture with them and permitting free water to enter as

Figure 3-7 The vines in this photograph must be removed immediately if damage to the window is to be prevented.

well. Such water can aid the growth of fungi and bacteria in wood and eventually rot it away. Severe corrosion can result when water reaches metal surfaces. In addition, coatings may blister or peel, and mildew and moss may grow on wet surfaces. Even in mild cases, water in a wood substrate can set up a chemical reaction between the water and the natural extractives in the wood and stain the coating.

Damage should be promptly repaired, to prevent more serious problems later.

Improperly cleaning anodized aluminum can damage its finish. The cleaners used must be neutral, because both alkaline and acidic materials will damage coatings. Ordinary wax cleaners and mild soaps and detergents are the strongest materials that should be used without consulting with the metal and finish producers (see section 3.13.3).

Failing to reoil statuary bronze will result in a deterioration of the appearance of the finish (see section 3.13.3).

Metal finishes may be damaged severely if they are cleaned using sandpaper, steel wool, or abrasive or caustic cleaners.

The improper or premature cleaning of copper alloys can lead to unwarranted damage to them. Patinas and statuary bronze finishes may be removed or at least severely damaged by improper cleaning.

Applying unnecessary coatings in an attempt to protect copper alloys often damages them in the process.

3.12.9 Natural Aging

All metal and wood finishes age and lose their properties over time. Product manufacturers, finish producers, or associations representing them are the best sources of data about the expected life of a finish. It should be kept in mind, however, that the manufacturer's or finish producer's projections may assume the best of conditions.

Failing to recoat or refinish wood or metal at proper intervals may contribute to premature aging of the finish and eventually lead to its complete failure (Fig. 3-8). Most chemical finishes and both inorganic and organic coatings that are exposed to the sun's ultraviolet rays and weathering will fail faster than those used in other locations. They will all fail eventually if not maintained and recoated or refinished at the proper intervals. Exterior wood with a transparent finish is particularly susceptible to damage from sunshine because its transparency permits the sun's rays to reach the wood. In addition, ultraviolet rays may attack the finish itself. The result can be delamination of the finish from the surface and flaking and peeling. Exterior transparent finishes containing alkyd resins may fail in as little as six months; urethane resins may last eighteen months. Even the best materials,

Figure 3-8 Years of neglect have led to the condition shown in this photo.

the phenolic resins with ultraviolet absorbers, may last only as long as two years.

One form of failure is loss of elasticity. An organic coating that has lost its elasticity becomes hard and no longer able to respond to the expansion and contraction of the material to which it is applied. The result is crazing and eventual alligatoring, flaking, or peeling. Laminated coatings are also susceptible to a loss of elasticity, and may crack or delaminate as a result.

Unprotected wood exposed to weather will turn dark over time as its water-soluble impurities bleed out. The appearance of bleeding on transparent-finished exterior wood surfaces indicates that the finish has failed and water has reached the substrate. The worst woods for bleeding are redwood and cedar.

Another problem is mildew growth due to a failed finish that permits water to reach the substrate.

Natural erosion is yet another problem. Weathering will slowly wear away finishes until their appearance alters drastically and they can no longer protect the substrates properly. For this reason anodic coatings on aluminum must be carefully selected so that the surfaces that will be exposed are heavier and more resistant than those to be used in less-exposed locations.

3.13 Cleaning and Repairing of Woods and Metals and Their Finishes

Existing woods and metals and their finishes that are damaged or in an unsightly condition but still usable may often be repaired and restored to a usable condition.

When working with existing materials and finishes, it is best when possible to use materials and methods that match those of the original installation.

3.13.1 General Requirements

This section contains some suggestions for repairing existing woods and metals and their finishes. Because these suggestions are meant to apply in many situations, they might not apply to a specific case. In addition, there are many possible cases that are not specifically covered here. When a condition arises in the field that is not addressed here, advice should be sought from the additional data sources mentioned in this book. Often, consulting with the manufacturer of the item being repaired will help, though sometimes it is necessary to obtain professional help (see section 1.4). Under no circumstances should the specific recommendations in this book be followed without careful investigation and the application of professional expertise and judgment.

Before an attempt is made to repair metal or wood or their finishes, the manufacturer's and finisher's recommendations for making the repairs should be obtained. It is necessary to be sure that their recommended precautions against materials and methods that may be detrimental to the materials and finishes requiring repair are followed. Preventing future failure requires that the repairs not be undertaken without careful, knowledgeable investigation and the application of professional expertise and judgment. The repairs should be carried out by experienced workers under competent supervision.

3.13.2 Repairing Specific Types of Damage

In addition to the general requirements for repairs already discussed, the following recommendations apply to specific problems.

Corroded or Fractured Metals. Corroded metals can often be cleaned and refinished. Metals corroded beyond repair must of course be removed and new materials provided. The new materials should be properly protected to prevent their failing for the same reason the replaced items failed.

Fractured metals are often not directly repairable, because fractures often involve more damage than is apparent. The grain structure of the material may be damaged to a considerable distance beyond the actual crack. Sometimes a strengthening plate may be added to support a fractured metal item, but such repairs are often unsatisfactory. Usually, the fractured component must be removed and a new one substituted. The new item may have to be of a different material or thickness to prevent a reoccurrence of the fracture.

Before any metal item is repaired, the advice of its manufacturer should be sought and followed. Where there is doubt about the manufacturer's advice, professional help should be sought (see section 1.4). The prevention of future failure, particularly when the failure is one involving fracture, requires that the repairs not be undertaken without careful, knowledgeable investigation and the application of professional expertise and judgment. The repair should be carried out by experienced workers under competent supervision.

The method of repair chosen will depend on the type and extent of the damage. The source of corrosion must first be removed and the corroded area sanded or the corroded material otherwise removed completely. Such cleaning and other needed repairs to the metal must be done before the finish is repaired. Cleaning and repairing should be done, substrates be prepared, and repairs made strictly in accordance with the material manufacturer's recommendations. Repairs should be made only by experienced workers.

Damaged Finish. The method and extent of repairs needed depend on the type of material and the type and extent of the damage. Repairs needed to substrate material must be done before finishes are repaired. Cleaning and repairing should be done, substrates prepared, and repairs made strictly in accordance with the material manufacturer's recommendations. Repairs should be made only by experienced workers.

Damaged finishes on members that are not exposed should be repaired using the same type material used in the original finish. Painted finishes should be repainted. Galvanized surfaces should be recoated using galvanizing repair paint.

Scratched and abraded areas in opaque organic coatings may be touched up, using any paint type that is recommended for such use and compatible with the existing finish. Preparation should be done as recommended by the paint's manufacturer.

Materials used in making touch-ups must be compatible with the existing finish. If there is doubt about the compatibility of materials, it is best to contact the coating's manufacturer for advice.

Except for minor touch up of scratches, it is usually best to refinish an entire door or frame or window, rather than just the damaged area, so that the repairs will be less apparent (Fig. 3-9). Refer to section 3.15 for a discussion about recoating existing surfaces.

Figure 3-9 A touch-up is definitely not the solution to the peeling paint problem shown in this photo. The window should be completely refinished.

3.13 Cleaning and Repairing of Woods and Metals and Their Finishes 113

The repair of factory finishes should be made in accordance with the manufacturer's and finisher's recommendations. Techniques that work well with one finish may be harmful to a similar one.

Finishes to be repaired should first be prepared as discussed in section 3.15.3. The requirements suggested there for major surfaces to be refinished generally apply.

Transparent coatings on aluminum can usually be repaired using the same coating material that was originally used, but the repairs may be visible. Transparent coatings on aluminum are usually, but not always, lacquer. Transparent coatings on other metals may not be easily repairable.

Most paint, laminates, and high-performance coatings can be touched up using the materials recommended, and often produced, by the manufacturer of the coating.

The repair of mechanical, chemical, most inorganic and some organic finishes with the finished item left in place is usually restricted to minor damage only. It is difficult, and sometimes impossible, to repair mechanical or inorganic finishes in the field in the true sense of the word. Small patches and repairs are usually obvious and are often not acceptable for that reason. Since most types of finishes used on stainless steel, aluminum, and copper alloys can only be applied in a shop, not with the metal installed in a building, major damage often requires that the metal item with the damaged finish be removed and refinished in the shop. Alternatively, of course, the damaged finish may be removed in the field, or properly treated there and a new finish applied. Such new finishes are usually paint.

The repair of mechanical or chemical finishes, anodized aluminum, plastic coatings, fluoropolymer finishes, and baking-finish enamels should be made in accordance with the manufacturer's and finisher's recommendations. Similar coatings may not respond alike to the same refinishing method; techniques that work well with one finish may be harmful to another, similar one. In the absence of manufacturer's or finisher's recommendations, the same advice should be sought from the association representing the metals' manufacturers or finishers. Repairs should then be made in strict accordance with the recommendations received. Only experienced workers should be used to repair mechanical or inorganic metal finishes.

Areas where repairs are to be made should be inspected carefully to verify that the existing materials are as expected and that the repair can be satisfactorily made. This inspection should preferably be made by a representative of the metal's manufacturer, finisher, or both.

When a metal's mechanical or inorganic finish is in need of repair, it is usually necessary first to remove applied wax, oil, and organic coatings. An anodized aluminum item, for example, may have been coated with a lacquer that must be removed before the aluminum or its anodizing can be repaired. Such coatings are often reapplied after the repairs have been completed. Waxes and oil coatings are discussed in sections 3.13.3 and 3.15.

The repair of procelain enamel is sometimes possible using a synthetic porcelain glaze developed specifically for that purpose, but this is a job for a professional. Porcelain enamel can also be reenameled, but it is often more economical simply to remove a damaged porcelain-enamel panel and provide a new one. See section 3.16 for sources of additional information.

Dents and Holes in Metals. If the damage is not too severe, it may be possible to remove dents and fill holes in metals. A slight dent in a thin metal can often simply be pushed back into place if the area is accessible from the opposite side. When the back side is accessible but the dent resists hand pressure, it can sometimes be beat back into shape with a leather or wooden mallet. Holding a sandbag against the finished side of the area being bent back into shape will usually help make the rebent metal smoother. Sometimes, dents in thin metal can be pulled back into place using simple suction tools such as a plumber's helper.

Small dents in heavier metals can often be filled with epoxy or another filler, then ground down flush with the surrounding surface. Even large dents in thick metals can sometimes be filled with epoxy. It may be possible to pull out large dents in thick metals by inserting screws into the metal and pulling on the screws.

Holes in metal can be filled by applying an epoxy mixture against a screen wire held fast to the back side of the metal.

Most metal repairs, however, especially those involving bent or twisted metal, are not do-it-yourself projects. With such problems expert knowledge and equipment are needed. Often such damage cannot be repaired until the metal item has been removed. It is frequently necessary to send the item to a shop for repairs to be made. Rebending metal may set up stresses that can only be relieved by heat treatment of some sort, which is usually impracticable to do in the field. The necessity of removing a door or window to a shop for repairs sometimes makes the process so expensive that it is cheaper to discard the damaged item and provide a new one.

Dented or Punctured Wood. If the damage is not too severe, dents in wood can usually be repaired. After the finish is removed, dents can be filled using plastic wood, wood filler, or putty. Slight dents can be removed using steam.

Small drill holes and other round holes in painted wood can be filled with plastic wood. Such holes in wood with a transparent finish can be filled with dowels or plugs that match the adjacent wood. Large holes in, and missing portions of, painted wood doors and frames can be satisfactorily repaired with wood patches and fillers. Large holes in doors with a transparent finish probably cannot be repaired without damage that would be objectionable to the appearance of the door. Usually, it is best to remove severely damaged frame members and trim and replace them with new components (Fig. 3-10).

3.13.3 Cleaning Existing Wood and Metal Finishes That Do Not Need Repair

All types of woods and metals and their finishes must be cleaned and maintained periodically if their failure is to be prevented. Black marks and other soiling, mortar, sealants, stains, dirt, dust, grease, oil, efflorescence, salts, unwanted paint, and other unsightly contaminants should be removed completely. The materials and methods suggested in this section are general in nature and do not necessarily apply in every situation. The soaps, detergents, cleaning compounds, and methods used should always be com-

Figure 3-10 The rotted part of the frame shown in this photo was cut away and a new piece of wood inserted because the door was merely an entrance to a warehouse and not accessible to the public. This solution was unsightly, however, and would not have been acceptable in a more visible location.

patible with the surfaces to be cleaned and with adjacent surfaces and not damage either in any way. Abrasives and caustic or acidic cleaning agents should not be used. The materials and methods used should be those recommended by the material's producer, the fabricator, finisher, and manufacturer of the finishing material, and applicable industry standards. In no case should the suggestions made here supersede their recommendations. Finishes damaged from using incorrect products or improper cleaning agents should be refinished. When a condition arises that is not addressed here, advice should be sought from the additional sources of data mentioned in this book. Often, consultation with the manufacturer of the materials being cleaned will help. Sometimes it is necessary to obtain professional help (see section 1.4). Under no circumstances should the specific recommendations in this book be followed without careful investigation and the application of professional expertise and judgment.

In some cities there are organizations established specifically to clean buildings. When a metal to be cleaned consists of major surfaces on an entire building and such an organization is available, it might be advisable to at least talk to them about it. The organization's references should be carefully checked, of course, and the methods they propose to use should be ascertained and verified against the recommendations of the materials' producers.

Materials' producers publish their own recommendations about cleaning the products they manufacture. The Kawneer Company, for example, has a publication called "Architectural Finishes" that includes a description of the various finishes available on Kawneer products with their recommendations for maintaining and cleaning those finishes.

There are also industry standards and producers' associations' recommendations for cleaning and maintaining metals. AAMA 609.1-1985, "Voluntary Guide Specifications for Cleaning and Maintenance of Architectural Anodized Aluminum" is the standard for cleaning anodized aluminum. AAMA 610.1-1979, "Voluntary Guide Specification for Cleaning and Maintenance of Painted Aluminum Extrusions and Curtain Wall Panels" is the standard for cleaning painted aluminum. The Copper Development Association's (CDA) publication *Copper, Brass, Bronze Design Handbook, Architectural Applications* contains advice about maintaining copper alloys. The CDA publication "Taking Care of the Metal in Your Building" also contains advice about maintaining copper alloys.

Before an attempt is made to clean existing metal or wood finishes, their manufacturer's and finisher's recommendations for cleaning should be available and referenced. It is necessary to be sure that the manufacturer's recommended precautions against materials and methods that may be detrimental to finishes are followed.

To prevent streaking, metal should not be cleaned when there is bright sunlight on its surface. Cleaning in the shade or on cloudy days is best. Cleaning should always proceed from top to bottom so that runoff is removed during subsequent cleaning.

Dirt, soot, pollution, cobwebs, insect cocoons, and the like can often be removed from exterior surfaces using a water spray followed by scrubbing with a mild household detergent mixed with water. The cleaned surfaces should be rinsed thoroughly with clean water and permitted to dry.

Areas of soiling can be washed from most organic coatings on metal or wood with a mild soap and a damp cloth or sponge. Some types of finishes can withstand more scrubbing than others without damage, but each should first be tested in a small, inconspicuous area. It is not always prudent to accept blindly the claims of a manufacturer. Detergents, abrasives, compounds containing acids or caustic cleaning agents, and abrasive devices should not be used.

Surfaces with mechanical, chemical, inorganic, or transparent organic finishes offer different cleaning problems than those with opaque organic coatings. Such finishes should be cleaned only according to their manufacturer's instructions. For example, stainless steel, aluminum, and copper alloys should first be cleaned with clear water only. If that does not remove foreign deposits or soiling, they may then be washed with clear water and a mild soap. When such mild methods do not work, these materials may be washed with mineral spirits, then with mild soap and clean water, and finally rinsed with water and permitted to dry. Detergents, abrasives, and caustic or acidic cleaning agents should not be used. If these methods still do not satisfactorily clean the surfaces, it may be necessary to use more drastic cleaning methods. No drastic method should be used, however, until it has been reviewed by the door or window manufacturer and finisher or by a nationally recognized association representing them. Drastic methods should be used only by professionals.

Particularly difficult to remove deposits on stainless steel can be removed using phosphoric- or oxalic-acid compounds. The directions of the manufacturers of the cleaning agent and metal material should be followed strictly when using such materials.

With the concurrence of the metal and finish producers, solvent- or emulsion-based cleaners may be used on aluminum to remove oil and grease. Again, when the manufacturer and finisher agree, abrasive cleaners may be used on aluminum to remove materials that the milder methods do not remove. Such cleaners often contain oil, wax, and silicones. The cleaning agents in them may be soaps, acids, alkalies, and abrasives. Professionals may also use other cleaning methods on severely damaged or corroded aluminum finishes including etching cleaners, steel-wool polishing, power-driven wire-brush cleaners, and steam cleaners. Most of these methods remove all or part of the existing finish, so they should be used only when absolutely necessary. Finishes damaged by these cleaning operations must be renewed. The cleaning materials must also be removed. Because steel particles left on aluminum will rust and cause stains, it is best to use stainless steel when steel wool is appropriate. In some cases a finish can be renewed in the field, but damaged anodic surfaces can be reanodized only in an anodizing plant.

Aluminum surfaces are sometimes waxed, to provide weather protection and make them easier to clean. Such wax coatings tend to wear away and turn yellow with time and must be renewed periodically. It is often necessary to remove old wax buildup before rewaxing. There are also some specialty sealers on the market that are used to seal anodized surfaces, to restore their original luster and make them easier to clean. These materials are usually used on older surfaces that have begun to lose their luster or become porous.

Clear anodized aluminum is often protected with temporary tape or strippable coatings or by a permanent transparent organic coating, usually lacquer. Temporary coatings and tape should be removed immediately after the danger of damage has passed.

Copper alloys are frequently oiled or waxed to enhance their appearance and help protect them by excluding moisture. The oil or wax will dissipate over time and must be renewed periodically. Linseed oil has been traditionally used to coat copper alloys. Unfortunately, boiled linseed oil contains varnish, which weathers unevenly and may yellow or become mottled in appearance. Raw linseed oil dries slowly and tends to pick up dust and other contaminants during the drying process. It also reacts with the copper, which partially defeats the purpose of oiling. Lemon, lemon grass, castor, and other oils are also used, with differing degrees of success. The Copper Development Association's current literature suggests that high-grade paraffin oils may be the best materials to use in many situations.

The types of waxes used on copper alloys include high-quality commercial paste waxes and mixtures of Carnauba wax or beeswax and wood turpentine. These materials are used mostly on architectural components that are accessible to people. Bronze sculptures displayed where people can touch or climb on them are often waxed, for example, as are handrails and similar components.

Bronze is sometimes cleaned using a process called glass-bead peening, which is similar to other abrasive-blasting procedures except that the abrasive used is small glass beads.

Anyone responsible for maintaining copper-alloy surfaces should contact the Copper Development Association at the address listed in the Data Sources Appendix for recommendations about the particular situation and copper alloy involved.

Except for simple washing with mild soap and water, porcelain enamel cleaning should be done only in accordance with the enamel producer's recommendations (see section 3.1).

Surfaces adjacent to those being cleaned should be protected. If they become soiled during the cleaning process, they too should be cleaned, as recommended by their manufacturers or by the associations representing their manufacturers. Cleaning should not harm the surfaces in any way.

3.14 Finishes That Cannot Be Refinished in the Field

Items with baking-finish coatings so severely damaged that they must be recoated are usually refinished in place, using paint. It is possible to return such items to the shop for refinishing, but the procedure is so expensive that it is seldom done, and the methods used are so specialized and depend on so many variables, they are not usually subject to control by building owners, architects, engineers, or general building contractors.

Removal to the factory or shop is necessary when a mechanical or chemical finish on stainless steel needs refinishing, a color anodic coating on aluminum needs reanodizing, or a patina or statuary bronze finish on a copper alloy must be refinished. None of those finishes can be refinished in the field. Some inorganic coatings, and a few organic ones, cannot be refinished in the field, either, at least not with the same material. Some transparent finishes are difficult to remove without damaging their underlying mechanical or chemical finishes. They must often either be just patched, which may prove unsightly, or the entire item must be removed and shipped to a shop for refinishing. When there is any doubt about whether needed refinishing can be done in the field, the manufacturer's advice should be obtained. Shop or factory refinishing must be done under the supervision of someone knowledgeable and expert in such work. When a refinished item is returned to the owner's building, it should have all the characteristics of a new product.

Since removal and return to the factory is an expensive solution to finish failures, such problems are sometimes solved by removing old protective coatings and painting the items in place. This option is seldom used, however, on copper alloy or extruded aluminum items.

3.15 Refinishing Existing Wood and Metal in the Field

Organic coatings are used extensively to refinish wood and metal in the field, even when an existing finish is not an organic one. Transparent coatings on wood and metal are frequently refinished using the same or a compatible transparent coating material. Opaque coatings on wood or metal that have been damaged beyond the range of normal repair are often field painted. New paint may be applied over existing paint, baking finish coatings, and even high-performance coatings when high-performance repair materials are not available. Paint is also sometimes applied where transparent finishes were originally used and over metals that have a mill or other mechanical finish or an inorganic coating such as galvanizing or another metallic coating. Paint may also be applied over copper-alloy doors and frames, but doing so is unusual.

Existing wood and metal finishes can be damaged in many ways and at many levels, from minor scratches to complete delamination of an applied finish. Unfortunately, while there is no easy way to determine the exact cause of some finish failures, others will be obvious at a glance. A stain from a water leak will be obvious, for example, but a cracked applied finish may have resulted from causes as diverse as bent metal to errors in applying the finish.

Discolorations may either be stains or simply the result of selecting the wrong finish. Some apparent discolorations may not be discolorations at all but rather the result of using the same finish on different alloys of the same metal.

Before an attempt is made to clean or repair an existing wood or metal finish that is to receive a new organic coating, the existing materials manufacturers' brochures for products, installation details, and recommendations for cleaning and repairing should be studied. It is necessary to follow the manufacturers' recommended precautions against materials and methods that may be detrimental to the materials. It is impossible to emphasize strongly enough the need to have access to such data when existing coatings are of the baking finish or high-performance types or are transparent coatings over decorative finishes, such as anodized aluminum or copper alloys.

This section contains some suggestions for determining the causes of some types of failures (see, for example, section 3.15.3) and for applying new coatings on existing wood and metal surfaces. Because these suggestions are meant to apply to many situations, they may not apply to a specific case. In addition, there are many possible cases that are not specifically covered here. When a condition arises in the field that is not addressed here, advice should be sought from the additional data sources mentioned in this book. Often, consulting with the manufacturer of the materials being installed will help. Sometimes it may be necessary to obtain professional help (see section 1.4). Under no circumstances should these recommendations be followed without careful investigation and the application of professional expertise and judgment.

The discussion in section 3.11 about new organic coating materials and their installations on new surfaces, also apply generally to organic coatings applied on existing surfaces. There are, however, a few additional considerations to address when new material is applied on an existing surface.

Existing finishes should be removed, substrates prepared, and materials applied in accordance with the organic coating materials manufacturer's recommendations. Only experienced workers should be used to prepare for and apply organic coatings.

Adequate ventilation should be provided while work associated with preparing and applying organic coating materials is in progress. Providing ventilation may be somewhat more difficult when operating in an existing building, especially where occupants are involved. The organic coating materials manufacturer's recommended safety precautions should be followed.

The moisture content of the substrates should be within the range recommended by the organic coating materials manufacturer, especially if excess moisture has contributed to the failure.

Areas where repairs will be made should be inspected carefully, to verify that existing materials that should be removed have been and that the substrates and structure are as expected and are not damaged. Sometimes substrate or structure materials, systems, or conditions are encountered that differ considerably from those expected. Sometimes unexpected damage is discovered. Both damage that was previously known and damage found later should be repaired before new organic coating materials are applied. Substrates damaged during the removal of an organic coating should also be repaired before a new coating is applied.

3.15.1 Materials and Manufacturers

Organic coating materials should be as listed in section 3.11, and the discussion in section 3.11.2 also applies here.

In addition to having materials be compatible with each other within each organic coating system, as is necessary in new buildings, the materials for use on existing finished surfaces must be compatible with existing materials that will remain in place. Tests should be conducted in the building to ensure compatibility. Incompatible materials should not be used. For example, applying water-based paints over solvent-based ones can create serious problems. The water-based paint may lift the oil-based one from the substrate or refuse to bond with it. Paint applied before 1950 is likely to have an oil base, since water-based paints were unheard of before then. Even today, except in residences, most paint applied on metal is either oil- or alkyd-based.

Some materials are limited in their usability for repainting or refinishing existing painted or finished surfaces. Some transparent finishes containing urethane resins, for example, are not applicable over other finishes or even over a previous coating of the same material without extensive preparation. Lacquer cannot be used over solvent-based paints, because it will act as a paint remover.

Until relatively recently, fluorocarbon polymers were difficult to repair successfully in the field, but now there is at least one air-drying, fluorocarbon polymer resin-based repair material available.

Air-drying materials are available to repair or recoat most transparent finishes, but such repairs may be apparent. Some transparent coating materials dissolve themselves even after they have dried, making small repairs easy to accomplish. Large repairs, however, may be difficult, and may force the complete removal of an existing transparent coating before a new one can be successfully applied.

3.15.2 Organic Coating Systems

The discussion in section 3.11.5 applies also to systems to be used in existing buildings, but with several differences. The materials themselves are basically the same, but some coats may be eliminated when the existing coating is sound. It is also necessary to ensure that the selected new organic coating systems will be appropriate to recoat every existing surface under each condition that will be encountered.

3.15.3 Preparation for Applying Organic Coatings

The discussion in section 3.11.10 applies as well to the preparation of existing surfaces. This section contains additional requirements that are applicable when the surfaces to be coated are not new.

Areas to be recoated and the conditions under which a coating will be applied should be inspected carefully. It is not unusual to find conditions other than those expected. Unsatisfactory conditions should be corrected before work related to organic coatings is begun.

Before a decision can be made about whether to repair a damaged surface or completely recoat it, it is necessary to inspect the surface and probably perform some tests. Some painting contractors are able to conduct such examinations and tests but many are not, especially when the coating is a baking finish or a transparent coating over aluminum or a copper alloy. Most national paint manufacturers employ personnel trained and equipped to make inspections and tests and to recommend methods for dealing with paint failures. Some of them are also knowledgeable about failures in baking-finish organic coatings. But, it will probably be necessary to contact the metal producer for assistance with a transparent coating over aluminum or a copper alloy.

Tests that may be required include those for compatibility, existing-coating thickness, and moisture content. The results will often point directly to the cause of a failure and can also help prevent further failures caused by misdiagnosis or using the wrong methods to make repairs or recoat failed surfaces. Examination may show, for example, whether just a surface coat is peeling or the entire thickness is affected. Microscopic examination may be necessary to determine the number of coats already applied. Chemical testing may be needed to determine the type of organic coating used in the existing coats.

Even before tests are made, it may be possible to determine the cause of some organic coating failures. Where paint has peeled or chipped, for example, a piece of paint from the failed surface or the substrate from which the paint peeled can be examined. Slick surfaces might mean that the substrate was too smooth and must be sanded or otherwise roughened before new paint will adhere properly. The back of the paint chip may also be slick if the substrate was coated with oil or grease when the paint was applied. A chalky residue on the back of paint that was on metal may be rust. White rust will form even on zinc-coated metal. A paint chip that was over a galvanized surface and which is coated with a waxy film may indicate that an oil-based primer was used.

When blistering has occurred, a diagnosis can be made by slitting a blister. If bare substrate is visible, the blister was probably caused by entrapped moisture vapor. If there is coating material beneath the blister, the problem was caused by solvent vapor that could not escape into the atmosphere because the coating's surface skin formed too rapidly.

After diagnosing the problem by examination and testing, and before removing existing materials or applying a new coating to the overall existing surfaces, the technique and materials to be used should be applied in a small, inconspicuous area. Doing so will test the suitability of the removal method, the compatibility of the materials, and how well the new color and finish match the existing.

Existing organic coatings should be properly cleaned before they are painted or recoated. Paint, for example, should be scrubbed using household cleaner and water, rinsed thoroughly with clean water, and allowed to dry before repainting. Mildew, stains, and other blemishes should be eliminated.

The extent of the recoating done may be just enough to cover the damage itself, but more normally it will include the entire door, frame, or window. Metal items with a baking finish cannot be recoated in the field. If complete recoating of such items is desired, they must be sent back to the shop. Touch-up of small areas in such surfaces is often possible, however, using air-drying repair coatings that match the original coating successfully. Damaged baking-finish coatings are most often simply painted over. Usually, existing transparent finishes are refinished completely. Surfaces that cannot be refinished without affecting their edges and projections, are usually also refinished.

Surfaces to be coated should be prepared in accordance with the coating manufacturer's recommendations. Removal of transparent coatings should be done in strict accordance with the instructions of the coating manufacturer and the wood or metal producer.

Hardware, accessories, machined surfaces, plates, fixtures, and similar items that will not be painted or finished should be removed before surface preparation for recoating begins. Alternatively, such items may be protected by surface-applied tape or another type of protection. Even when such items will be painted, they are often removed to make painting them and adjacent surfaces easier. The removed items can be reinstalled after the new coating has been applied.

Paint that is not slick and does not require sanding should be scrubbed using household cleaner and water, then rinsed thoroughly with clean water and allowed to dry before it is repainted. Mildew, stains, and other blemishes should be eliminated.

Surfaces should be cleaned before the first coat is applied and between coats if necessary, to remove oil, grease, dirt, dust, and other contaminants. Activities should be scheduled so that contaminants do not fall onto wet, newly painted or coated surfaces. Organic coatings should not be applied over dirt, rust, scale, oil, grease, moisture, scuffed surfaces, or other conditions detrimental to the formation of a durable coating film. Care should be exercised when using a solvent, of course, to ensure that it does not remove the existing coating.

When surfaces that will be coated cannot be put in proper condition by customary cleaning, sanding, and puttying, proper conditions must be achieved using whatever extraordinary methods are necessary.

Activities should be scheduled so that contaminants do not fall onto wet, newly coated surfaces.

Existing surfaces containing natural salts that have not been washed away because rain does not strike their surfaces should be washed with a solution that will remove the salts. Trisodium phosphate is recommended by many sources, but its use is not permitted in some jurisdictions.

Glossy existing surfaces to be recoated that do not require complete removal of the existing coating should be roughened by sanding or another appropriate method that will produce surface tooth sufficient to properly accept the new coating.

The removal of existing finishes should be complete, including removing undercoats and other materials that would affect or show through new materials. Such cleaned substrates should be completely free of films, coatings, dust, dirt, and other contaminants.

When existing organic coatings have been completely removed, the substrate should be prepared as if it were new.

Shoulders at the edges of sound existing organic coatings should be ground smooth and sanded to ensure that flaws do not telegraph through the new coating. When so recommended by the coating's manufacturer, such edges may be feathered, using drywall joint compound or another method recommended by the coating's manufacturer.

Abrasions in previously painted surfaces should be touched up before calking and application of the first field coat of paint. The touch-up paint should be the same material as the first field coat. Surfaces that are to be covered should be touched up before they are concealed. Touch-ups in exposed locations should be sanded smooth.

Before application of organic coatings is started in an area, that space should be swept clean with brooms and excessive dust removed. After coating has begun, the area should not be swept with brooms. Then the necessary cleaning should be done using commercial vacuum-cleaning equipment.

Surfaces to be painted or finished should be kept clean, dry, smooth, and free from dust and foreign matter that would adversely affect the adhesion or appearance of the finish.

Substrates damaged during the removal of an existing coating should be repaired before a new coating is applied.

Additional Requirements for Preparation of Wood Surfaces. Before wood is prepared for recoating, deteriorated material should be removed, damaged members repaired, and required new wood items installed. It may be necessary to remove the finish in some locations to determine the state of underlying wood members.

Before it is recoated, existing wood finished with an organic coating that does not need to be removed should be properly cleaned and prepared using scrapers, mineral spirits, sandpaper, fillers, and sealants as necessary. Under no circumstances should organic coatings be applied over abraded surfaces, moisture, wax, oil, grease, dirt, dust, or other contaminants that would adversely affect the adhesion or appearance of the new coating.

When the surfaces of existing transparent finished wood to be refinished are in good condition, old wax and polish should be removed by wiping the surfaces with mineral spirits. Rags should be changed frequently. Surfaces should be sanded lightly until they are dull. Dust should be removed using a tack rag.

When the surfaces of existing transparent-finished wood are porous or in poor condition or are severely discolored or stained, and where the existing finish is scratched or otherwise damaged, the old finish should be completely removed. The surface should then be cleaned and prepared as discussed in section 3.11.10 for new wood. Extraordinary cleaning and bleaching may be necessary to remove some stains and discolorations. Application of a new wood stain may be necessary to achieve an acceptable appearance.

Nails fastening wood to be repainted or refinished should be set and similar screws should be countersunk where this was not previously done. After the prime coat has been applied and has dried, the nail and screw holes and cracks, open joints, and other defects should be filled with putty or wood filler. Where a transparent finish will be used, the filler should be tinted to match the color of the wood.

Wood that is weathered and displays an open, fuzzy grain will usually not hold paint. It must first be sanded smooth and coated. One possible coating mixture contains two parts turpentine and one part linseed oil, but severely weathered wood may require a one-to-one mix. One coat may be enough, but as many as three are sometimes necessary. After the coating has thoroughly dried, the surfaces should be primed using an oil-alkyd primer.

Additional Requirements for Preparation of Metal Surfaces. To prepare it to receive paint, bare existing aluminum should be washed with mineral spirits, turpentine, or lacquer thinner, as is appropriate. It should then be steam cleaned and wirebrushed to remove loose coatings, oil, dirt, grease, and other substances that would reduce bond or harm new paint, and to produce a surface suitable to receive new paint.

Bare existing stainless steel should be washed with solvent and prepared according to the paint material manufacturer's recommendations.

Bare existing copper should have its stains and mill scale removed by sanding. Then dirt, grease, and oil should be removed, using mineral spirits. If a white alkyd primer was used, the copper should be wiped clean using a clean rag wetted with Varsol.

It is necessary to remove dirt, oil, grease, and defective paint from previously painted ferrous-metal surfaces using a combination of scraping, sanding, wirebrushing, sandblasting, and a cleaning agent recommended by the paint manufacturer. Surfaces should be wiped clean before they are painted. Shoulders should be sanded, if necessary, to prevent telegraphing. Where necessary to produce a smooth finish, existing paint should be stripped down to the bare metal.

Removing rust and other corrosion before painting is essential. The methods used will depend on the amount of rust or other corrosion present and the severity of the problem. Methods include sandblasting, power-tool cleaning, hand cleaning, and phosphoric-acid wash cleaning. In each case the recommendations of the Steel Structures Painting Council should be followed for ferrous metals and those of the manufacturer for nonferrous metals.

Just as welds, abrasions, factory- or shop-applied prime coats, and shop-painted surfaces in new materials should be touched up after erection and before calking and application of the first field coat of paint, so should previously painted surfaces also be touched up. On previously painted surfaces the touch-up paint should be the same material as the first field coat. Surfaces that will be covered should be touched up before they are concealed.

Specific Existing Conditions. In this section are some recommendations for dealing with a few specific types of failures in common organic coatings applied over wood or metal. They should, of course, be adapted to suit actual conditions. In every case, the advice of the coating's manufacturer and of the wood- or metal-item producer should be sought before attempting to repair organic coatings.

Stains and Discolorations. When an opaque coating has become stained, the source of the stain should be located and the conditions that produced it should be corrected.

When a stain results from metal corrosion, the metal should be uncovered, hand sanded, and coated with a corrosion-inhibiting primer or other coating to prevent further corrosion.

Nailhead stains can be removed from painted or transparent-coated surfaces by removing the coating from the nails using a wire brush, countersinking the nails, priming and filling the holes with wood filler or putty, and recoating the entire surface. Spot repairs will probably be noticeable.

Discolorations caused by chemical reactions between natural extractives in wood and water can often be cleaned using equal parts of denatured alcohol and water. The cleaned area should then be rinsed, permitted to dry, and painted with two coats of a stain-blocking primer specifically formulated for the purpose, after which it can be repainted to match the adjoining surfaces.

Water marks and other stubborn stains can be coated with primers specifically formulated for that purpose. The primer should be in addition to the number of coats recommended by the coating's manufacturer for the coating system that will be applied.

Brown spots in a painted surface that have resulted from the use of a latex paint in an area where high humidity is present can usually be cleaned away with a damp sponge. The use of abrasive cleaners should not be necessary, and repainting is not usually needed.

Brown spots inherent in a coating may resist removal and necessitate complete recoating of the surface. Tests should be made first, however, to ensure that the spots are not a chemical problem that will extend also through the new finish. If so, it may be necessary to remove the existing coating down to bare substrate and start all over again.

When mildew is suspected, it is first necessary to establish that the observed stain is actually mildew. A simple test is to place a drop of household bleach on the stain. Mildew will turn white; dirt will not. Mildew can be removed using the formulation and methods mentioned in section 3.11.13. Prevention of reoccurrence may be more difficult, however. Mildew-resistant paints may help. The most effective prevention technique is to remove the cause of the shade and dampness that permitted the growth in the first place. Vegetation may be trimmed to let sunlight strike the surface, for example. Sources of water should be controlled or removed. Adding a rain gutter and downspout system will sometimes help.

Telegraphing or Shadowing. When stains or soiling beneath a paint are visible through the paint, adding another coat will often solve the problem. When additional coats are unsuccessful, it is necessary to remove the paint and the source of the stain.

When discolorations are visible through a transparent coating, the coating and the discoloration must be removed and the wood recoated. Sometimes it is necessary to bleach the wood to remove stains. Refer to section 3.11.10 for more discussion about eliminating bleeding and other stains and covering knots.

Chalking. Although not harmful to the underlying material, light chalking may prevent a new coat of paint from adhering. Often, light chalking can be removed by gently brushing the surface. Sometimes it will be necessary to wash the chalking away with a household detergent and water, using soft brushes.

Excessive chalking, though, can weaken a coating as it becomes thinner, eventually resulting in a film that is too thin to protect the underlying metal or wood. Heavy chalking must also be removed unless, of course, the coating material itself is unsound and must be removed. A mild method for removing excess chalking is to wash the surface using a solution of one-half cup household detergent to a gallon of water and a medium-soft bristle brush. If that does not work, the surface can be scrubbed with a solution of trisodium phosphate and water, using a stiff brush. Where the use of trisodium phosphate is prohibited, a commercial solution made for the purpose must be substituted.

After cleaning, a previously chalked surface should be flushed with clean water and allowed to dry. Care should be taken to prevent damage to the metal or wood. The new paint should be applied before the chalking begins again.

Soft, Gummy Coatings. Organic coatings will sometimes become soft and gummy if the substrate becomes wet. If failure of new coatings is to be prevented, existing soft coatings must be removed completely, the wood dried, and further wetting prevented.

Wrinkles. When wrinkles in an opaque organic coating affect only the surface coat and are not extensive, it may be possible to scrape and sand down the wrinkled coat and touch up the sanded areas, or recoat the entire surface, using paint. When wrinkles affect undercoats or cover a large area, the entire coating must usually be removed, to expose bare metal or wood, before new paint is applied.

Wrinkles in baking-finish coatings will probably require completely refinishing the surface. If an air-drying coating is available that matches a wrinkled baking-finish coating, it may be possible to treat that surface as described above for wrinkled paint.

In a transparent coating, when only the top coat is affected it is sometimes possible to sand away wrinkles, but usually such a condition requires that the coating be removed and the surface be recoated.

Crazing. In paint the early stage of alligatoring called crazing can be sanded and repainted without completely removing the paint. The tiny cracks will be apparent on close examination, but will be filled with new paint and will, therefore, protect the substrate. If such an appearance is not acceptable, complete removal of the crazed paint will be necessary.

The same approach applies to a baking-finish coating if an air-drying patching material is available that will match the baking finish. If not, the item must be refinished, which will probably require returning it to the shop, or it may be sanded down and painted.

Most crazed transparent finishes must be completely refinished, but some can be repaired.

Checking and Alligatoring. Most checking and alligatoring are caused by the top coat of an organic coating being unable to adhere properly to underlying coats. Because only the surface coat is usually affected, a few sources say that in opaque coatings it is appropriate to sand or scrape away only the affected coats before recoating. Most sources, however, maintain that complete removal to the bare substrate of checked or alligatored organic coatings is always necessary before a new coating is applied. All agree that complete removal is necessary when the cracks extend completely through the coating.

Complete removal of alligatored or checked transparent finishes is usually necessary.

Blistering. Blisters may appear either as pimples, pinholes, or pits. When the blisters were caused by moisture on the substrates, the organic coating must be removed completely down to the bare substrates, and the substrates allowed to dry before they are refinished.

When the blisters were caused by solvent vapors between coats it is necessary to remove only the affected coats.

Blistered transparent coatings and most high-performance coatings must usually be removed completely, down to the bare substrates, and the substrates allowed to dry before recoating.

Cracking of a Transparent Coating. The appropriate cure for cracks in transparent coatings depends on the materials used, the extent of the cracking, and the age of the finish. Regardless of the materials used, the first step is to completely remove wax and foreign materials from the finish. Mineral spirits will often clean the surface satisfactorily without harming the finish, but any cleaning agent should be first tested in an inconspicuous location to ensure that it will do no harm.

The next step is to test the finish to determine which finish material was used by applying small patches of denatured alcohol and lacquer thinner in inconspicuous locations. Denatured alcohol will soften shellac, but not varnish or lacquer. Lacquer thinner will soften lacquer, but not others. Neither will soften varnish.

Damaged varnish cannot be repaired and must be completely removed and the surface refinished.

Shellac and lacquer can be melted and smoothed out in place. To do so, the appropriate solvent is spread onto the surface, softening the finish material, which can then be brushed out to a smooth finish. It may be best to apply an additional coat of the finish after the softened material has again dried.

If the damage is extensive or the material has been restored several times, it will probably be necessary to completely remove the existing coating materials and apply a new coating.

Other Cracking, Peeling, and Scaling. Cracks in opaque organic coatings, and the peeling and scaling of all organic coatings, usually affect the entire thickness of the coating. When they do it is necessary to remove existing coating materials completely before recoating the underlying metal or wood. Because the defect may have been caused by the coating material being too brittle for the surface, the new paint should be more flexible than the failed material.

When the surface coat of a paint has peeled away from its undercoats, it may be possible to remove only the unsound paint. When the peeling was caused by the use of incompatible paints, however, the incompatible top coat or coats should be removed completely.

When an organic coating has peeled away from its substrates, complete removal and recoating are necessary. If the peeling coating is a baking-finish material it will be necessary to return the affected item to the shop for refinishing, unless an air-drying coating is acceptable. Alternatively, the baking-finish coating may be removed and the item painted.

When a transparent or high-performance organic coating has peeled away from its substrates, complete removal and recoating are necessary.

Small Worn Spots in Transparent Finishes. It is often possible to refinish small areas in a transparent finish so that the patch is not objectionable. Covering wax and other topping materials must first be removed. Then rough areas in a wood substrate should be given a coat of penetrating sealer. The new finish should be applied carefully so that lap marks do not show.

Failure of an Organic Coating to Dry. When an organic coating does not dry as quickly as expected, the proper approach is to wait—good materials will eventually dry.

If the organic coating does not eventually dry, the only solution is to remove the coating completely down to the bare substrate and recoat it with good materials.

Runs and Sags. Proper application is the best cure, but runs and sags in paint that have been permitted to dry can be removed using sandpaper. Another coat of paint may then be necessary to achieve the desired finish.

Runs and sags in baking-finish, high-performance, and transparent coatings may not be repairable without refinishing.

Excessive Paint Buildup. Existing paint with a total film thickness of more than one sixteenth inch should be removed completely.

Coating Wash-Off. When rain strikes an organic coating that has not dried or has not dried properly, the outer layer can be washed away. Such a surface must be sanded and recoated.

Naturally Aged Organic Coatings. When an organic coating has reached its expected life span, recoating is necessary. After a coating has been permitted to begin to fail, recoating is much more difficult. At this stage complete removal of the existing finish may be necessary.

Damaged Transparent Coatings over Other Metal Finishes. Transparent organic coatings are often applied over anodized aluminum, and sometimes over copper alloys, to protect their finishes. When the underlying metal or its mechanical or chemical finish has become damaged, it is necessary to remove the transparent coating to repair the underlying material or finish.

Mill-finished or clear anodized aluminum may begin to oxidize and turn dark. Before such aluminum can be cleaned, covering organic coatings must be removed. If reanodizing is required, the item must be shipped to an anodizing plant for the operation to be performed. After the aluminum has been cleaned, and reanodized if desired, a transparent organic coating, usually methacrylate lacquer, may be applied to protect the new finish. For reasons mentioned earlier, however, color anodized aluminum should not have a methacrylate lacquer applied, regardless of the extent of the cleaning or repairs. Another possible solution is to skip the lacquer and give the newly finished aluminum a coat of wax or oil. The wax or oil will need renewal more frequently than lacquer would.

Copper alloys with a lacquer or other transparent coating may begin to appear dull over time. Their luster can often be restored by removing the transparent coating and refinishing the copper alloy. If, for example, the copper alloy is a lacquered brass, the lacquer can be removed using amyl nitrate (banana oil), denatured alcohol, or acetone. Other transparent coating types will require different removal methods. The finish may also need restoration. The method used will depend on the finish. No attempt should be made to restore the finish on a copper alloy until the exact identity of the existing finish is known. Then the restoration should be done in accordance with the metal manufacturer's and finisher's recommendations. Restoration may be simple, as in the case of tarnished brass, which can be restored with a common brass cleaner, or complicated, as in the case of a faded patina on bronze. After the finish has been restored, a new transparent organic coating of either the original or a different type will probably be suggested by the metal's finisher. Sometimes, however, a wax or oil coating will be recommended.

Organic Coating Removal. There may be a compelling reason to remove an existing coating that has nothing to do with the coating itself. Perhaps, for instance, an existing coating contains lead and the building's use has changed so that the coated item will now be accessible to children. Removal is also necessary when the thickness of a paint's coating reaches one sixteenth inch or thicker. Under most circumstances, however, when an existing organic coating is essentially sound it is necessary to remove only the portion that is cracked, loose, peeling, or otherwise damaged. It is also necessary to remove rust, oil, grease, dust, soiling, and other substances that would affect the bond or appearance of new paint. It is not necessary to remove organic coatings from surfaces that have only minor blemishes, dirt, soot, pollution, cobwebs, insect cocoons, and the like, unless removal is necessary so that repairs can be made to the substrates. Such foreign matter can be removed using a water spray, followed by scrubbing with a mild household detergent and water solution. The cleaned surfaces should be rinsed thoroughly with clean water and permitted to dry.

When an existing paint displays crazing, blistering, peeling, or cracking of its top layer or layers only, it may be possible to remove it only down to sound paint.

When an existing paint displays excessive chalking or blistering, peeling, flaking, cracking, or scaling through its entire thickness, it must be removed completely, exposing the substrates, and the underlying problem corrected. It may also be necessary to remove an existing paint when introduction of a new substrate material prohibits making a smooth transition from the existing paint to the new one.

When complete removal of an organic coating is necessary, it should be accomplished by one of the methods described in this section. Some limitations apply, however. For example, transparent coatings can be removed by sanding or scraping, but they are usually best removed with chemical removers. On open-grain woods it may be necessary to use steel wool dipped in remover to loosen an embedded finish from the wood's pores. After a finish has been removed, the chemicals used must be neutralized according to their manufacturers' instructions.

Removing some finishes is difficult, if not impossible. It might not be possible, for instance, to completely remove an existing penetrating oil stain from wood, although bleaching might lighten it. When removal of such a stain is not possible, the surface must either be stained again, using the same material as originally used, or covered with an opaque paint.

Some of the methods included here are not applicable for every type of item. Abrasives and harsh chemicals that might severely damage glass, glazing sealants, and plastic gaskets, for example, should not be used to remove paint from windows.

Limited Paint Removal. Where paint removal is limited to small areas, abrasive methods such as scraping or sanding should be used. Hand tools work best in most cases. Gouging of the substrates should be avoided. Mechanical abrasive methods, such as ones using orbital sanders and belt sanders, may be appropriate in some circumstances, but such devices should be used carefully to prevent their removing too much material. Rotary sanders, sandblasting, and water blasting are inappropriate and should not be used for limited paint removal.

Total Paint Removal Using Heat. For total paint removal, thermal methods such as an electric heat plate or electric heat gun should be used.

Blowtorches and other flame-producing methods are dangerous and should not be used for paint removal. Fire-insurance requirements should be checked and removal methods cleared with the insurance provider before any hot-process removal is undertaken.

Total Paint Removal Using Chemicals. Chemicals formulated for the purpose, such as solvent-based strippers and caustic strippers, may also be used for total paint removal. The manufacturer's directions and applicable legal requirements should be followed. Precautions are necessary to protect workers and others against harm from inhaling vapors, from fire, eye damage, and chemical poisoning resulting from skin contact with chemicals, and from other dangers associated with using chemicals. Lead residue and other harmful substances should be disposed of properly. Chemical strippers must be removed completely before paint is applied, because the residue left by some strippers will impede the adhesion of new paint.

Total Paint Removal Using Abrasives. Abrasive methods of paint removal include scraping, sanding, water blasting, sandblasting, and blasting with other abrasives. One abrasive product that has been used successfully in some conditions is made from corncobs. Hand tools should be used where appropriate. Gouging of the substrates should be avoided. Using mechanical abrasive methods, such as orbital sanders and belt sanders, may be appropriate in some circumstances, but such devices should be used carefully, to prevent their removing too much material. Sandblasting is probably not an appropriate means of removing paint from galvanized metal, because preventing damage to the metal while doing it requires more expertise than is commonly available and because it removes the galvanizing.

Hazardous Materials. Improper removal and disposal of existing chromium- and lead-based paints can create serious health and legal problems. The handling of such materials is regulated at the national level and by many state and local jurisdictions. For example, material containing more than 5 parts per million of lead or chromium is classified by the Environmental Protection Agency (EPA) as hazardous waste and is subject to all applicable regulations.

Serious problems can also occur when the materials to be removed contain asbestos, arsenic, or other toxic substances. Even methylene chloride, which is used in many commercial paint removers, is classified as a toxic waste. Removal and disposal of such materials must be done in strict accordance with laws and recognized safety precautions. Stiff fines and penalties may be imposed on owners, architects, and contractors if such materials are handled or disposed of improperly, regardless of who is at fault. A complete discussion of hazardous materials removal and disposal is beyond the scope of this book. When faced with such a problem, obtain copies of the applicable rules and regulations and follow them explicitly. The Steel Structures Painting Council, at the address listed in the Appendix, is a source for current recommendations about cleaning hazardous materials from metals. Their recommendations may also be applicable to other materials or at least point to other sources. The actual removal of hazardous wastes is best done by professionals who are expert in such work.

3.15.4 Applying Paint

Except when they are modified in this section, the suggestions for applying organic coatings in section 3.11.15 apply as well to paint applied over existing surfaces.

The extent of new paint applied on a previously coated surface may be over the entire surface or on only part of it, depending on the requirements of the project and the recommendations of the paint's manufacturer. When only partial repainting is required, it is usually best to extend the repainting to a natural break, such as the complete face of a door, its entire frame, or a complete window.

The painting of existing surfaces should not be started until patchwork, extensions, repair work, and new work in the space have been completed.

When a flat paint finish is applied over a previously painted or coated surface, it is usually best to apply all the coats recommended by the paint manufacturer for the selected paint system on new work.

When a semigloss or gloss paint finish is applied over a previously painted or coated surface, the first coat (primer) should be applied in accordance with the paint manufacturer's specifications and the final coats should be applied as they would be on a new surface.

Before applying new paint over existing paint, other organic coatings, or another existing finish, it is best to test for compatibility by applying the paint in small, inconspicuous locations representative of each condition that will be encountered.

Primers may be omitted from previously painted or coated surfaces where the existing coating is sound, but all other coats recommended by the manufacturer should be applied to every surface, including previously painted or coated surfaces, even when the existing coating is sound, unless the paint manufacturer specifically recommends otherwise for the particular situation. When properly applied, the required number of coats should produce a fully covered, workmanlike, presentable job. Each coat must be applied in heavy body, without improper thinning. When stains, dirt, or undercoats show through the final coat of paint, the defects should be corrected and the surface covered with additional coats until the paint film is of uniform finish, color, appearance, and coverage.

It is necessary to sand lightly between each succeeding paint coat. High-gloss paint should be sanded between coats using very fine-grit sandpaper. Dust should be removed after each sanding, to produce a smooth, even finish.

The tops, bottoms, and side edges of exterior doors should be finished the same as the exterior faces of the doors.

The first coat of new paint should be applied to surfaces that have been prepared for recoating as soon as practicable after preparation and before subsequent surface deterioration occurs. Sufficient time should be allowed between successive coatings to permit proper drying.

At the completion of recoating, surfaces should be examined, touched up and restored where damaged, and left in proper condition.

3.15.5 Applying Other Organic Coatings

Except when they are modified in this section, the suggestions in section 3.11.15 apply as well to coatings applied over existing surfaces.

Organic coatings other than paint include baking-finish coatings, high-performance coatings, and transparent coatings. Of those, only transparent coatings can be applied in the field. Some baking-finish coatings can be painted with air-drying material of the same composition. Some high-performance coatings are repairable using air-drying materials similar to those in the original coating. They should be used strictly in accordance with their manufacturer's directions.

Patches in transparent coatings are usually obvious and are seldom acceptable for that reason. Except for the repair of minor cracking mentioned earlier, when transparent coatings have been damaged, the existing materials should be removed completely, the substrates prepared, and new materials applied in accordance with the coating material manufacturer's recommendations. Only experienced workers should be used to prepare for and apply transparent organic coating materials. Where a surface cannot be refinished without affecting its edges and projections, the edges and projections are usually also refinished.

Refinishing of existing surfaces should not be started until patchwork, extensions, repair work, and new work in that space have been completed.

The first coat of a new organic coating material should be applied to surfaces that have been prepared for recoating as soon as practicable after preparation and before subsequent surface deterioration occurs. Sufficient time should be allowed between successive coatings to permit proper drying.

At the completion of recoating, surfaces should be examined, touched up and restored where damaged, and left in proper condition.

3.16 Where to Get More Information

The book in this series entitled *Repairing and Extending Nonstructural Metals* contains a more detailed discussion of the sources, production, and finishing of metals than is included here. Similarly, the book in this series entitled *Repairing and Extending Finishes, Part II* contains additional information about finishes used on wood. All of the information in those books has not been repeated here. The bibliographical references there that are applicable to the finishing and repair of finishes discussed in this chapter, however, have been repeated in this book's Bibliography. The applicable listings under "Where to Get More Information" in those books have also been repeated here.

Many national and regional coatings manufacturers offer quite good information about the coating systems they manufacture. As has been mentioned in the text, a reputable manufacturer may be the best source of advice about repairing a failed organic coating.

The following AIA Service Corporation's *Masterspec*, Basic Version sections contain helpful information about the coatings addressed in this chapter. Anyone involved in selecting paint or other organic coating systems should obtain a copy. The author assumes that *Masterspec*'s later editions will contain similar data.

- The February 1988 edition of Section 09800, "Special Coatings" includes discussions of epoxy, polyurethane, chlorinated rubber, heat-resistant enamel, black enamel, silicone-alkyd enamel, zinc rich, and aluminum-pigmented coatings.
- The November 1988 edition of Section 09900, "Painting" includes a discussion and detailed listing of the national and regional paint and organic coatings manufacturers operating at the time of publication. The regional manufacturers are broken down into those in the Northeast region, the southern states and Gulf Coasts, the north-central states, the Southwest region and the Northwest states.

    This section also includes a discussion of the requirements for selecting organic coating products and descriptions of the components of organic coating materials that are more extensive than those in this chapter.

The following American Architectural Manufacturers Association (AAMA) publications contain valuable information for anyone interested in finishes on aluminum:

- *Aluminum Store Front and Entrance Manual.*
- *Metal Curtain Wall, Window, Store Front and Entrance Guide Specifications Manual.*
- AAMA 603.8-1985, "Voluntary Performance Requirements and Test Procedures for Pigmented Organic Coatings on Extruded Aluminum."
- AAMA 604.2-1977, "Voluntary Specification for Residential Color Anodic Finishes."
- AAMA 605.2-1985, "Voluntary Specification for High Performance Organic Coatings on Architectural Extrusions and Panels."
- AAMA 606.1-1976, "Voluntary Guide Specifications and Inspection Methods for Integral Color Anodic Finishes for Architectural Aluminum."
- AAMA 607.1-1977, "Voluntary Guide Specifications and Inspection Methods for Clear Anodic Finishes for Architectural Aluminum."
- AAMA 608.1-1977, "Voluntary Guide Specification and Inspection Methods for Electrolytically Deposited Color Anodic Finishes for Architectural Aluminum."
- AAMA 609.1-1985, "Voluntary Guide Specification for Cleaning and Maintenance of Architectural Anodized Aluminum."
- AAMA 610.1-1979, "Voluntary Guide Specification for Cleaning and Maintenance of Painted Aluminum Extrusions and Curtain Wall Panels." Describes recommended procedures for cleaning and maintaining painted aluminum after it has been installed.

The June 1988 article "Direct to Rust Coatings" in *American Paint Contractor* contains a good description of the various materials available for painting rusted metals and their application.

The 1982 American Society for Metals' publication *Metals Handbook*, 9th edition, Vol. 5: Surface Cleaning, Finishing, and Coating contains detailed, excellent discussions of materials and methods for the preparation for and application of organic coatings on metals. It also discusses the reasons for failures. It is an excellent source of data for anyone responsible for dealing with the finishes used on metals.

Section 1500 of the Architectural Woodwork Institute's *Architectural Woodwork Quality Standards, Guide Specifications and Quality Certification Program* is the definitive guide on factory finishing of wood doors.

Every designer should have the full complement of applicable ASTM Standards available for reference, of course, and anyone who needs to understand the types of metal and wood materials and their finishes discussed in this chapter should have access to a copy of the ASTM Standards marked with a [3] in the Bibliography.

Most of the more than six hundred ASTM standards related to organic coating materials are testing requirements for the chemicals in those products and are therefore not too helpful in selecting new organic coating materials or determining the types of organic coating materials that might be found in an existing building. The following list contains some ASTM standards that might be helpful:

- Standard D 16, "Standard Definitions of Terms Relating to Paint, Varnish, Lacquer, and Related Products."
- Standard D 1730, "Recommended Practices for Preparation of Aluminum and Aluminum-Alloy Surfaces for Painting."
- Standard D 1731, "Recommended Practices for Preparation of Hot-Dip Aluminum Surfaces for Painting."
- Standard D 2833, "Standard Index of Methods for Testing Architectural Paints and Coatings."
- Standard D 3276, "Standard Guide for Painting Inspectors (Metal Substrates)."
- Standard D 3927, "Standard Guide for State and Institutional Purchasing of Paint."

C. R. Bennett's 1987 article "Paints and Coatings: Getting Beneath the Surface" is as complete a general introduction to organic coatings as one would expect to find in a magazine article. Mike Bauer in a 1987 letter to the editor takes issue with some of Mr. Bennett's statements, however, so both pieces should be read together.

The Construction Specifications Institute's 1988 Monograph 07M411, "Precoated Metal Building Panels," contains basic information about the coil-coating process and gives the sources of the information it contains.

The publications of the Copper Development Association (CDA) that are marked with a [3] in the Bibliography contain very helpful information related to the subjects in this chapter.

Mario J. Catani's 1985 article "Protection of Embedded Steel in Masonry" contains valuable information about why steel corrodes and how to prevent it. While the article is specifically aimed at embedded steel, the principles in it apply to all steel corrosion.

The Sheldon W. Dean and T. S. Lee edited *Degradation of Metals in the Atmosphere* published in 1988 heads a list of ASTM books about metal corrosion. It includes articles about the corrosion of stainless steel, copper, wrought-aluminum alloys, zinc, and other metals. There are also discussions in it of materials performance, environmental characterizations, and test methods. Other ASTM books cover related aspects of the same subjects and methods for testing and monitoring corrosion using nondestructive and other methods. Anyone concerned with corrosion in existing metals or preventing such corrosion may want to contact ASTM and ask for a list and description of their books on the subject.

J. Scott Howell's 1987 article "Architectural Cast Iron" contains a table showing the various types of alloys generally used in iron, steel, stainless steel, aluminum, and copper alloys.

Larry Jones's 1984 article "Painting Galvanized Metal" is a good discussion of the problems associated with painting galvanized metal, which is one of the more difficult surfaces to paint successfully. Some of the recommendations in that article are not universally accepted, however.

Clem Labine's 1982 article "Restoring Clear Finishes" is an excellent practical discussion of the methods for restoring an existing transparent finish. It covers cleaning, reviving, and stripping such finishes. Mr. Labine includes a method for determining which finish material has been used and a logical process to follow to obtain the best results with the least effort. Obtaining a copy is well worth the effort for anyone who must deal with existing transparent finishes on wood.

Timothy D. McDonald's 1987 article "Technical Tips: Coatings That Protect Against the Corrosion of Steel" contains the criteria for selecting organic coatings for steel and other important recommendations regarding primers, curing, and what to do about already corroded surfaces.

Probably the best source of information about metal finishes available today is the National Association of Architectural Metal Manufacturers' (NAAMM) *Metal Finishes Manual for Architectural and Metal Products*. Anyone interested in designing, applying, or maintaining metals or their finishes should have access to a copy of the latest edition.

The National Decorating Products Association's 1988 *Paint Problem Solver* is a definitive source that specifically addresses paint problems and their solutions. It discusses such adhesion problems as alligatoring, blistering, checking, cracking, flaking, and excessive chalking; peeling from galvanized metal, metal doors and garages, wood doors and frames, and other wood; sagging, wrinkling, and peeling of a latex top coat from previously painted hard, slick surfaces, and more. It also addresses application problems, including the applicator not holding enough paint, brush marks, cratering, excessive shedding of bristles onto a painted surface, and excessive splatter from rollers. Discolorations, including fading, mildew, rusted nail and other fastener heads, and staining from flashings are also covered. Other problems covered include lap marks, poor hiding, and uneven gloss. Everyone responsible for paint maintenance and for painting over existing materials should have a copy. The *Paint Problem Solver* is also available from the Painting and Decorating Contractors of America.

Harold B. Olin, John L. Schmidt, and Walter H. Lewis's 1983 edition of *Construction Principles, Materials, and Methods* is a good source of background data about iron, steel, stainless steel, aluminum, and wood production and finishing. Unfortunately, it contains little that will help specifically with troubleshooting failed metals or wood or their finishes. This book is slightly out of date, but an updated version will soon be published.

The Old-House Journal's 1982 article "Stripping Paint" is an excellent discussion of the available methods for completely removing existing paint.

The Old-House Journal's 1983 article "48 Paint Stripping Tips" contains practical tips for removing existing paints. Most of the data there are most useful to those actually doing the stripping, however, and the article is not too helpful in determining whether to strip existing paint or deciding on the general method to be used.

The following Painting and Decorating Contractors of America (PDCA) publications are of varying value to those having to deal with the maintenance of paint or painting existing surfaces. Some critics have rightly pointed out that much of the data in the PDCA publications are also available from some major national paint manufacturers, but there is a twofold advantage in having the PDCA data. First, it is a source of needed data that are always available even when the selected manufacturer does not publish similar recommendations, as is often the case. And second, it serves as a second opinion to compare with data printed in other sources.

- The 1975 edition of *Painting and Decorating Craftsman's Manual and Textbook, Fifth Edition* is used in training craftspersons.
- The 1982 *Painting and Decorating Encyclopedia* is a good resource for designers, decorators, and craftspersons. It is not specific to dealing with existing conditions.
- The 1986 *Architectural Specification Manual, Painting, Repainting, Wallcovering and Gypsum Wallboard Finishing, Third Edition* was actually published by Specifications Services of the Washington State Council of the PDCA in Kent, Washington. It should be on the shelf of every person responsible for designing, maintaining, and repainting painted surfaces. It includes an evaluation of finishing systems, a discussion of new- and existing-surface preparation, general information and finish schedules for interior and exterior paint finishes, and guide specifications. The guide specifications are not as definitive as some other available ones and are not based on the current CSI format. The parts of the book about repainting include a discussion on how to handle existing sound, slightly deteriorated, or severely deteriorated paint on metal and wood, including preparation methods for repainting and also for dealing with ferrous metal, galvanized metal, and aluminum. There is a discussion of cleaning and removal of existing paint by hand, solvent, steam, power tool, burn off, chemicals, and sandblasting. It also includes removal methods for efflorescence and mildew,

and acid-etching methods. Other subjects include methods for handling extractive bleeding from cedar, redwood, mahogany, and Douglas fir.
- The 1988 *The Master Painters Glossary* is an excellent glossary of paint industry terminology that is well worth having.

Maurice R. Petersen's 1984–1985 article series "Finishes on Metals: A View from the Field" discusses mechanical, chemical, and other inorganic finishes, and organic coatings used on metals and compares them. It includes recommendations about repairing coatings.

Patricia Poore's 1985 article "Stripping Paint from Exterior Wood" is an excellent discussion on the subject.

The Porcelain Enamel Institute (PEI) offers no specific advice about cleaning, repairing, or reenameling porcelain enamel, but it will furnish a list of enameling companies that can clean, repair, or reenamel damaged porcelain enamel surfaces. It may be more convenient to call a local contractor who handles porcelain-enamel products or a local member of the PEI and ask for the name of a nearby organization that can clean or repair such surfaces.

Jack C. Rich's book *The Materials and Methods of Sculpture* contains a great deal of data relating to metals, especially castings. It discusses the physical characteristics of various metals and their finishing and includes data about cleaning and retouching bronzes and other copper alloys.

Ramsey/Sleeper's *Architectural Graphic Standards* has extensive information about stainless steel and aluminum shapes, bars, and wire, including information about shapes, sizes, and thicknesses.

Thomas R. Scharfe's 1988 article "New Metal Coating Technologies Enhance Design Opportunities" includes a list of coatings and some of their characteristics.

The following publications of the Steel Structures Painting Council should be available to everyone responsible for painting steel:

- The 1983 *Steel Structures Painting Manual,* Vol. 1, *Good Paint Practice,* 2nd ed.
- The 1983 *Steel Structures Painting Manual,* Vol. 2, *Systems and Specifications,* 2nd ed.

The editions listed in the Bibliography of the U.S. Department of the Navy's Guide Specification NFGS-09910, "Painting of Buildings (Field Painting)," the U.S. Department of the Army's CEGS-09910, "Painting, General," and the U.S. General Services Administration's PBS: 3-0990, "Painting and Finishing, Renovation, Repair, and Improvement" contain excellent guidance for dealing with conditions where new paint is to be applied over an existing surface. Later editions of the Navy and Army guides should be equally helpful. Current editions of the GSA guides, however, are based on AIA's *Masterspec* and may not be as specifically helpful for dealing with existing surfaces.

The U.S. General Services Administration Specifications Unit's federal specifications applicable to organic coatings are sometimes out of date or not available, and sometimes the quality level they require is substandard. Unfortunately, however, they are often the only applicable standard. Some of the many federal specifications applicable to materials usable on metal and wood are more often referred to than are others. Some of the more-often-referenced ones are included in the following list, which is by no means complete. Complying with federal specifications does not necessarily guarantee a satisfactory product for any particular condition.

- TT-C-535D, "Coating, Epoxy, Two-Component, for Interior Use on Metal, Wood, Wallboard, Painted Surfaces, Concrete and Masonry."
- TT-C-542E, "Coating, Polyurethane, Oil-free, Moisture Curing."
- TT-E-489G, "Enamel, Alkyd, Gloss (For Exterior and Interior Surfaces)."
- TT-E-496B, "Enamel, Heat Resisting (400 deg F), Black."
- TT-E-505A, "Enamel, Odorless, Alkyd, Interior, High Gloss, White and Light Tints."
- TT-E-506K, "Enamel, Alkyd, Gloss, Tints and White (For Exterior and Interior Surfaces)."
- TT-E-508C, "Enamel, Interior, Semigloss, Tints and White."
- TT-E-509B, "Enamel, Odorless, Alkyd, Interior, Semigloss, White and Tints."
- TT-E-527C, "Enamel, Lusterless."
- TT-E-543A, "Enamel, Interior, Undercoat, Tints and White."
- TT-E-545B, "Enamel, Odorless, Alkyd, Interior-undercoat, Flat, Tints and White."
- TT-E-1593B, "Enamel, Silicone Alkyd Copolymer, Gloss (For Exterior and Interior Use)."
- TT-F-322D, "Filler, Two-component Type: For Dents, Cracks, Small-holes, And Blow Holes." Materials meeting this standard are used as a filler and for patching cracks and seams in wood, metal, concrete, and cement mortar.
- TT-F-336E, "Filler, Wood, Paste."
- TT-F-340C, "Filler, Wood, Plastic."
- TT-L-58E, "Lacquer, Spraying, Clear and Pigmented For Interior Use."
- TT-L-190D, "Linseed Oil, Boiled, (For Use in Organic Coatings)."

- TT-L-201A, "Linseed Oil, Heat Polymerized."
- TT-P-25E, "Primer Coating, Exterior (Undercoat For Wood, Ready-mixed, White And Tints)."
- TT-P-28F, "Paint, Aluminum, Heat Resisting (1200 F)."
- TT-P-29J, "Paint, Latex Base, Interior, Flat, White and Tints."
- TT-P-30E, "Paint, Alkyd, Odorless, Interior, Flat White and Tints."
- TT-P-37D, "Paint, Alkyd Resin; Exterior Trim, Deep Colors."
- TT-P-47F, "Paint, Oil, Nonpenetrating-flat, Ready-mixed Tints And White (for Interior Use)."
- TT-P-52D, "Paint, Oil (Alkyd Oil) Wood Shakes and Rough Siding."
- TT-P-55B, "Paint, Polyvinyl Acetate Emulsion, Exterior."
- TT-P-81E, "Paint, Oil, Alkyd, Ready Mixed Exterior, Medium Shades."
- TT-P-86G, "Paint, Red-lead-base, Ready-mixed."
- TT-P-615D, "Primer Coating, Basic Lead Chromate, Ready Mixed."
- TT-P-636D, "Primer Coating, Alkyd, Wood And Ferrous Metal."
- TT-P-641G, "Primer Coating; Zinc Dust–zinc Oxide (For Galvanized Surfaces)."
- TT-P-645A, "Primer, Paint, Zinc Chromate, Alkyd Type."
- TT-P-650C, "Primer Coating, Latex Base, Interior, White (For Gypsum Wallboard)."
- TT-P-664C, "Primer Coating, Synthetic, Rust-inhibiting, Lacquer-resisting."
- TT-P-791A, "Putty, Pure-Linseed-Oil (For) Wood-sash-glazing."
- TT-P-1511A, "Paint, Latex-base, Gloss and Semi-gloss, Tints and White (for Interior Use)."
- TT-S-176E, "Sealer, Surface, Varnish Type, Floor, Wood and Cork."
- TT-S-300A, "Shellac, Cut."
- TT-S-708A, "Stain, Oil; Semi-transparent, Wood, Exterior."
- TT-S-711C, "Stain, Oil Type, Wood, Interior."
- TT-T-291F, "Thinner, Paint, Mineral Spirits, Regular And Odorless."
- TT-V-86C, "Varnish, Oil, Rubbing (For Metal and Wood Furniture)."

Martin E. Weaver's 1989 article series "Fighting Rust" is a good reference about the causes and cures for rusted iron and steel.

Weaver's July 1989 article "Caring for Bronze" is an excellent discussion of the corrosion, cleaning, and protection of copper alloys. Anyone interested in caring for bronze should obtain a copy. Contact *The Construction Specifier* (see Appendix: Data Sources).

Kay D. Weeks's and David W. Look's 1982 publication "Exterior Paint Problems on Historic Woodwork" is a fine discussion of exterior paint failures that is applicable to all conditions, not just historic preservation projects. It gives causes, recommended treatments, methods of paint removal, and some general recommendations for selecting paint for repainting exterior wood.

The Western Lath, Plaster, and Drywall Contractors Association (formerly the California Lathing and Plastering Contractors Association) is responsible for a 1981 publication called *Plaster/Metal Framing System/Lath Manual* that is a good source of information about metal framing for partitions. A new edition was published in late 1988 by McGraw-Hill.

Forrest Wilson's 1984 book *Building Materials Evaluation Handbook* offers sound advice about cleaning metal surfaces.

Until recently the Zinc Institute was a source for data about zinc. A recent telephone call to them, however, netted an answering-machine message stating that the Zinc Institute was no longer in operation and that questions should be addressed to zinc suppliers. In addition, the American Anodizers Council (AAC) has produced standards for anodizing that it will make available upon request. Alice Koller's 1981 article "Hot-Dip Galvanizing: How and When to Use It" is also an excellent article on the subject.

In addition, the following publications may be of some value:

- Kenneth Abate's 1989 article series, "Metal Coatings, Fighting the Elements with Superior Paint Systems."
- *Architectural Technology*'s 1986 article "Technical Tips: Paints and Coatings Primer."
- Able Banov's 1973 book *Paints and Coatings Handbook for Contractors, Architects, and Builders.*
- David R. Black's 1987 article "Dealing with Peeling Paint."
- The *Canadian Heritage*'s 1985 article "Take It All Off? Advice on When and How to Strip Interior Paintwork."
- Sarah B. Chase's 1984 article "Home Work: The ABC's of House Painting."
- The Construction Specifications Institute's 1988 Monograph 07M411, "Precoated Metal Building Panels" and its 1988 *Specguide* 09900, "Painting."
- Dan Elswick's 1987 article "Preparing Historic Woodwork for Repainting, Part 2—Thermal and Chemical Cleaning."
- Caleb Hornbostel's 1978 book *Construction Materials, Types, Uses, and Applications.*
- Larry Jones's 1984 article "Don't Overlook the Heat Plate."

- Mary Kincaid's 1982 article "What Paint Experts Say."
- Robert Lowes's May/June 1988 article "Abrasive Blasting" in *PWC Magazine*.
- Dave Mahowald's 1988 article "Specifying Paint Coatings for Harsh Environments."
- Charles R. Martens's 1974 book *Technology of Paints and Lacquers*.
- Ambrose F. Moormann, Jr.'s, 1982–83 article series "Paint and the Prudent Specifier."
- *The Old-House Journal*'s 1983 articles "Our Opinion of 'Peel Away'" and "Paint On Paint"; the 1987 article "Exterior Painting: Problems and Solutions"; and the 1988 article "Commercial Paint Stripping."
- The Reader's Digest Association's *Complete Do-it-yourself Manual*.
- The U.S. Department of the Army's 1980 *Painting: New Construction and Maintenance (EM 1110-2-3400)*.
- The U.S. Department of the Army's 1981 *Corps of Engineers Guide Specifications, Military Construction,* CEGS-09910, "Painting, General."
- The U.S. Departments of the Army, the Navy, and the Air Force's 1969 *Technical Manual* TM 5-618, "Paints and Protective Coatings."
- The U.S. Department of Commerce, National Bureau of Standards, 1968 *Organic Coatings BSS 7*.
- The Guy E. Weismantel edited 1981 *Paint Handbook*.

See also the other entries in the Bibliography marked with a [3].

CHAPTER

4

Metal Doors, Frames, and Store Fronts

This chapter covers interior and exterior carbon-steel doors and frames and similar frames used for sidelights and other glazed exterior conditions, interior cased openings, and borrowed-light openings. It also includes interior and exterior formed stainless steel, aluminum, and copper-alloy doors and frames, and extruded aluminum doors, frames, entrances, and store fronts. The door types covered include swinging and revolving doors. The discussion also includes some special door and frame conditions such as lead-lined doors and frames, reinforced frames, structural frames, and automatic doors. The discussion covers the applicable door hardware.

Wood doors, frames, and store fronts, including those wood doors that might be mounted in the types of metal frames discussed in this chapter, are addressed in Chapter 5.

Metal glazed sliding doors are covered in Chapter 6.

Curtain walls are in Chapter 7. Refer to the Glossary for a description of the difference between curtain walls and store fronts.

Glazing for the doors and frames addressed in this chapter is discussed in Chapter 8.

Beyond the scope of this book are grilles; rolling, coiling, and overhead doors; doors for compartments, cubicles, cabinets, casework, elevator entrances, and coolers and freezers; access doors; and roof and floor hatches.

4.1 General Requirements for Metal Doors and Frames

There are two kinds of steel door and frame manufacturers. One group mass produces what are called standard doors and frames. The other makes so-called custom doors and frames to fit an architect's or owner's specifications. Standard and custom doors and frames should both meet the same requirements. Because of the unusual conditions found in many existing buildings, many of the steel doors and frames used in them are custom-made. Standard steel doors and frames are often less costly than custom ones, but some kinds of special hardware cannot be used on them, and the number of designs available in standard doors and frames is somewhat more limited.

Stainless steel and aluminum doors and frames are also available either as standard or custom units. Copper-alloy doors and frames are custom units.

4.1.1 Standards

Carbon-steel doors and frames should comply with the Steel Door Institute's publication SD-100-85, "Recommended Specifications for Standard Steel Doors and Frames." All those in the same project should be manufactured by a single firm that specializes in producing the types of doors and frames required.

Stainless steel, aluminum, and copper-alloy doors and frames and their hardware and finishes should conform with the applicable standards and recommendations of the representative industry associations, including those of the American Architectural Manufacturers Association (AAMA), the Aluminum Association (AA), ASTM, the National Association of Metal Manufacturers (NAAMM), the American National Standards Institute (ANSI), the Copper Development Association (CDA), and the American Iron and Steel Institute (AISI). Specific standards are mentioned in the body of this text, more are listed in section 4.14. All are listed in the Bibliography and followed by a [4].

Except where industry standards say otherwise, stainless steel and copper-alloy doors and frames with construction similar to that of carbon-steel doors should also comply with the standards applicable to carbon-steel doors.

The design of extruded-aluminum components should be in accordance with the recommendations in the AAMA publications *Metal Curtain Wall, Window, Store Front and Entrance Guide Specifications Manual* and the *Aluminum Store Front and Entrance Manual*.

Door and frame manufacturers should furnish written certification that the metal surfaces of their products have been finished as required.

A door and frame erector should have at least three but preferably five years' experience in installing such work. Stainless steel, aluminum, and copper-alloy doors and frames are often installed by their manufacturer or by a firm licensed by the manufacturer, especially when handicapped or security hardware, or revolving doors are involved.

Fire-rated doors and frames should have been tested by Underwriters Laboratories, Factory Mutual, or another recognized testing laboratory, as fire-door assemblies complete with the type of hardware to be used. Each fire door that falls within the size limitations imposed by the tests should be identified with the testing laboratory's labels, indicating the applicable fire rating of both the doors and the frames. Whether labeled or not, fire door and frame assemblies should be fabricated and installed in compliance with NFPA Standard No. 80 or ASTM Standard E 152.

Thermal-rated (insulated) door and frame assemblies should have been fabricated and tested as assemblies in accordance with ASTM Standard C 236.

Sound-rated (acoustical) assemblies should have been fabricated and tested as assemblies in accordance with ASTM Standard E 90 and classified in accordance with ASTM Standard E 413.

Finish hardware and the metals and finishes used for it should comply with the recommendations of the American National Standards Institute's standards A156.1 through A156.21. These current ANSI standards replace the Builder Hardware Manufacturers Association (BHMA) standards that were used for many years.

4.1.2 Controls

The general requirements discussed in sections 1.6.1 and 1.7.1 apply also to doors installed in new construction. The controls discussed in this section are in addition to those discussed in sections 1.6.1 and 1.7.1.

The door and frame manufacturer's specifications for fabrication, shop painting, and installation should be submitted for the owner's approval.

The manufacturer should submit for the owner's approval shop drawings for the fabrication and installation of metal doors and frames. The shop drawings should include details of each frame type, the elevation of each door type, the conditions at each opening, details of construction, location and installation requirements of finish hardware and reinforcements, and details of joints and connections. They should show anchorage and accessory items and describe anchorage systems. They should include an opening schedule, using reference numbers for details and openings. When the owner has engaged an architect, the references on the shop drawings should be the same as those used on the architect's drawings. The shop drawings should include new doors and frames; existing doors and frames that are to be removed and discarded or turned over to the owner; existing doors and frames that are to be removed and altered or refinished and reinstalled; existing doors and frames that are to be relocated; and existing frames that are to receive new doors.

The manufacturer or fabricator should also submit for the owner's approval sample corner sections of the required metal frames and doors showing construction details.

The best way for an owner to control the quality and appropriateness of finish hardware is to employ a certified Architectural Hardware Consultant (AHC). Refer to section 1.4 for additional information.

4.1.3 Materials

Refer to sections 3.2 "Carbon Steel and Stainless Steel," 3.3 "Aluminum," 3.4 "Copper Alloys," and 3.5 "Other Metals" for general discussions about those metals.

Carbon steel for interior door facings and frames should be either hot-rolled commercial-quality steel sheet and strip that comply with ASTM standards A 568 and A 569, with stretcher-level standard of flatness, or cold-rolled, commercial-quality carbon-steel sheets that comply with ASTM standards A 366 and A 568, with stretcher-level standard of flatness.

Carbon steel for exterior door facings and frames should be hot-dip zinc-coated (galvanized), commercial-quality carbon-steel sheets complying with ASTM standards A 525 and A 526, with G60 zinc coating. Sheets should be mill phosphatized and comply with stretcher-level standard of flatness.

Concealed stiffeners, reinforcements, edge channels, and moldings associated with carbon-steel doors and frames may be fabricated either from cold-rolled or hot-rolled steel, usually at the fabricator's option. Other steel products may also be used for framing and bracing that will be concealed within carbon-steel doors and frames, and for anchors and hardware.

Frames, subframes, concealed stiffeners, reinforcements, edge channels, moldings, clips, and louvers in stainless steel doors and frames should be fabricated from stainless steel; in aluminum doors and frames from aluminum; and in copper-alloy doors and frames, from copper alloy. The metal's thickness should be in accordance with the applicable industry standard.

Stainless steel is used for formed doors and their frames, for both interior and exterior installations. Most door facings on stainless steel doors are flat, stainless steel sheets, but patterned stainless steel sheets, plastic laminates, wood veneer, and other materials are also used. Sidelights, borrowed-light frames, and store-front frames are also made from stainless steel. Formed stainless steel is also used to make revolving doors.

Stainless steel is usually polished, but it may also be given an organic coating in the factory.

Aluminum is used to make both interior and exterior swinging doors and their frames. Sidelight, borrowed-light, and store-front frames are also made from aluminum. Frames and door stiles and rails are usually made from extruded aluminum. Most door panels are flat aluminum plate, but plastic laminate, wood veneer, and other materials are also used. Aluminum revolving doors are also usually made from extruded aluminum components.

A variety of alloys and tempers are used to make aluminum doors and frames. A common aluminum type used for extrusions has the properties of ASTM Standard B 221, alloy 6061, 6063, or 6463, with a temper that is suitable for the finish to be applied. Temper T5 is often used.

Sheet aluminum and bent plate materials used in doors and frames are typically required to have the properties described in ASTM Standard B 209 for alloys 5005, 5086, or 6061, as suitable for the finish to be applied.

Concealed aluminum is usually left with its mill (as fabricated) finish. Exposed aluminum is usually either clear or color anodized, but may be given an organic finish. A particularly popular organic finish today is fluorocarbon polymer, but various baked enamel and other opaque organic and inorganic coatings are also used.

Copper-alloy materials are used to make formed doors and frames for either interior or exterior locations. Most door facings on copper-alloy doors are copper-alloy sheets. Copper-alloy materials are also used to make formed frames for sidelights, borrowed lights, and store fronts, and the various components of revolving doors.

Copper-alloy materials are usually given a patina or architectural-bronze finish.

Exposed fasteners should have the same finish as the material being fastened. Painted fasteners should not be used unless the metal being fastened is also painted. If they are, the difference will be obvious.

Bolts used to attach stainless steel or aluminum frames to a supporting structure should be cadmium-plated steel, non-magnetic stainless steel, or zinc-coated steel. Bolts for use with copper-alloy frames should be of a copper alloy. The fasteners for frame assemblies should be stainless steel for stainless steel assemblies, non-magnetic stainless steel or aluminum for aluminum assemblies, and copper alloy for copper-alloy assemblies. Exposed fasteners should be countersunk.

Inserts for use in concrete or masonry may be of cast iron, malleable iron, or hot-dip galvanized steel.

Two types of sealants are used in conjunction with metal doors and frames. The first, which is almost always an elastomeric type, is used between two pieces of metal in a door or frame. It should be formulated to remain plastic permanently.

The second type is used between the frame and adjacent construction. In exterior locations, elastomeric sealants are usually used. Interior sealants may either be elastomeric types or an acrylic-latex type.

4.2 Design and Fabrication of Carbon-Steel Doors and Flush Panels

Carbon-steel doors and panels, which are often called hollow metal, should be rigid, neat in appearance, and free from defects, warp, and buckle.

The tops and bottoms of interior doors should either be welded flush or closed with a recessed, spot-welded enclosure. The top and bottom edges of exterior doors should be formed as an integral part of the door construction or by adding an inverted steel channel. Joints should be fully welded.

The cores of carbon-steel doors may be hollow, in which case interior steel bracing in one of several designs is used to stiffen the door. Many hollow steel doors are filled with a fireproof, rotproof, sound-deadening material. In thermal-rated (insulated) doors, the filler is either a glass fiber or foam insulation. In lieu of interior steel bracing, some hollow doors have a honeycomb core.

The maximum top, bottom, and edge clearances between doors and their frames and between the leaves of a pair of doors should be those recommended in the Steel Door Institute's publication SDI-100. In no case, however, should the clearances for fire-rated doors exceed those required by the authorities having jurisdiction.

4.2.1 Carbon-Steel Door Types, Styles, and Designs

There are many carbon-steel door types, styles, and designs available today, and even more have been used in the past. Most of them fit into the classifications and are used as explained by the Steel Door Institute in SDI-100 and SDI-108. References produced before 1983 will show different classifications. SDI-100 contains a chart comparing the old designations and those it currently recommends. The following is a summary only of the most-used types, styles, and designs. SDI-100 and SDI-108 contain much more complete descriptions.

Each of the major SDI grades (I, II, and III) includes four models (1, 2, 3, and 4). Grade I, Standard-Duty, includes doors that are either 1-3/8

4.2 Design and Fabrication of Carbon-Steel Doors and Flush Panels 153

or 1-3/4 inches thick. They are usually used as apartment, model, dormitory, and office-building doors into such spaces as closets, bathrooms, bedrooms, and offices. Grade I doors are not usually used in hospitals, nursing homes, or schools. Grade I door facings are 20 gage.

Grade II, Heavy-Duty, and Grade III, Extra Heavy-Duty, doors are 1-3/4 inches thick. Grade II doors are used as apartment-unit entrances, on public toilets, equipment rooms, and stairwells, and as fire doors. Grade III doors are used for main entrances; stairwells in apartment buildings, motels, and dormitories; and in industrial plants and other heavy-use locations. Grade II door facings are 18 gage; Grade III, 16 gage.

Models 1 and 2 are full flush doors in which seams are permitted on door edges but not on faces. Model 1 is of hollow steel construction; Model 2 has a solid core.

No seams are permitted on Model 3 or 4 doors. Model 3 doors are of hollow steel construction. Model 4 doors have solid cores.

Grade III also contains a Model 5 door, which includes flush panel style and rail doors with 16-gage steel face sheets.

The more common door designs include flush doors (Fig. 4-1), which have no applied moldings, panels, or openings. They may be of any grade except III and any model except 5.

Other designs are paneled doors, which have moldings on their faces. The moldings may either be applied over flush doors or else the face panels may have the moldings embossed in them. Various designs are available (Fig. 4-2).

Carbon-steel doors may also be fully glazed (Fig. 4-3) or only partly glazed (Fig. 4-4). In each case the glazing panels are held in place using moldings of not less than 20 gage that are made of the same metal as the door. The moldings may be a narrow rectangular flush type, with shaped

Figure 4-1 A flush door.

Figure 4-2 A few paneled-door designs.

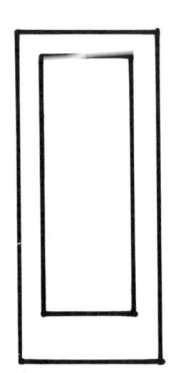

Figure 4-3 A fully glazed door.

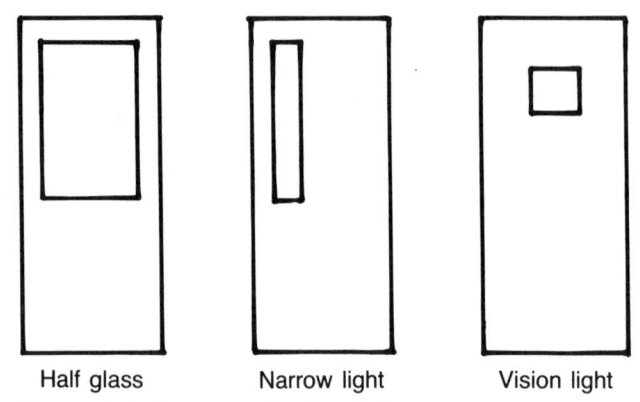

Half glass   Narrow light   Vision light
Figure 4-4 Some partly glazed door types.

4.2 Design and Fabrication of Carbon-Steel Doors and Flush Panels

Figure 4-5 A fully louvered door.

types, or half rounds. They all should be factory fit to avoid gaps in their corners.

Both interior and exterior steel doors may be either fully louvered (Fig. 4-5) or simply contain inserted louvers (Fig. 4-6). Refer to section 4.6 for a discussion of louver types and materials.

Steel doors may be a combination of design types. Louvers may be mounted in a paneled door, for example, or in a door that also has a glazed panel. Some of the panels in a paneled door may be glazed.

Dutch doors are a special design type that may be made up of any of the other designs. Dutch doors may have shelves but typically do not.

Lead-lined doors should be constructed and lined with sheet lead to comply with the recommendations of the National Council on Radiation Protection and Measurement (NCRPM).

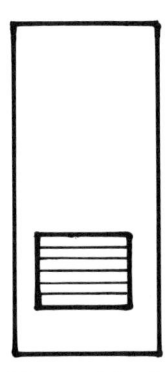

Figure 4-6 A door with an inserted louver.

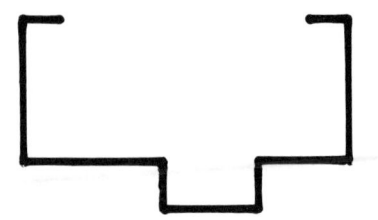

Figure 4-7 A typical frame shape.

4.2.2 Carbon-Steel Panels

Both insulated and uninsulated panels are used in transoms and in the bottom halves of sidelights or other glazed frames. The panels may be either of hollow steel construction or have solid cores. They are usually of the same construction as associated doors,

4.3 Design and Fabrication of Carbon-Steel Frames

Carbon-steel frames are formed from steel sheets (Fig. 4-7). They should be fabricated to be rigid, neat in appearance, and free from defects, warp, and buckle. Where practicable, units should be fitted and assembled in the manufacturer's plant. Units that cannot be permanently factory assembled before shipment should be clearly identified and marked to assure their proper assembly at the project site.

Exterior frames should be at least 14 gage, but thicker metals are also satisfactory and may be required for large frames and other demanding conditions.

Interior frames should be of not less than 14-gage steel for double door openings and single door openings not more than three feet wide or seven feet high. Other frames may be as light as 16 gage, although some conditions may require heavier steel.

Frames should be delivered to their installation location complete with fasteners, anchors, floor clips, reinforcements, and accessories. Fasteners should be concealed whenever possible. Those that must be exposed should be countersunk, flat Phillips-head screws and bolts.

Welded door frames should be prepared for shipment by welding temporary steel spreaders to the bottoms of the frames. Before they are shipped, frames should be stenciled, labeled, or tagged with instructions necessary for installation.

Except on weather-stripped frames and the frames of double-acting doors, smoke doors, and fire-rated doors, the stops should be drilled to

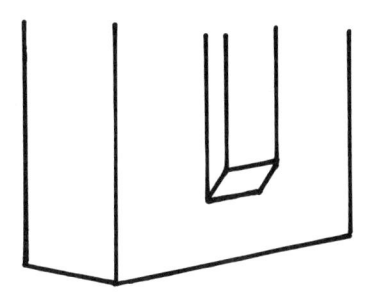

Figure 4-8 A terminated stop.

receive three silencers on the strike jambs of single-swing frames and two silencers on the heads of double-door frames.

Except for fire-labeled doors, lead-lined doors, and doors in smoke-barrier partitions, many jurisdictions require that interior doors in hospitals, nursing homes, and some other medical buildings have stops terminated six inches above the floor (Fig. 4-8). The bottoms of terminated stops should have filler plates welded in place at either 45 or 90 degrees, as required by the authorities having jurisdiction.

Plaster guards or mortar boxes of at least 26-gage steel should be welded to the frame at the back of finish-hardware cutouts where mortar or other materials might obstruct hardware operation.

Glazing beads should be removable and fabricated from at least 20-gage steel. They may be square or rectangular and either mitered or butted at the corners, but they should be factory fitted in full lengths to avoid gaps at the corner joints and joints other than in corners. The removable beads should be located on the side of the glazing material that falls in the space to be protected.

Cover and closure plates, uninsulated panels, and trim necessary to complete the installation of each frame should be fabricated from at least 14-gage steel sheet. The same steel should be used that is in the adjoining frame.

Loose beads and trim should be secured to the frame using oval-head, countersunk, self-tapping screws spaced approximately nine inches on center. Stainless steel screws should be used in removable stops.

Transom bars and mullions should be of tubular construction (Fig. 4-9). Sills should be of the desired shape, and may or may not be tubular. Their shape is usually similar to or identical with the other parts of the frame. All should be fabricated from the same steel and be of the same gage as the adjacent frame.

Subframes (Fig. 4-10) should be formed from not less than 12-gage steel.

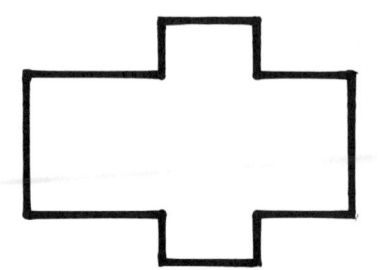

Figure 4-9 A tubular mullion or transom bar.

Most carbon-steel frames are not designed to support loads other than themselves and therefore require separate lintels to support loads imposed by a wall or partition above them. Sometimes, however, frames are reinforced by structural steel sections so that they carry imposed loads, and some frames are actually fabricated from channels, angles, beams, and other structural-steel sections with applied stops.

Even when they do not support superimposed loads, the heads of frames that are more than 49 inches wide should be reinforced with a continuous steel channel that is at least 14 gage. In addition, frames for power-operated doors and lead-lined doors should be reinforced as necessary to properly support the doors. Other frames may also require additional reinforcement under some circumstances, such as when they are required to carry a superimposed load. The forces, weights, equipment, and sizes to be encountered must be taken into account when sizing reinforcements. The

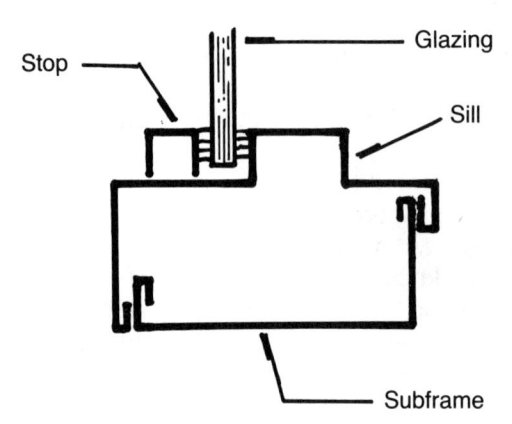

Figure 4-10 A sill with a subframe. Similar subframes are also used at jambs.

recommendations of the door-frame and automatic-door equipment manufacturers should be followed.

Frame tolerances should comply with the applicable portions of the ANSI A115 Series standards.

4.3.1 Carbon-Steel Frame Types

There are two standard carbon-steel frame types: welded buck and trim frames, and knockdown (K.D.) or drywall (DW) frames. The first type are used with steel or wood doors and as glazed frames. They are also used for sidelights, cased openings, borrowed-light frames, as frames for operating pass windows, and as steel store fronts. Knockdown (K.D.) frames, which are also called drywall (DW) frames, are designed to be installed after the partitions have been erected. Special types of frames are also available specifically for use in remodeling. These frames are often two-piece or half-frames designed to fit over an existing wood or steel frame. Adjustable two-piece frames are also available, as either welded or knockdown types. Some remodeling-type frames permit prehanging of the doors at the shop.

In every frame, molded members should be fabricated straight and true, with corner joints well formed, in true alignment, and with fastenings concealed where practicable. Metallic fillers should not be used to conceal manufacturing defects.

Welded Buck and Trim Frames. Frames in the welded buck and trim category should be fully welded, combination buck-and-trim, pressed-steel-type frames, with mitered corners.

Head, jamb, mullion, and sidelight intersections should be accurately mitered and welded to produce completely rigid assemblies. Joints should be continuously arc welded for the full depth and width of the frame, and the welds should be dressed smooth and flush on exposed surfaces. There should be no visible joints.

When possible, welded construction should also be used for cased opening frames, corner posts, counters, mullions, sills, removable heads, trim, and members for use at other special conditions. Short lengths should be avoided. Exposed ends should be closed with fillers. Concealed reinforcing should be used for rigidity and concealed connections should be used when details permit.

Knockdown Frames. Knockdown frames with mechanical joints are available in several types and styles. Those with mitered joints are usually the best, but this is not always true.

Knockdown frames are available with similar stop, trim, and glazing-bead features as those found in welded frames.

Store Front–Type Frames. Carbon steel frames of the welded type are sometimes used to produce glazed ground-floor interior and exterior walls and entrances. Frame sections that are too large to be shipped to the installation site in one piece must be designed and fabricated for field assembly. These frames must be designed, fabricated, and erected to prevent the passage of wind and water into the building. Many of the applicable requirements in section 4.5 apply to carbon-steel store front-type frames.

Lead-lined Frames. Frames can be specially fabricated or modified to receive lead linings when they will be used in diagnostic x-ray rooms, radiation laboratories, radiation treatment areas, and in other locations where radiation is a hazard. Lead linings should comply with the applicable recommendations of the National Council on Radiation Protection and Measurement (NCRPM). Frames should be lined with a single, continuous, unpierced sheet of lead.

Usually, the contractor is permitted to provide steel frames with factory-installed lead linings or to install the lead in the field.

4.4 Design and Fabrication of Formed Stainless Steel and Copper-Alloy Swinging Doors and Frames

Many of the designs available in carbon-steel doors (see section 4.2.1) are also available in standard doors manufactured from stainless steel. Most copper-alloy doors and frames and many stainless steel doors and frames are custom made, so full-flush, paneled, glazed, louvered, or almost any other desirable door design is possible. Consideration must be given to stability and strength, of course.

In most cases the Steel Door Institute's grades and models discussed in section 4.2.1 can also be made applicable to formed stainless steel and copper-alloy doors.

As is true for steel panels used in steel frames, panels for use in stainless steel or copper-alloy frames are usually made in the same way, from the same materials, as the associated doors.

Spreaders should not be welded to stainless steel or copper-alloy door frames. Frames should be kept in alignment, using temporary braces that will not mar the frame's finish.

Exterior door and frame assemblies should be required to meet the performance standards discussed in section 4.5 for extruded aluminum entrances and store fronts.

4.5 Design and Fabrication of Extruded Aluminum Swinging Doors, Frames, and Store Fronts

Although extruded stainless steel and copper-alloy products often appear in finish hardware, they are seldom used to make swinging doors or frames. Therefore, this section refers exclusively to extruded aluminum items.

Extruded aluminum components are used to make most entrances and store fronts, and some interior doors and frames. Flush doors, however, usually have aluminum face panels made from sheet or strip materials.

4.5.1 Design and Fabrication of Extruded Aluminum Swinging Doors

Each manufacturer's doors are fabricated according to that manufacturer's standards. Doors should, however, be designed and fabricated in accordance with the standards of the organizations mentioned in section 4.1.1.

Flush-Panel Doors. Aluminum flush-panel doors are generally manufactured using extruded tubular or channel frame and bracing members and aluminum sheet facing panels.

Joints in the framing for a door can either be welded or just held together by metal reinforcements.

Within the framing, most flush-panel aluminum doors are filled with a core material. This can be either a honeycomb made from resin-impregnated kraft paper, a rigid, closed-cell polyurethane insulation, or a rigid, noncombustible mineral insulation board.

The faces of most flush-panel doors are either flat or embossed aluminum panels made from sheet material. The panels should be not less than 0.064 inches thick. They should be fastened to or interlock with the framing and be bonded to the core with an adhesive.

Flush-panel doors may have glazed openings similar to those in formed stainless steel and copper-alloy doors. They may also have louvers.

Stile-and-Rail Doors. The tubular frame assemblies that make up the frames of stile-and-rail doors should be joined by welding or with concealed mechanical attachments. Their frames are often stabilized with tie rods.

Extruded aluminum stile-and-rail doors are identified according to their jamb-stile widths. There are four in general use. Thin stiles are less than 1-3/4 inches wide. Narrow stiles are nominally 2 inches in width. Medium stiles are nominally 4 inches wide. Wide stiles are 5 or more inches in width. Head and sill stiles also vary with the jamb stile but are not the same width as the jambs. Their thickness is usually 1-3/4 inches.

Doors may be either standard or a thermal-break type. In the latter, exterior metals are separated from interior ones by a nonconductive material.

Standard doors are usually glazed with 1/4-inch-thick glass. Thermal-break doors are usually glazed with 1-inch-thick insulating glass, but other glass thicknesses are also used.

Doors are sometimes required to comply with structural performance requirements based on applying certain specific loads to the top edge of the door on the side opposite the hinges and measuring the resulting deflection.

4.5.2 Design and Fabrication of Extruded Aluminum Door, Entrance, and Store Front Framing

In 1989 the technical director of the AAMA stated that about 75 percent of all store fronts are made using standard off-the-shelf components, as opposed to custom ones. The author suspects that a similar percentage would hold for all extruded aluminum doors and frames.

General Requirements for Framing. Frames should be furnished complete with glazing accessories, gaskets, glazing stops, doorstops, clips, fins, concealed flashings, anchors, and fasteners necessary for installation.

Incidental aluminum tubing, structural angles, channels, flats, and other supporting members, reinforcements, fasteners, trim, sleeves, drips, closer angles, and other components necessary to complete their installation should be furnished with frames. Aluminum or steel reinforcements for frames should be provided as necessary to meet the performance requirements of the framing.

Aluminum frame and opening assemblies should be fabricated in accordance with the industry standards mentioned earlier and with the standards of their manufacturer, so long as those standards do not conflict with industry ones.

Mechanical joints in frames and trim should be accurately milled and fitted to make hairline joints. Joints in exterior frames should be watertight. Concealed reinforcements should be used to make rigid connections. Metal-to-metal joints should be filled with sealant.

Anchor brackets at joints and reinforcements for hardware should be completely concealed and fastened in place invisibly.

Welds should be deep-penetration type made on the insides of shapes, and should be pickled and washed. No weld discolorations should appear on finished surfaces.

Trim, glazing beads, and gaskets should provide the minimal grip on the glazing recommended in the industry standards mentioned in Chapter 8 and should comply with the performance requirements of the frame. Exterior frames are usually glazed with 1/4-inch-thick single-pane glass or 1-inch-thick insulating glazing. Other thicknesses of single-pane or insulating glass may be used, however.

Anchorages should be designed to permit horizontal and vertical expansion of the frames.

Specific Requirements for Entrance and Store Front Framing. Extruded aluminum entrance and store front frames are usually tubes or channels, or variations of those shapes (Fig. 4-11). Frames for entrances and store fronts are typically 1-3/4 to 2 inches wide, although other widths are also used. Some are as thin as 1 inch. Mullions are usually 4 or 4-1/2 inches deep, but 5-1/2 inch, 6-inch, 7-inch, and other depths are also used. Base members may be the same width as the mullions but in many cases are wider. The typical extrusion thickness is .125 (1/8) inch.

Frames are available either standard (see Fig. 4-11) or with a thermal barrier between the metal that is exposed to the elements and the interior portion of the framing (Fig. 4-12).

Frames may be designed to be flush glazed or with applied glazing stops (see Fig. 4-11). The glazing channel or beads may be in the center of the frame, off-center, or close to the exterior face. Sometimes the jamb or head members are set so that the metal is flush with the surrounding material.

Store fronts are also sometimes glazed with gaskets mounted on the exterior face of the frame. Gasket glazing is more frequently used in glazed curtain walls, however, and is therefore addressed in Chapters 7 and 8. Framing associated with structural glazed conditions is also addressed in those chapters.

Recessed glazing   Applied glazing stops

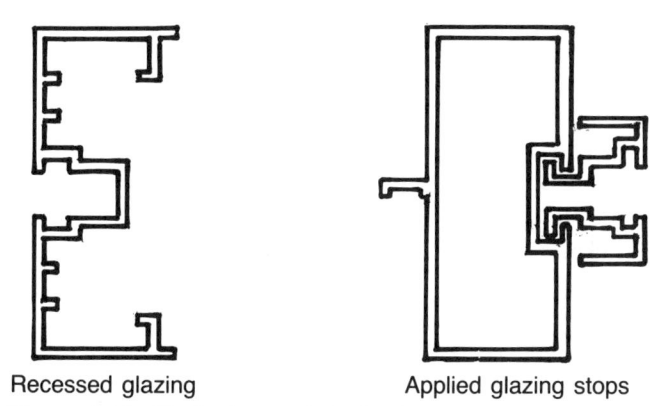

Figure 4-11 Typical extruded aluminum entrance and store front frames. Many other types are also available.

164   Metal Doors, Frames, and Store Fronts

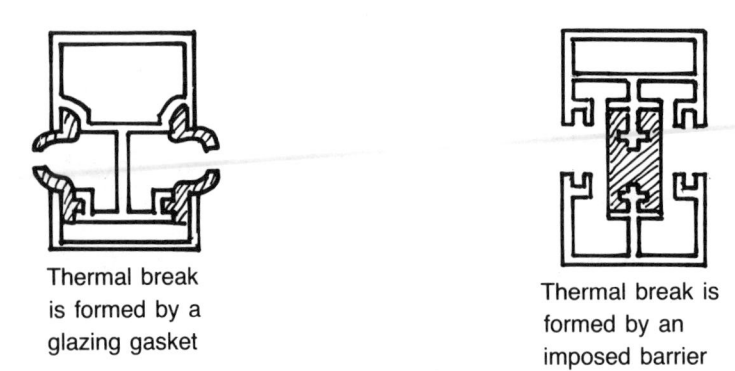

Thermal break is formed by a glazing gasket

Thermal break is formed by an imposed barrier

Figure 4-12 Two typical frames with a thermal barrier. Many other configurations are also available.

Entrances and store fronts should be required to meet performance requirements based on their location and use. The following list shows a typical group of requirements, which are examples only. After each category is the industry standard that the tests for that requirement are usually based on. Following the industry standard is the performance standard. The requirements for a particular project may vary. When a failure related to the listed types of performance is discovered, it is necessary to determine the limits that should have been required for the particular project and compare them with the actual installation. Failing to meet necessary performance requirements is a major cause of failure in entrances and store fronts.

Air infiltration: ASTM Standard E 283; .06 cubic feet per minute for each square foot of fixed area.

Water penetration: ASTM Standard E 331; no penetration at a test pressure of 8 pounds per square foot.

Condensation resistance: AAMA 1503.1; Condensation Resistance Factor (CRF) should be 56.

Thermal resistance: AAMA 1503.1; thermal transmittance (U-value) should be less than .6 BTU per hour per square foot per degree Fahrenheit.

Structural strength: ASTM Standard E 330; under a wind load of 20 pounds per square foot, the system should not deflect more than 1/175 of its span, with a safety factor of 1.65.

Specific Requirements for Interior Extruded Frames. Extruded aluminum frames for interior doors are available in a variety of sizes to fit the conditions. These frames often wrap around interior-partition construc-

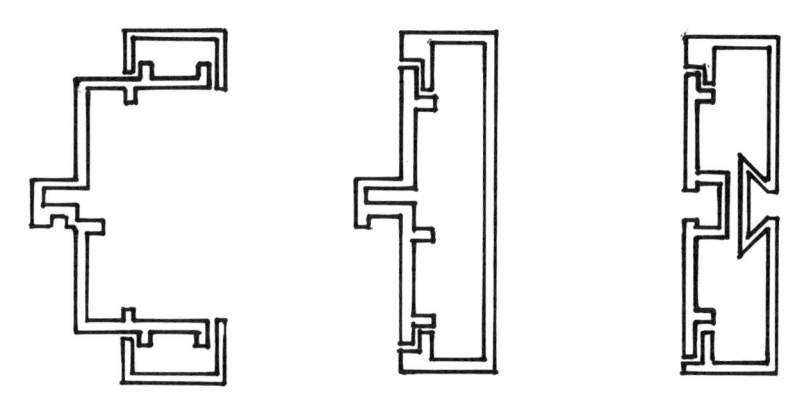

Figure 4-13 Some typical interior extruded-frame shapes. Many others are also available.

tion and are designed and sized accordingly (Fig. 4-13). In general, the requirements for this type of frame are similar to those discussed earlier in this chapter for formed frames.

4.5.3 Design and Fabrication of Packaged Entrances

Packaged entrances are factory-assembled units complete with doors, frames, and hardware. Their main advantage is that little field assembly is needed. They should, however, meet the requirements suggested in this chapter for aluminum entrances and store fronts.

4.6 Fabrication of Louvers

Doors may be partially or fully louvered (see Figs. 4-5 and 4-6). Louvers are also often used in carbon-steel frames and may occur occasionally in stainless steel or aluminum frames.

Louvers in carbon-steel doors and frames may be either carbon steel, stainless steel, or aluminum. Carbon-steel louvers in exterior doors or frames should be hot-dip galvanized. Louvers in stainless steel, aluminum, or copper-alloy doors or frames should be of the same material and finish as the doors and frames. Louvers are usually finished to match the adjacent door or frame.

Many properietary louver-blade shapes are available for interior louvers. Where vision blocking is unnecessary, louver blades may be of the straight type, but most interior louvers, especially those in doors, have sightproof, inverted Y-type (Fig. 4-14) or V-type blades. Door louvers are

166   Metal Doors, Frames, and Store Fronts

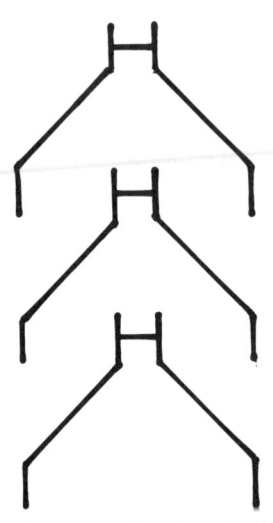

Figure 4-14 A typical inverted Y-shaped louver blade.

usually of the stationary type. Interior-frame louvers are also usually fixed, but may be backed up by operating louvers.

Exterior-door louvers should be weatherproof, as is the Z-shaped one in Figure 4-15. Exterior louvers require insect screens, which should be mounted on the interior face of the louver in a removable frame. Most exterior door and frame louvers are fixed, but some are operable. The frames of exterior louvers should be designed so that water entering the louver drains away. Wide louvers may require bracing. The most satisfactory type of bracing is concealed mullions fastened to the interior of the louver.

The blades of interior louvers may be secured to their surrounds by welding or some other method. Exterior louver blades should be welded in place, although some are not.

Fixed louvers are usually required to have at least 40 percent free air before they are screened, but other percentages are also often required.

The best door louvers are welded into the door flush with the door face. Overlapping moldings are sometimes used, but are undesirable, especially on exterior doors, where they are subject to leaks.

Special louvers, such as fusible-link or light-proof ones, are occasionally needed.

Door louvers are usually finished to match the doors.

Where ducts are connected to louvers in exterior-wall frames and the ducts are smaller than the louvers, the unused portions of the louver should be blanked out on its inside face. Blank-off panels are manufactured in

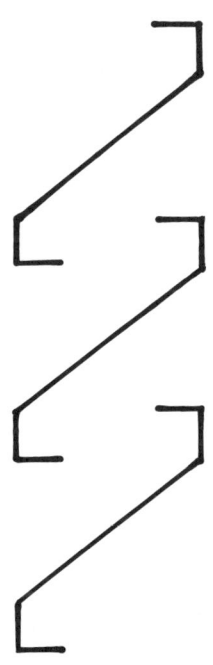

Figure 4-15 A Z-shaped louver blade.

several configurations. One of the most used types consists of one-half-inch-thick glass-fiber insulation or its equivalent installed between two sheets of metal matching the louver.

4.7 Finish Hardware for Swinging Doors

Finish hardware for carbon-steel, formed stainless steel, copper-alloy, and interior extruded-aluminum swinging doors is usually furnished by a hardware supplier. The finish hardware for extruded-aluminum entrances and store fronts is usually furnished by the door and framing manufacturer, except that lock cylinders are often furnished by the supplier of the building's other finish hardware.

Unless the owner or one of the owner's consultants is a finish-hardware expert, it is a good idea to engage the assistance of someone who is. Certified Door Consultants (CDCs) and Architectural Hardware Consultants (AHCs) both qualify. Refer to section 1.1.4 for more information about CDCs and AHCs.

4.7.1 Hardware Types

This section discusses four types of hardware: standard, handicapped-person, automatic door, and security.

Standard hardware includes butts, hinges, lock and latch sets and their trim, exit devices, door controls, closers, holders, auxiliary locks, architectural door trim, weather stripping, and thresholds.

Handicapped-person hardware is designed to permit easier operation of doors by handicapped persons. Such hardware ranges from counterbalanced closers that allow opening a door with little force to fully automatic door operators.

The many available kinds of automatic operators include electromechanical, hydraulic, and pneumatic-operated devices. There are also several types of actuating devices, including floor treadles, motion sensors, sonic sensors, photoelectric cells, manual push buttons or plates, and remote switches.

Probably the simplest of automatic hardware is the fire- or smoke-actuated closer in which a detector senses smoke or fire and releases a hold-open device that permits the closer to close the door. The detector might be just a fusible link built into the hold-open device. When the temperature reaches a preset level, the link melts, and the door closes.

Building security may require simply replacing the standard components of a bored or mortise lock with high-security cylinders and other components. It may, however, require one of a variety of special control devices, such as electric (including electromagnetic) or electronic locks or three-way key locks in which a single key operation drives bolts into the jamb, head, and sill. Security locks operate by means of keys, push buttons, or card readers.

4.7.2 Hardware Materials and Finishes

Finish hardware comes in many different materials and finishes. Materials include cast iron, steel, stainless steel, aluminum, copper alloys, and zinc. Metal-plated metal is also used. Refer to sections 3.5 and 3.10.1 for discussions about other metal and metallic-coated metals. The finish hardware on stainless steel, aluminum, and copper-alloy doors is often of the same material and finish as the doors.

The finishes used on hardware items include all those discussed in Chapter 3 that are applicable to the basic metal used to make the hardware in question. Currently, finish hardware finish designations are addressed in ANSI A156.18, "Materials and Finishes," which replaces older U.S. Department of Commerce standards (US26D, for example) that were used for many years.

4.7.3 Preparation of Doors and Frames for Finish Hardware

Doors and frames should be prepared to receive mortised and concealed finish hardware in accordance with the hardware supplier's templets and the owner's instructions as expressed in a finish-hardware schedule. Formed carbon-steel, stainless steel, and copper-alloy doors and frames, and extruded-aluminum interior doors and frames, should be prepared in compliance with ANSI A115 Series standards. Finish hardware should be located in accordance with *Recommended Locations for Builder's Hardware for Standard Steel Doors and Frames,* published by the Door and Hardware Institute (DHI).

Extruded-aluminum entrances and store fronts should be prepared at the factory to receive finish hardware in accordance with its manufacturer's instructions and the owner's hardware schedule.

Doors and frames should be reinforced at the factory to receive mortised hardware, other concealed hardware, and surface-applied hardware. They should also be drilled and tapped at the factory for mortised and other concealed hardware. Extruded-aluminum entrances and store front components should be drilled and tapped at the factory for all hardware. Drilling and tapping of other doors and frames for surface-applied finish hardware, however, can be done at the installation site. Templets provided by the hardware supplier should be followed. Welded-in-place reinforcement should be provided for hinges and pivots, but hinges should not be welded to frames. Reinforcement should also be provided for locks, closers, and other mortised or surface-mounted hardware.

Where the surrounding walls or partitions will be erected or finished after the doors have been installed, cover boxes should be placed in back of hardware cutouts to prevent intrusion of mortar, plaster, gypsum board joint compounds, and other materials into the hardware.

4.8 Frame Supports and Anchors

Supports and anchors for metal frames should be fabricated from heavy-gage metals. Anchorage for fire-rated assemblies should be UL or Factory Mutual–approved types, as required for the conditions.

Inserts, bolts, and fasteners should be the door and frame manufacturer's standard units. Steel units to be built into exterior walls should be hot-dip galvanized in compliance with ASTM Standard A 153, Class C or D, as applicable.

Anchors should be of the shapes and sizes required for the adjoining type of wall construction. Floor clips should be provided at the bottom of each jamb member. They may be the fixed type or the extension type, as required by conditions. Each clip should be drilled to receive two 3/8-inch-diameter anchor bolts.

Jamb anchors should be located near the tops and bottoms of each frame and at intermediate points not more than twenty-four inches apart. At least four clips per jamb should be provided for single doors in stud partitions.

Head anchors should be of the same type as jamb anchors. In stud partitions, two should be provided in the head of single doors, and four per head of double doors.

Anchors for frames set in masonry should be at least eight inches long. They should be adjustable and corrugated or have another deformed shape.

Anchors for frames set in studs should be clips designed for the purpose. They should be welded in the frame jamb and head.

Frames set in previously placed concrete, stone, or unit masonry walls may be anchored by screwing or bolting them to subframes of the same material as the frame, which are in turn bolted to the wall. Alternatively, such frames may either be bolted to the wall through the stop or anchored with another type of device of a design suitable for the location and conditions.

Knockdown and renovation-type frames are usually screwed or toggle bolted to existing wood or metal framing, or bolted to expansion shields in concrete, stone, or unit masonry.

In every case, countersunk holes should be provided in the frames for exposed anchorage.

Anchorage for fire-rated frames must be UL approved types as required for the condition.

4.9 Installation of Metal Doors and Frames in New Construction

Metal doors and frames should be delivered to the installation site cartoned or crated to protect them during transit and storage at the site. Items with minor damage may be repaired and installed, but those with major damage should be discarded and new, undamaged items used.

Doors and frames should be stored at the building site under cover. The units should be placed on wood sills that are at least four inches high or stored on floors in a manner that will prevent corrosion and damage. The use of unvented plastic or canvas shelters that could create a humidity chamber should be avoided. If the cardboard wrapper on a door becomes wet, it should be removed immediately. At least 1/4-inch-wide spaces should be provided between stacked doors, to promote air circulation.

4.9.1 Preparation for Installation

The installer should examine the substrates, supports, and conditions under which metal doors and frames are to be installed and ascertain conditions detrimental to proper and timely completion of the work. Installation of frames or doors should not proceed until unsatisfactory conditions have been corrected.

Where a frame would otherwise be insufficiently supported, framing should be built in to provide proper support. Steel angles, channels, or other shaped members are usually used to construct an overhead support for free-standing doors and frames.

4.9.2 Installation of Formed Frames and Doors

Formed doors and frames are made from carbon steel (hollow metal), stainless steel, and copper alloys. Such frames should be installed in accordance with the provisions of the Steel Door Institute's publication SDI-105, *Recommended Erection Instructions for Steel Frames*. Care must be exercised so that the frames are not bent, twisted, or racked. Damaged frames should not be installed. Frames should be set plumb, aligned properly, and anchored securely in place.

The installation of frames for power-operated doors and lead-lined frames must be coordinated with the installation of electrical work, the lead lining, and other associated work.

Except for frames in already placed concrete, stone, or masonry walls and knock down-type frames in wood or metal stud walls or partitions, frames should be placed before the enclosing walls and ceilings are erected. Frames should be set accurately in position, plumb, in the proper alignment, and securely braced until permanent anchors are set. After wall construction has been completed, temporary braces and spreaders should be removed, leaving surfaces smooth and undamaged.

Exposed connections should fit accurately together to form tight, hair-line joints. Connections that will not be left as open joints or screwed, bolted, or riveted together should be welded, in the shop when possible. Joints can be welded in the field when shipping-size limitations preclude shop-welding. In both cases, exposed welds should be ground smooth. Shop-paint coats should be touched up after welding has been completed, using the same paint used for the shop coat. Exterior frames that have been hot-dip galvanized after fabrication, or that will have bolted or screwed field connections, or that have been factory prefinished should not be welded, cut, or abraded.

As discussed earlier in this chapter, frames should be reinforced when necessary, with structural sections of the same material used in the rest of the frame where necessary. Members used as reinforcement should be anchored securely to the structure.

In new masonry construction, three wall anchors should be installed in each jamb, at the hinge and strike levels.

At in-place concrete, stone, or masonry construction, frames should be set and secured to the adjacent construction with not less than three machine screws and masonry anchorage devices at each jamb. At existing openings having wood subbucks, the subbucks should be removed and the jambs repaired, using masonry. The new mortar should be allowed to set properly before the frame is secured to the new masonry or loads are applied to embedded anchors.

In metal stud partitions, the number of wall and head anchors mentioned earlier in this chapter should be attached to the studs, using self-tapping screws.

Fire-rated frames should be installed in accordance with NFPA Standard No. 80. Labels should neither be removed nor painted over.

The bottoms of frames should be anchored to the floor with expansion bolts. Powder-actuated fasteners will also work, but they are generally too dangerous to use if the building is occupied.

Carbon-steel and stainless steel frames in masonry walls should be filled solid with mortar as the walls are laid. Corrosive additives should not be used in the mortar. Copper-alloy frames in masonry should not, however, be filled with mortar unless the metal is protected as discussed in section 3.11.7. Most copper-alloy frames are not designed to wrap around adjacent masonry construction and are supported using concealed bolts or screws into sleeves or subbucks rather than by masonry anchors.

Frames in metal stud partitions should be fastened to the studs as the partitions are erected. Carbon-steel and stainless steel frames are usually spot grouted, but they should be fully grouted when either the frame manufacturer or the partition manufacturer or supplier recommends doing so. Copper-alloy frames in metal stud partitions should not be spot grouted unless they are protected from the grouting material.

Except under extraordinary circumstances, metal frames should not be used as lintels to carry stone, unit masonry, or wood or metal partition framing. Instead, a separate structural lintel should be installed over each frame.

Frames that have been installed out of square or out of plumb should be removed and reset, to provide a properly operating installation with an acceptable appearance.

The perimeters of frames in concrete, stone, or unit masonry and of other frames where joints would otherwise be left open or unsatisfactory for painting should be filled with a sealant.

Metal doors should be fitted accurately in their frames within the clearances specified in SDI-100. Fire-rated doors should be placed with the clearances specified in NFPA Standard No. 80. Fire-rating labels must not be altered, removed, or painted over.

4.9.3 Installation of Extruded Aluminum Swinging Doors, Frames, and Store Fronts

Interior aluminum doors and frames of other than entrance and store front types should be installed in accordance with the requirements in section 4.9.2, even though the frames are extruded.

Extruded-aluminum entrance doors and associated frames and store front frames should be installed in accordance with the recommended procedures of the AAMA's *Aluminum Store Front and Entrance Manual*, the manufacturer's recommendations, and the approved shop drawings.

Units should be erected plumb, level, true, and in proper alignment and should be securely fastened in place. Sealants should be used where necessary to make the installation weather- and watertight.

Frames should be securely anchored to the building's frame or construction. The necessary anchors and inserts should be set properly in concrete and masonry. Units should be anchored securely, with concealed fasteners where practicable.

Accessories and trim should be installed rigidly, plumb, square, level, and true in the proper locations, in alignment with other work, and properly scribed to fit where necessary.

Exposed fasteners should be used sparingly. Those that are necessary should be countersunk Phillips-head screws finished to match the metal being fastened.

Sealants should be installed within frame and trim assemblies and at the perimeter, to make the installation weathertight.

4.9.4 Installation of Glazing and Louvers

Glazing and its installation are discussed in detail in Chapter 8.

Door louvers and their screens are usually installed in metal doors at the factory. They should be set accurately and securely anchored in place.

Louvers and blank-off panels should be installed in metal frames in accordance with applicable industry standards and the instructions of the manufacturer.

Removal louvers should be installed using removable stops set with glazing compound. Fasteners should be either cadmium-plated or stainless steel countersunk screws.

Screens with metal frames should be attached to the insides of exterior wall louvers. Where ducts are not present, the screening is usually 14 by

18 mesh, or some other size of insect screen. Where ducts are present, the screening is often one-half-inch-mesh bird screening. Screening is made from many materials, including aluminum, steel, copper, and glass-fiber wire. Some metal screen wire is coated with plastic.

4.9.5 Adjustments, Cleaning, and Protection

Immediately after they have been erected, corroded or damaged areas of shop-applied prime-paint coats on carbon-steel doors and frames should be sanded smooth and touched up using a compatible air-drying primer.

In the few instances when stainless steel, aluminum, or copper-alloy doors or frames come to an installation site with a factory-applied prime coat for finish painting in the field, immediately after erection, corroded and damaged areas of the prime coat should be sanded smooth and touched up using a compatible air-drying primer.

Where aluminum or a copper alloy is to be placed in contact with dissimilar metals, pressure-treated wood, concrete, masonry, mortar, or plaster, concealed surfaces should be painted with a heavy coat of alkali-resistant bituminous or zinc-chromate paint, or the metal should be separated from the incompatible materials using another acceptable method.

Exterior-finish-system materials, cementitious materials, and sealants should be removed immediately from metal surfaces. Such materials must not be allowed to harden or set or to damage the metal's finish.

Metal surfaces should be cleaned after completion, without damage to the finish. Strippable coatings and other protection should be removed.

Damaged and broken door and frame components should be properly repaired, or removed and new, undamaged components installed. Temporary protection should be removed.

After an installation has been completed, the finish hardware should be checked and adjusted to ensure its proper operation.

Metal doors and frames should be left undamaged, complete, properly operating, and free from dirt, excess sealant, and glazing compounds.

4.10 Design, Fabrication, and Installation of Revolving Doors

Revolving doors are proprietary items that come as complete packages including operating hardware. They are available in stainless steel, aluminum, and copper alloys, in almost any type of finish those materials are capable of receiving.

Most revolving doors vary in diameter from six to ten feet, but some are even larger. They come in three- or four-wing designs. Many options are available, from special hardware to glass ceilings. Fully customized installations are frequent.

Revolving doors operate either manually or with a power assist. Both must have speed controls, to prevent damage to the door and harm to the people using it. The optimal speed of a revolving door is ten revolutions per minute.

Some special applications may be found for revolving doors. For example, doors with larger than normal wings and slower than usual operating speeds may be used in entrances to hospitals, airports, and train terminals. Revolving doors sometimes have access-control devices so that entry can be restricted.

4.11 Why Metal Doors and Frames Fail

Metal doors and frames and their finishes fail for many reasons, most of which can be grouped into three categories. The first consists of problems with the surrounding or supporting construction. It includes excess structural movement, failed steel or concrete structure, failed wood structure or wood wall or partition framing, failed metal wall or partition framing, failed concrete, stone, or masonry walls or partitions, and other building element problems. These causes are discussed in Chapter 2. While they may not be the most probable cause of the failure of doors or frames or their finishes, they are usually more serious and costly to fix than the other possible causes. Consequently, the possibility that they are responsible for a failure should be investigated.

The second category concerns the materials and finishes used in doors and frames. It includes bad materials, inappropriate finish selection, improper preparation for and application of finishes, failure of the immediate substrate, failure to protect materials and finishes, failure to properly maintain applied finishes, and natural aging. Those failure causes and the failures they can cause are discussed in Chapter 3.

The third category includes improper design, inappropriate door or frame type selection, inappropriate hardware selection, improper fabrication, improper installation, unforeseen trauma, and natural aging. These types of failure are discussed in this section.

4.11.1 Improper Design

Designing a door or frame that is too weak for the loads that will be applied to it can lead to its failure. This error includes using metals that are too thin (Fig. 4-16). A related error is selecting fasteners or anchors that are

Figure 4-16 An inadvertently applied load on this store front trim caused it to buckle at its weakest point.

too weak for the application, too few in number, placed too far apart, or of the wrong type. Any one of these factors can lead to failure of the anchor or fastener.

Placing a door or frame in a location where it is likely to be damaged will often lead to its failure (Figs. 4-17 and 4-18).

Selecting poor shapes for frames can produce sections that do not fit together well. There are limits on size and configuration that prevent some designs from being made at all, but the kinds of problems that find their way into existing construction are more subtle, such as misfitting joints and surfaces that do not align.

Failure to provide sufficient and appropriate means of fastening the components of a frame together can lead to joint separation and both air and water leaks.

Requiring that incompatible fasteners be used will lead to problems. Steel fastened with copper-alloy bolts, for example, will corrode. Requiring that incompatible anchors and attachments be used is a similar problem.

Requiring the wrong type of joining method is another error. Insisting that aluminum frame components be welded or brazed in the field is one

Figure 4-17 The store front corner shown in this photo suffers from constant battering because of its exposed location.

example. Specifying that a thin metal be welded to a thick one is another. Calling for highly polished metals to be joined by brazing will almost surely lead to a failed joint. Welding most copper alloys will produce unsightly joints. Designing open joints where a unit must exclude water will not necessarily lead to failure of the joint, but the unit will not do its job.

Requiring that butt joints be used in anodized aluminum or copper-alloy panels will produce an unsatisfactory appearance. No two anodized panels or finished copper-alloy panels will ever exactly match in color and shade.

There are some sizes and configurations that prevent a given design from being made at all. Others can be made but not given a final finish. But the kinds of problems that find their way into existing construction are usually more subtle, such as misfitting joints and surfaces that do not align.

4.11.2 Inappropriate Door or Frame Type or Design Selection

Selecting a flush door design that includes a long, narrow vision light for a door that must have a louver (Fig. 4-19) can result in a door that is too weak.

Figure 4-18 The store front corner in this photo fared no better than the one in Figure 4-17, in spite of its wood sill.

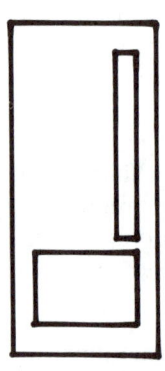

Figure 4-19 Most door manufacturers will not warrant doors with configurations such as the one shown in this drawing, because such doors often fail.

A similar weakness occurs when vision panels or louvers are selected that are too large for the door, which leaves the portions of the door surrounding the openings too small. This problem can exist even in stile and rail doors.

Selecting door sizes without taking into account adjacent finishes or door operation can lead to misfitting doors and damaged finishes. A door that strikes a carpet will surely damage the carpet, for example. Not using the recommended tolerances is a similar error that can lead to problems with door or weather-stripping operation.

Selecting the wrong type of louver can lead to problems. A Y-shaped louver may be fine for restricting vision, for example, but it will not prevent light from penetrating.

Frame-design errors can also lead to failures. Designing too shallow a rabbet, for example, can result in losing glass in a storm.

4.11.3 Inappropriate Hardware Selection

A problem frequently associated with door failure is that of using the wrong hardware. Failing to coordinate hardware selection with door design is one example. Another is selecting surface-mounted weather stripping for a door with provisions for recessed weather stripping. This will only create a problem, of course, if the door installer uses the wrong weather stripping anyway.

Selecting hinges, butts, or pivots that are too small for the door's weight can lead to sagging or even inoperative doors. Using too few hinges or butts can have a similar effect, as can placing hinges, butts, or pivots in the wrong locations.

Selecting standard closers for entrances for the handicapped will result in doors that are too hard for such a person to open.

Failing to require metal receivers for dead bolts can make unauthorized entry as easy as a kick.

Selecting the wrong type of hardware for a revolving door can make it operate too fast or too slow or be too hard to open.

4.11.4 Improper Fabrication

Improperly fabricated metal doors or frames are a sure precursor of failure. Perhaps the most frequent cause of improper fabrication is failure to follow the requirements of the design, including the shop drawings and specifications and the standards of the industry.

Many fabrication errors involve joints. Errors include improperly made welded joints. Welding without proper preheating can lead to stress-corrosion cracking, for example. Using the wrong weld metal can result in cracking in either the weld itself or the adjacent metal.

The welding or brazing of metals, especially thin ones, from the visible side will usually lead to unsightly joints.

Failing to properly clean a joint to be welded, brazed, or soldered will almost always result in a failed joint.

Not using flux before brazing or soldering a joint will probably result in failure of the brazing metal or solder to adhere to the metal, and the joint will fail.

Another common error in producing metal doors and frames is joining the components in such a way that corners or intersections are not square, plumb, and properly aligned and the joints are not flush, tight, and correctly fitted.

Using damaged materials in a door or frame, or damaging them during or after fabrication, must be avoided. Damaged components and materials should be repaired or removed and new ones installed.

The damaging of coil coatings and other preapplied finishes during fabrication is a problem. Such damage should be touched up before the item loses enough protection to permit it to suffer damage from rust or other corrosion.

4.11.5 Improper Installation

Correct preparation and installation are essential if failure in metal doors and frames or their finishes is to be prevented.

Probably the number-one cause of failure due to installation errors is failing to follow the design and recommendations of the manufacturer and recognized authorities, such as the ASTM, ANSI, SDI, DHI, or NAAMM.

Failing to follow the design can mean using anchors, fasteners, and other materials that will rust or otherwise corrode where high humidity will occur or where they will be set in or in direct contact with masonry or concrete. When materials that will corrode must be used for anchors, an often-made error is that of failing to protect them properly with an applied bituminous or other appropriate coating.

Not following the design and recognized standards can also lead to using anchors, fasteners, clips, and other devices that are of the wrong type for the installation, are too small or weak to perform the necessary function, or are not in the proper number or location.

Failures can also occur if the area where a metal frame is to be installed is not properly prepared or ready to receive the frame.

An allied error is to install a damaged frame or door. While the damage itself may or may not be the fault of the installer, installing a damaged door or frame is.

4.11 Why Metal Doors and Frames Fail

Installing a frame in such a way that building loads that the frame is not designed to support are transmitted into it will lead to problems later. Failure to provide a lintel over a metal frame in a masonry wall is an example.

The installer must be careful to install doors and frames so that their various components finish in the proper plane, within acceptable tolerances, so that the unit is in the proper location, and is plumb, level, and in alignment. Improperly aligning a component may not only cause the unit to present an unpleasing appearance but can force loads to be applied eccentrically, which can cause damage to the door or frame.

Failing to seal around the perimeter of a metal frame will leave an ill-appearing installation that may result in air or water leaks (Fig. 4-20).

A common error with aluminum store front sections is failing to remove anodizing acids from the aluminum. If the acid used to anodize the aluminum is not properly removed, it will eventually wash off and damage the aluminum and adjacent surfaces. Acids trapped in hollow sections are a particular problem, as are acids trapped in joints between assembled sections.

Figure 4-20 Calking did not make a thing of beauty out of the conditions shown in this photo, but it did prevent further leaking.

Weep holes are usually needed in hollow sections and assemblies to ensure proper removal of acids.

Another error in installing aluminum doors and frames is failing to remove strippable coatings and tape soon after the installation has been completed. Such materials may become difficult or impossible to remove after prolonged exposure to sunlight.

4.11.6 Unforeseen Trauma

There is little that can be done to prevent natural disasters from causing damage to doors and frames. A tornado, hurricane, or earthquake can damage a building's structure or cause it to move in unanticipated ways, thereby damaging associated doors and frames. Such an event can also damage doors and frames even when it does no harm to other portions of the building (Fig. 4-21).

4.11.7 Natural Aging

Aging affects all aspects of doors and frames. Hardware may eventually simply wear out, for example. Most aging problems can be delayed for many years by performing proper maintenance. Refer to section 3.12.9 for a discussion about the effects of aging on materials and finishes.

4.12 Repairing Metal Doors and Frames

Existing doors and frames that are damaged or in an unsightly condition but still usable may often be repaired and restored. Sometimes it is necessary to remove a door or frame to make repairs, but many can be made without removing the unit. Refer to section 3.13 for a discussion about cleaning finishes and repairing wood and metal materials.

When working with existing doors and frames, it is best when possible to use materials and methods that match those of the original installation. For instance, when a portion of a frame has been damaged, it might be possible to remove and discard only that part and provide a new replacement section. When fasteners or attachments have failed, it will be necessary to provide new attachments. It will also probably be necessary to remove damaged finish hardware and install new items. Doors and frames that are damaged too severely to repair must, of course, be discarded and new units provided.

Sometimes an apparent failure in a door or frame will actually be a failure in an adjacent or supporting material or structure. Each condition must be examined in light of the data included in this book and in the

4.12 Repairing Metal Doors and Frames   183

Figure 4-21 The damage to the doors shown in this photo was caused by a tree limb being driven by tornado-force winds.

referenced sources of additional information. The condition of the damaged door or frame should then be brought as close as possible to the condition usually required for that type of item when new. When damage has been observed, apply the specific data in this book, those in the listed sources of additional information, and the expert knowledge of someone knowledgeable about the type of problem that exists, to solve the problem.

Where existing doors or frames are to be repaired in place or removed, repaired, and reinstalled, there are additional requirements beyond those discussed earlier. At least some selective demolition, dismantling, cutting, patching, and removal of existing construction is usually necessary. The comments related to such matters in sections 1.6 and 1.7 are applicable when existing metal doors or frames must be repaired.

4.12.1 General Requirements

This section contains suggestions for repairing existing metal doors and frames. Because these suggestions are meant to apply to many situations, they might not apply to a specific one. In addition, there are many other cases that are not specifically covered here. When a condition arises in the field that is not addressed here, advice should be sought from the additional data sources mentioned in this book. Consultation with the manufacturer of the item being repaired will often help, but, sometimes it is necessary to obtain professional help (see section 1.4). Under no circumstances should the recommendations here be followed without careful investigation and the application of professional expertise and judgment.

Before an attempt is made to repair an existing metal door or frame, the manufacturer's and finisher's recommendations for making the repairs should be obtained. When finish hardware is involved, the participation of a certified Architectural Hardware Consultant (AHC) is called for. It is necessary to follow the recommended precautions against using materials and methods that may be detrimental to the items requiring repair. Prevention of future failure requires that repairs not be undertaken without careful, knowledgeable investigation and the application of professional expertise and judgment. The repair should be carried out by experienced workers under competent supervision.

Unless existing doors or frames must be removed to facilitate repairs to them or other portions of the building, or they are scheduled to be replaced with new materials, only sufficient materials should be removed to do the cleaning and repairing. The more materials removed, the greater the possibility of additional damage to them during handling and storing. Materials that are loose, sagging, or damaged beyond in-place repair should be removed. Components that are sound, adequate, and suitable for reuse may be left in place or be removed, cleaned, and stored for reuse later.

After doors and frames have been removed, concealed damage to supporting elements such as framing or substrates may become apparent. Such damage should be repaired before repaired or new doors or frames are installed.

Existing doors, frames, and supports that will be removed should be taken out carefully, and adjacent surfaces should be protected so that the process does no damage to the surrounding area.

Unless a decision is made to discard them, damaged metal doors and frames and support system components that can be satisfactorily repaired should be repaired, whether they have been removed or left in place. Failing to repair known damage may lead to additional failure later on. The methods recommended by the manufacturer of the doors and frames should be followed carefully when making repairs. Doors, frames, and associated items that cannot be satisfactorily repaired should be discarded and new, matching items installed.

Areas where repairs will be made should be inspected carefully to verify that existing components that are to have been removed have been, and that the substrates and structure are as expected and not damaged. Sometimes, substrate or structure materials, systems, or conditions are encountered that differ considerably from those expected, or unexpected damage is discovered. Whether the damage was previously known or was found later, it should be repaired before door or frame components are reinstalled. Door and frame reinstallation should not proceed until unsatisfactory conditions have been corrected.

4.12.2 Repairing Specific Types of Damage

In addition to the general requirements for repairs already discussed, the following specific recommendations apply to specific problems.

Bent, Twisted, or Deformed Doors or Frames. If the damage is not too severe, it may be possible to straighten twisted or deformed door or frame components.

Most repairs to metal, especially those involving bent or twisted materials, are not do-it-yourself projects. Expert knowledge and equipment are needed. The types of repairs addressed here will usually damage the finish. An opaque coating can usually be restored or painted over, although doing so sometimes makes it necessary to paint every similar door and frame. Damage to the common types of mechanical and chemical finishes and coatings used on stainless steel, aluminum, and copper alloys is not usually repairable in the field. When an item must be removed and shipped back to its manufacturer or finisher it is usually best to have the repairs made in the shop, where working conditions can be controlled more easily.

It is usually not possible to straighten bent, twisted, or warped doors or frame members without removing them. Often it is necessary to send the item to a shop for the repairs to be made. Rebending metal may set up stresses in it that can be relieved only by a heat treatment of some sort, which is usually impracticable to do in the field.

The necessity of shipping a door or frame to a shop for repairs is sometimes so expensive that it is cheaper to discard the damaged item and provide a new one.

Often, straightening a badly damaged metal door or frame is impracticable, even when cost is no object. In that case, removing the damaged door or frame and providing a new one may be the only alternative.

Loose, Sagging, or Out-of-Line Doors or Frames. A metal door or frame may become loose or sag or appear to be out of line because it was poorly designed or fabricated. Sometimes the metal used will simply be too thin for the item to support itself. It is very difficult to fix a badly designed door or frame. It is usually impracticable to repair doors or frames that have failed because they were made from metal that was thin.

Most of the time, when a metal frame is loose, sagging, or out of line or falls away from its support, the fault lies either in undue movement of the supporting wall, partition, or structure or in the anchoring system. Sometimes such problems are relatively easy to solve. New anchors in the right locations will often do wonders. However, the appearance of the finished installation may not be good, especially if existing anchors' being in the wrong places or inadequate in size or number has made it necessary to install new anchors where they will be visible.

Some of the reasons for doors being loose or sagging include loose hinges and frames that have changed in shape due to settlement or other movement in the supporting wall, partition, or structure. Loose hinges can be repaired easily, of course, simply by tightening their screws. Sometimes a binding door can be made to operate properly by shimming its hinges with thin metal or cardboard strips placed between the hinge and the frame, but this technique will work only if the sagging is slight. To correct severe sagging it may be necessary to square up the frame, which might require removing it. Broken welds will require rewelding, of course, which might require the frame to be sent to a shop.

Missing Hardware. It is usually possible to find or make replacements for missing hardware. Often the new hardware can be installed quite easily in the field.

Repairing Materials Adjacent to Metal Doors and Frames. When damage to a door or frame has also affected adjacent construction, repairing that construction is also required.

4.13 Metal Doors and Frames in Existing Construction

The requirements discussed earlier in this chapter for new doors in new construction apply equally when doors and frames are to be installed in existing construction. There are, however, some additional considerations that must be taken into account when doors and frames are installed into an older building. Many of those considerations are addressed in this section.

Since the number of possible conditions is large, it is not practicable to try to cover every one of them in this book. The following discussion is therefore generic in concept and should be viewed in that light. The specific suggestions here may not apply to a particular condition, but the principles should be applicable to most. In the end there is no substitute for professional judgment, which must be applied when using the following data.

4.13.1 Door, Frame, and Hardware Materials, Finishes, Types, and Fabrication

There are no essential differences in the doors and frames or associated hardware, glazing, or louvers installed in all new buildings and those installed in existing ones, or the metals or finishes used. For details refer to discussions earlier in this chapter.

There are, however, a few differences in frames and their anchors. Most frame types are the same as those used in new buildings, but two-piece steel and aluminum frames and other types of remodeling frames and knockdown frames are used more often. In addition, frames with subbucks may be used more frequently, and some installation details and anchoring methods may differ.

There is little difference between the types, materials, and finishes for hardware used on new doors placed in existing construction and those used on new doors in new construction. When the doors themselves or their frames already exist, however, the hardware must be selected for its compatibility with existing cutouts and penetrations. Often the same hardware manufacturer must be used to ensure compatibility.

When the walls and partitions where a new metal frame is to be installed are already in place, the attachments and anchors may not be of the same types as would be used if frames were installed before the surrounding construction. Refer to sections 4.8 and 4.9 for an indication of the types of anchorages and anchor devices that might be used. When the supporting structure is already in place it is usually not necessary to supply inserts and anchorage devices before the doors and frames arrive at the site.

4.13.2 Standards, Controls, Delivery, and Handling

The standards and controls discussed in sections 1.6.1, 1.7.1, 4.1.1, and 4.1.2, and the delivery and handling requirements in section 4.9 apply as well to new doors and frames installed into existing construction.

While the work related to installing new doors and frames into existing construction is under way, existing doors and frames not scheduled to be removed and discarded should be protected from harm. Where such doors or frames become damaged, they should be repaired to a condition at least equal to that existing before the damage. Where making acceptable repairs is not possible, the damaged door or frame should be removed and a new unit installed.

Damaged doors and frames should not be installed without first repairing them. When satisfactory repairs are not possible, damaged doors and frames should be discarded and new units should be provided. The requirements indicated for installing new doors and frames apply also to reinstalling existing doors and frames.

4.13.3 Preparation for Installation

Just as preparations are necessary before new metal doors and frames and associated glazing and related items can be installed in a new space, so is it necessary to make preparations before installing such items into existing construction. In the latter case, however, the preparations are likely to be more extensive.

The requirements for demolition and removal of existing items and renovation work discussed in sections 1.6 and 1.7 apply to work related to metal doors and frames. The requirements in this section are in addition to those in sections 1.6 and 1.7. The requirements discussed in section 4.9.1 apply as well when the doors and frames will be installed in existing construction. Of course, it may be necessary to remove portions of the existing construction to permit installation of proper support for frames.

As is true when both the doors and frames and the space are new, the site where new doors and frames will be installed in an existing building should be examined carefully by the installer. When the doors and frames will be installed in an existing space, two examinations are usually best. The first should be made before demolition work begins. A report of this examination should be sent to the designer and owner so that modifications in the proposed demolition can be made, if necessary, to accommodate the door or frame design.

The second examination should be made after demolition in the space has been completed. This examination is necessary to verify that other construction to which the new frames are to be attached, and the areas and conditions where the frames will be installed, are in proper and satisfactory condition to receive the new frames. Preparatory installation work should not be started, and the frames should not be installed, until unsatisfactory conditions have been corrected.

4.13.4 Installation of Doors and Frames

In general, the requirements discussed in sections 4.9.2 and 4.9.3 related to installing doors and frames in new construction also apply to doors and frames installed in existing construction. This section contains modifications of the requirements in those sections and some additional ones.

Should conditions be encountered that differ significantly from those expected, the person discovering the difference should immediately notify the other concerned parties, especially the owner, so that alternative methods can be developed for handling the unexpected situation. Work should not proceed until such alternative methods have been approved by the owner and accepted by the contractor.

Regardless of the source of the damage, damaged doors and frames should not be installed. It is generally much easier to repair a damaged door or frame before it is installed. Often, damage will not be repairable after a door or frame has been installed, and the item must be removed to permit repair. Damaged doors and frames that have been installed, but cannot be repaired in place should be removed and repaired, or new, undamaged units installed. Installing already damaged doors and frames is a major cause of later problems with them.

Doors and frames should be installed in accordance with their manufacturer's latest published instructions, the approved shop drawings, and the latest editions of applicable industry standards and recommendations. There will be no difference in the frame design of extruded aluminum entrance and store front–type frames and, therefore, little difference in the installation or anchoring devices or methods for them, regardless of whether the construction is new or existing. Such frames are usually installed after the adjacent walls are in place anyway. There may be some differences in extruded aluminum interior-door frames and in formed carbon-steel, stainless steel, and copper-alloy frames, however. Wraparound frames will not be usable, for example, when the wall or partition is already in place. The same methods and designs used to install frames into already built concrete and masonry will still work, of course, but frames in stud partitions and walls are more likely to be of the knock down type or of another special remodeling-type design.

The types of anchors used will vary, of course, depending on when in the construction process the frame is installed. Frame anchors are usually built into concrete, stone, or masonry built after the frame has been erected. Of course, frames must be anchored to an existing concrete, stone, or unit masonry wall or partition using another method, such as exposed expansion bolts or screws into wood or metal subbucks, or expansion bolts placed into shields installed in drilled holes or in-place concrete or in masonry joints.

As a general rule, the anchoring of a metal frame is more complicated when it is installed in an existing building than when the construction is new.

4.13.5 Installation of Glazing and Louvers

Glazing and louvers are installed the same way whether the frame is in new or existing construction. Neither is there a difference in their installation whether the frame is itself new or existing, except that the method must be appropriate for the existing frame. It will usually be the same method as originally used.

4.14 Where to Get More Information

The following AIA Service Corporation's *Masterspec* Basic Version sections contain helpful information about carbon-steel, stainless steel, aluminum, and copper-alloy doors and frames and the materials and finishes from which they are made. Unfortunately, they contain little that will help with troubleshooting those installations.

- The May 1984 edition of Section 08110, "Steel Doors and Frames."
- The February 1987 edition of Section 08410, "Aluminum Entrances and Store Fronts."
- The February 1987 edition of Section 08460, "Automatic Entrance Doors."
- The May 1987 edition of Section 08470, "Revolving Doors."
- The May 1984 edition of Section 08710, "Finish Hardware."

The following American Architectural Manufacturers Association (AAMA) publications contain valuable information to anyone interested in finishes on aluminum:

- *Aluminum Store Front and Entrance Manual.*
- *Metal Curtain Wall, Window, Store Front and Entrance Guide Specifications Manual.*
- AAMA 701.2-1974, "Voluntary Specifications for Pile Weather Strip."
- AAMA 1102.7-1977, "Voluntary Guide Specifications for Aluminum Storm Doors."
- AAMA 1503.1-1988, "Voluntary Test Method for Thermal Transmittance and Condensation Resistance of Windows, Doors and Glazed Wall Sections."
- AAMA 1504-1988, "Voluntary Standard for Thermal Transmittance and Condensation Resistance of Windows, Doors and Glazed Wall Sections."

Every designer should have the full complement of applicable ASTM standards available for reference, of course, but anyone who needs to deal with problems concerning carbon-steel, stainless steel, aluminum, or copper-alloy doors and frames should obtain the standards referenced in this chapter. Additional applicable standards are listed in the Bibliography and marked with a [4].

The Copper Development Association publications marked with a [4] in the Bibliography contain helpful information related to the subjects in this chapter.

Ramsey/Sleeper's *Architectural Graphic Standards* has a glossary of terms used in steel door and frame work and many details about steel doors and frames.

Refer also to other items marked with a [4] in the Bibliography.

CHAPTER 5

Wood Doors and Frames

This chapter discusses wood swinging, interior sliding, and folding doors and frames and their finishes and hardware. Door types covered are architectural flush doors, including hollow-core, solid-core, and mineral-core doors, and stile and rail doors. The door facings covered are transparent finish and paint-grade wood and plastic. Door louvers and wood frames, including those used for doors, sidelights, borrowed-light frames, and store fronts, are also discussed.

Chapter 1 outlines problems that might cause wood doors and frames to fail and suggests steps that can be taken to solve them.

Chapter 2 addresses failures in structural systems, walls and partitions, and other building elements.

Wood materials and finishes for wood doors and frames are discussed in Chapter 3.

Metal doors and frames, including those used in conjunction with wood doors, are covered in Chapter 4.

Glazing for the doors and frames addressed in this chapter is discussed in Chapter 8.

5.1 Fabrication and Operation of Wood Doors and Frames

5.1.1 Standards

In general, wood doors and frames should comply with the requirements of the National Wood Window and Door Association's (NWWDA) ANSI/NWWDA I.S.1-1986 Series, "Industry Standard for Wood Flush Doors." They should also comply with sections 300, "Standing and Running Trim"; 900A, "Door Frames/Exterior or Interior"; 900B, "Interior Door Frames"; 1300, "Architectural Flush Doors"; and 1400, "Stile and Rail Doors," and the other applicable sections of *Architectural Woodwork Quality Standards, Guide Specifications and Quality Certification Program*, published by the Architectural Woodwork Institute (AWI). The Woodwork Institute of California (WIC) also has a set of standards, which differ in some respects from those of the NWWDA and AWI. Readers who have reason to believe that their doors were manufactured in accordance with WIC standards should request a copy from the WIC at the address listed in the Appendix.

Each door should bear the NWWDA Wood Flush Door Certification Hallmark certifying its compliance with applicable requirements of the ANSI/NWWDA I.S.1 Series documents. Alternatively, since all manufacturers do not participate in the NWWDA hallmark program, they are often permitted to substitute a certification of compliance in lieu of marking individual doors.

The cores of hollow-core doors should conform to the requirements of ANSI/NWWDA Industry Standard I.S.1. The particleboard used in cores should conform to ANSI Standard A208.1 for grade 1-L-1 particleboard.

Fire-rated doors should comply with governing regulations. Such doors should be identical to those used in assemblies that have been tested in compliance with ASTM Standard E 152 and should be labeled and listed by the UL or another nationally known testing laboratory. Installation should be in accordance with the National Fire Protection Association's publications *NFPA 80: Fire Doors and Windows* and *NFPA 101: Life Safety Code*.

Finish hardware and the metals and finishes used on it should comply with the recommendations of the American National Standards Association's standards ANSI A156.1 through A156.21. These current ANSI standards replace the Builder Hardware Manufacturers Association (BHMA) standards that were used for many years.

Hardware for fire-rated openings should comply with NFPA Standard No. 80. Such hardware should have been tested and listed by the UL, or another nationally known testing laboratory, for use on the types and sizes of doors required, and should comply with the requirements of the door and frame labels. Where emergency-exit devices are required on fire-rated doors, the door label should state that the door is to be equipped with fire-exit hardware. Such hardware should also contain a label stating that it is fire-exit hardware.

Sound-rated (acoustical) assemblies should have been fabricated and tested as assemblies in accordance with ASTM Standard E 90 and classified in accordance with ASTM Standard E 413.

5.1.2 Controls

General requirements related to controls are discussed in sections 1.6.1 and 1.7.1. The following controls are in addition to those discussed in those earlier sections.

The door and frame manufacturer's specifications for fabrication, factory finishing, shop painting, and installation should be submitted for the owner's approval.

The manufacturer should prepare for the owner's approval shop drawings for fabrication and installation of wood doors and frames. The shop drawings should show the location of each door and frame, the size and thickness of each door, details of each frame type, the elevation of each door type, the conditions at each opening, the details of construction, the location and installation requirements of finish hardware, blocking, fire ratings, details of joints and connections, the materials and methods to be used in factory priming, and the methods and materials to be used in factory or field finishing. They should show anchorage and accessory items and describe methods of joining (nailing, spline, screwing, doweling, dovetailing, mortising, etc.). For items to be repaired, the shop drawings should show the methods to be used in each case. They should also be accompanied by installation instructions for swinging, sliding bypass, sliding pocket, folding, and prehung doors and other pertinent data.

The shop drawings should include an opening schedule using reference numbers for details and openings. When the owner has engaged an architect, the references on the shop drawings should be the same as those on the architect's drawings. The shop drawings should include new doors and frames; existing doors and frames to be removed and discarded or turned over to the owner; existing doors and frames to be removed, altered, or refinished and reinstalled; existing doors and frames to be relocated; and existing frames to receive new doors.

The contractor should submit for the owner's approval sample sections of each door type, showing the finish veneer and edge, corner, and core construction. For doors with a transparent finish the submittals should include a veneer sheet from each available face-veneer flitch, with factory prefinishes applied. Such samples should include the entire range of available colors and flitches in the required materials and finishes. After the owner has selected the materials and finishes to be used, the contractor should submit twelve-inch-square samples of the selected flitch with the selected finish applied, and three-by-twelve-inch strips of solid wood of the species that will be used for exposed edges, trim, and other solid-wood components, also with the selected finish applied.

The contractor should also submit for the owner's approval sample corner sections of the required wood frames, showing construction details.

The door manufacturer should submit evidence that it is in business as an individual, company, firm, or corporation and has a trade name or mark recognized by the door-manufacturing industry.

Doors should be installed by a specialty subcontractor who has been regularly engaged in that business for at least two years, and by skilled workers under competent supervision.

Each door should have affixed to it labels or permanent markings identifying the manufacturer, core construction, adhesion bond type, face-veneer grade, and fire-rating label.

The shop drawings, templets, and schedules of doors, frames, hardware, glazing, and finishing should be coordinated and correlated to ensure that the doors are properly finished, machined for hardware, and ready to hang.

The owner or his representative should have the right to select at random as a test sample any one door from a shipment delivered to the job site for cutting and inspection.

The proposed manufacturer should be required to submit to the owner a certified statement of his qualifications.

The selected door manufacturer should submit to the owner the manufacturer's product data, specifications, and installation instructions for each type of wood door. This information should include details of core and edge construction, trim for openings and louvers, and similar components. It should also include certifications attesting the products' compliance with the requirements of the project. These certifications should be signed jointly by the contractor, door manufacturer, and door installer.

Where door warranties are required, the contractor should submit to the owner a written warranty on the door manufacturer's standard form signed by the manufacturer, installer, and contractor, agreeing to repair or replace defective doors that have warped, bowed, cupped, or twisted, that show telegraphing of the core construction in the face veneers, or do not conform to the tolerance limitations of the NWWDA and AWI. A warp in excess of one-quarter inch in the plane of the door or which adversely affects door operation or use should be considered a defect under the warranty. The warranty should also include the refinishing and reinstallation that may be required due to repair or replacement of defective doors. The warranty period may be anything from one year to the life of the installation.

The best way for an owner to control the quality and appropriateness of finish hardware is to employ a certified Architectural Hardware Consultant (AHC). Refer to section 1.4 for additional information.

Each kind of hardware—lock sets, hinges, closers, etc.—should be obtained from one manufacturer only. The hardware supplier should be a recognized finish-hardware supplier who has been furnishing hardware in the project's vicinity for projects of a comparable size and complexity for a period of not less than two years, and who is, or has in employment, an experienced hardware consultant who is available to the architect and contractor at reasonable times for consultation about the project's hardware requirements. The supplier should operate a fully stocked service facility within fifty miles of the project site.

The hardware supplier should submit the manufacturer's technical information for each item of hardware, including such information as may be necessary to show compliance with the project's requirements and instructions for installation and maintenance of operating parts and finishes.

The supplier should also submit for the owner's approval a hardware schedule that includes the following information:

The type, style, function, size, and finish of each hardware item.
The name and manufacturer of each item.
The fastenings required.
The location of each hardware set.
The mounting locations for the hardware.
Door and frame sizes and materials.
Keying information.

Hardware templates should be furnished to the fabricators of doors, frames, and other work to be factory prepared for the installation of the hardware.

Wood door frame and related finish carpentry work should be done by a firm that can demonstrate successful experience in work similar to that required. Only experienced personnel skilled in the processes and operations necessary should be used, working under competent supervision.

Samples of the hardwoods to be used in frames should be submitted for selection and approval, as should samples of laminated plastic door facings.

5.1.3 Materials and Finishes

General requirements for wood materials used in doors and frames are discussed in section 3.1. Finishes are discussed in section 3.11.

In some rare cases doors may have a different facing material on opposite sides, but usually have the same material on both faces. Doors with transparent finishes generally have the same facing material on their edges. The material used and the methods of cutting and matching it between adjacent leaves may be any of those discussed in section 3.1.

Finish-hardware items for use with wood doors are manufactured from many materials and given many different finishes. Materials used include but are not limited to cast iron, steel, stainless steel, aluminum, copper and its alloys brass and bronze, zinc, chromium-plated steel, and chromium-plated aluminum. Metal materials are discussed in sections 3.2, 3.3, 3.4, and 3.5.

Finishes for metal hardware items are discussed in sections 3.6 through 3.11. Current hardware finish designations are listed in ANSI A156.18, "Materials and Finishes," which has replaced the older U.S. Department of Commerce standards (US26D, for example).

Many miscellaneous materials are used in manufacturing and installing doors and frames. Among them are adhesives and sealants. To a certain extent, the adhesive used to manufacture a door dictates the quality of the door. Type I adhesives are considered to be waterproof. The AWI requires that they be used in exterior doors and in doors that will be installed in shower rooms and other high-humidity locations. Type II adhesives, which are considered water resistant, are used in interior doors that will not be installed in high-moisture areas.

Sealants are used between wood frames and adjacent construction to seal the joints there, for finishing and to fill unsightly cracks. In exterior locations, elastomeric sealants are usually used. Interior sealants may be elastomeric or acrylic-latex types.

5.1.4 Fabrication of Doors, Flush Panels, and Louvers

Wood doors and panels should be rigid, neat in appearance, and free from defects, warp, and buckle.

Most wood doors in commercial installations are 1-3/4 inches thick, but other thicknesses may also be used. Apartment closet doors, for example, may be 1-3/8 inches thick. Closet folding doors may be 1-1/8 inches thick.

Where exposed fasteners are used, they are usually screws or bolts with countersunk, flat Phillips heads.

The maximum clearances for doors should be those recommended in ANSI/NWWDA Industry Standard I.S.1. In no case, however, should the clearances for fire-rated doors exceed those required by the authorities having jurisdiction, or by NFPA 80.

Wood doors should be prefit, premachined, and prepared by the manufacturer to provide clearance at their edges, bevels, and radius stile edges. Mortising and cutting for locks, bolts, closers, hinges, louvers, vision panels, and other purposes should be provided, except for surface-applied hardware. Blocking should be provided within the door where hardware will not be otherwise fastened to solid wood. Sliding and folding doors should be prefitted to accommodate hardware, allow for proper operation and overlapping where appropriate, and fit closely at jambs and where doors meet.

Doors should not be later cut or machined to a size smaller than that for which they were originally manufactured. Prefinished doors should not be cut, machined, or otherwise damaged.

Many wood doors today come from the factory prehung in frames and ready for installation.

Doors to be field finished should be sanded smooth at the factory. Completed doors shall be free of defects and machine marks that will show through the finish. Doors should be dried again after the veneers have been applied, to remove moisture contained in the glue.

Wood doors fit either into the general category *flush*, also called *architectural flush*, or the category *stile and rail*.

Flush Doors. Flush doors are constructed of wood-framed cores faced on both sides with wood veneers or plastic-laminate facings. Fully flush doors (see Fig. 4-1) have no applied moldings, panels, or openings. Many flush doors, however, have openings in them for view panels or louvers. The openings in flush wood doors are similar to those in steel doors, as shown in Figures 4-2 through 4-6.

The cores of architectural flush doors may be hollow, in which case a ladder-frame construction, mesh construction, or cellular construction such as an impregnated-paper honeycomb, is used to stiffen the door. Side, bottom, and top rails and solid hardware (lock) blocks are also included. One or two solid-wood intermediate cross rails may also be used.

Cores may be solid. Solid cores may be either glued or nonglued staved wood blocks, particleboard, or a mineral composition material. Either particleboard or mineral-core doors may have a fire rating. Particle-board-core doors may have twenty- or thirty-minute ratings. Higher ratings of 3/4-, 1-, and 1-1/2 hours are obtained using mineral-core doors.

Special cores are used for sound-retardant, lead-lined, and electrostatic-shield doors.

Door faces are made in two layers on five-ply doors and three layers on seven-ply doors. The layers closest to the core, called the crossbands, are usually of 1/16-inch-thick hardwood veneer. The face layer can be any of the wood veneers or plastic laminates discussed in section 3.1.1.

Stiles and rails in paint-grade or plastic-laminate-faced flush doors may be either hardwood or softwood. Where the doors will be given a transparent finish, the stiles and rails should be of the same material as the facing. Finger joints, which are common in paint-grade doors, should not be permitted in doors that will receive a transparent finish. Stiles and rails in fire-labeled doors should be fire-retardant treated.

Openings should be factory cut and framed. The sizes and spacing between openings in the same door should comply with the door's warranty requirements. Figure 4-19 shows a case where two openings are too close together, which would void most door warranties. In doors with a transparent finish the opening trim should be made from prefinished wood that matches the door veneer, except where a fire label requires a metal frame. In painted doors, except where a UL label requires a metal frame the trim should be painted wood. Where labels require metal frames they should be factory-primed hollow-metal trim that is through-bolted to the door. Blocking must be provided in the door to accommodate the fasteners. Metal opening frames should be shipped with the doors. Wood beads should be installed in the factory on one side of the glazing material and loosely tacked in place on the other side, unless the louvers or glass are installed at the factory.

Stile-and-Rail Doors. Stile-and-rail doors are made up of wood stiles and rails with the spaces between them filled with either wood panels, glazing material, or louvers. Wood stile-and-rail doors may take on many configurations, such as those shown for steel doors in Figures 4-2 through 4-5 (Figs. 5-1 and 5-2).

Stiles and rails in paint-grade stile and rail doors may be either hardwood or softwood, and may be either solid wood or have a veneer facing. Finger joints, which are common in paint-grade doors, should not be permitted in doors that will receive a transparent finish.

The panels in stile-and-rail doors are either plywood or particleboard with a wood veneer facing.

Opening trim for these doors should be factory cut and fitted. In doors with a transparent finish, opening trim should be made from prefinished wood that matches the door's veneer. In painted doors the trim should be painted wood. Wood beads should be installed in the factory on one side of the glazing material and loosely tacked in place on the other side, unless louvers or glass are installed at the factory.

Figure 5-1 This photo shows a typical stile-and-rail door.

Special Doors. Sliding bypass, sliding pocket, and folding doors (Fig. 5-3) are all packaged proprietary units that come complete with doors, tracks, guides, rollers, pulls, and other hardware. The doors used may be either architectural flush or stile-and-rail doors, and may have solid panels or louvers. The louvers may be either in the entire door or only in the upper or lower panels. These types of doors are used mostly in residential construction but are often found in apartment and office buildings, and are occasionally found in other types of buildings as well.

Panels. Wood flush panels are frequently used in transoms. They should be made from the same materials and by the same construction methods as the associated doors.

5.1 Fabrication and Operation of Wood Doors and Frames 201

Figure 5-2 The style of this stile-and-rail door is the only thing typical about it. Note the size of the woman's shoe on its sill.

Louvers. Both interior and exterior wood doors may be either fully louvered, as shown in Figure 4-5, or simply contain inserted louvers, as in Figure 4-6.

Louvers in wood doors may be either wood or metal. Fire ratings require metal louvers, but wood louvers are used in most other cases.

Wood louver blades are usually a flat 45-degree wood blade or a sight-proof, inverted V-shaped slat-type with at least 35 percent free air. In painted doors the louver wood species is somewhat immaterial. In doors with a transparent finish, though, the wood in the louver should match the wood in the door veneer and be finished to match.

Metal louvers are usually formed from not less than 20-gage cold-rolled furniture steel. They are usually one inch thick with fixed, inverted V-shaped, 45-degree blades. Inverted Y-shaped blades may also be used. A

Sliding bypass

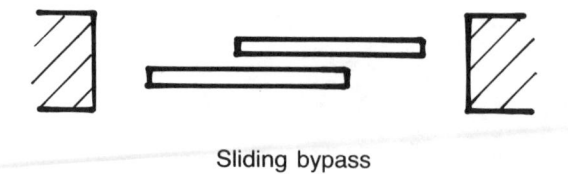

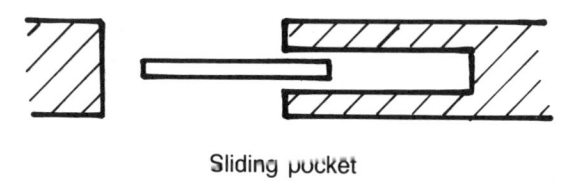

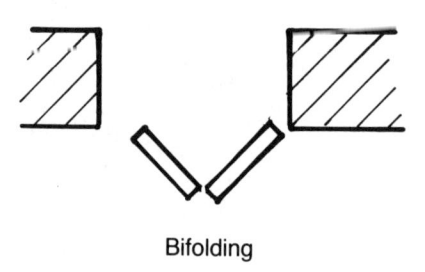

Sliding pocket

Bifolding

Figure 5-3 Special wood doors.

common finish for metal door louvers in wood doors is a bonderized, baked-on enamel. Colors are often available that will come close to matching the door finish.

Metal exterior door louvers should be a weatherproof design, such as the Z-type shown in Figure 4-15. Exterior louvers should be of not less than 18-gage steel blades set in a minimum 18-gage frame. Exterior louvers require insect screens, which should be mounted on the interior face of the louver in a removable frame.

In flush doors, adequate blocking must be installed in the door's core to provide solid wood to receive fasteners.

Figure 5-4 An early form of store front.

5.1.5 Wood Store Fronts

Early store fronts were combinations of wood windows and doors (Fig. 5-4). The windows were a bit larger than house windows, but were still not very large. Their glass panes were small, because the technology did not exist to manufacture glass in larger pieces (see Chapter 8).

Later store fronts were similar to those in Figure 5-4, but the window portions were larger (Fig. 5-5). The glass panes were still small, though.

When plate glass became widely available the modern store front was born, making the type of store front shown in Figure 5-6 possible. They had the general characteristics of modern store fronts. They had large glass panels without muntins and were separate display elements instead of just a way to look into the building.

Wood store fronts are seldom used in modern buildings, with extruded aluminum now being the material of choice. Most wood store fronts built today are either replacements for earlier ones or are in buildings designed to look old. Their construction is similar to that of earlier store fronts. Their vision panels are small and glazed in with putty, and they have lots of moldings and gingerbread around them.

Very few wood store fronts are built today with a modern look. Those

Figure 5-5 An early store front with a large window.

that are should be built in accordance with the discussion of wood frames in the following section.

5.1.6 Fabrication of Wood Frames

Wood frames should be fabricated at the mill to be rigid, neat in appearance, free from defects and warp, with wood surfaces free of machine and tool marks. Hardwood dowels should be provided to conceal screw heads where applicable.

Frames should be fabricated in accordance with the project's requirements, approved shop drawings, and applicable AWI and ANSI/NWWDA standards to the dimensions, profiles, and details required. Openings and mortises to receive hardware and other items should be precut when possible.

Joints should be made tight and should be constructed to conceal shrinkage. Trim, stops, and moldings should be mitered at exterior angles and coped at interior angles and returns. Materials should have no excessive warp. The backs of flat trim members should be routed or grooved and wide members should be kerfed, except when their ends will be exposed.

Figure 5-6 A plate-glass-and-wood store front.

The edges and corners of solid-wood members less than one inch thick should be eased to a 1/16-inch radius and thicker ones to a 1/8-inch radius.

When possible, fabrication, assembly, finishing, hardware application, and other work should be completed and doors should be hung before frames are shipped to the project site. Even when prehanging doors is impracticable, frames should be shipped preassembled when practicable. When prefabrication is not practicable, frames should be precut and shipped disassembled. Frames should be cut from stock in the field only on the rarest of occasions, because field cutting is inefficient and tends to be inaccurate. Where field assembly is necessary, ample allowance should be provided for scribing, trimming, and fitting.

Openings to receive glazing, louvers, and hardware should be located accurately and precut when possible. Templets or roughing-in diagrams should be used for hardware so that openings are the proper size and shape.

Frames should be delivered to the installation location complete with fasteners, anchors, clips, and accessories. Fasteners should be concealed whenever possible.

Prefabricated frames should be prepared for shipment by attaching temporary wood spreaders to the bottoms of the frames. Before they are shipped, the frames should be stenciled, labeled, or tagged with instructions for installation.

Glazing and louver-restraint beads should be removable. They may be square or rectangular and either mitered or butted at their corners, but they should be carefully fitted in full lengths to avoid gaps in the corners and joints other than at the corners. Removable beads should be located on the side of the glazing material that falls in the space that is to be protected.

Wood frames are not designed to support loads other than themselves and therefore require separate lintels to support the loads imposed by a wall or partition above the frame.

Frame tolerances should comply with the applicable portions of ANSI A115 Series standards.

Wood frames can be specially fabricated or modified to receive lead linings for use where radiation is a hazard. Lead linings should comply with the applicable recommendations of the National Council on Radiation Protection and Measurement (NCRPM). Frames should be lined with a single, continuous, unpierced lead sheet.

5.1.7 Finish Hardware

Unless the owner or one of the owner's consultants is a finish-hardware expert, it is a good idea to engage the assistance of someone who is. Certified Door Consultants (CDCs) and Architectural Hardware Consultants (AHCs) qualify. Refer to section 1.4.1 for more information about CDCs and AHCs.

Hardware Types. This section addresses five types of hardware: standard, handicapped-person, automatic door, security, and special door.

Standard Hardware. Standard hardware includes butts, hinges, lock and latch sets and their trim, exit devices, door controls, closers, holders, auxiliary locks, architectural door trim, weather stripping, thresholds, silencers, and stops.

Handicapped-Person Hardware. Some hardware is designed to permit easier operation of doors by the handicapped. Such hardware ranges from counterbalanced closers that allow opening doors with little force to fully automatic operators.

Automatic-Door Hardware. There are many kinds of automatic operators available, including electromechanical, hydraulic, and pneumatic-operated devices. There are also several types of actuating devices, including floor treadles, motion sensors, sonic sensors, photoelectric cells, manual push buttons or plates, and remote switches.

Probably the simplest of automatic hardware is the fire- or smoke-actuated closer in which a detector senses smoke or fire and releases a hold-open device that permits the closer to close the door.

Security-Type Hardware. Building security may require simply replacing the standard components of a bored or mortise lock with high-security cylinders and other components. It may, however, require one of a variety of special control devices such as electric (including electromagnetic) or electronic locks, or three-way-key locks in which a single key operation drives bolts into the jamb, head, and sill. Security locks may operate by means of keys, push buttons, or card readers.

Special Door Hardware. Sliding and folding door hardware is furnished as a part of the door assembly.

Preparation of Doors and Frames for Finish Hardware. Doors and frames should be prepared to receive finish hardware in accordance with the hardware supplier's templets and the owner's instructions, as expressed in a finish-hardware schedule. Door and frame preparation for hardware should comply with the requirements of the ANSI A115-W Series standards. Finish hardware should be located in accordance with DHI-WDHS-3, "Recommended Hardware Locations for Wood Flush Doors," published by the Door and Hardware Institute (DHI).

Doors should have wood blocks inserted at the factory to receive finish hardware. They should also be drilled and tapped at the factory for mortised and other concealed hardware. Drilling and tapping for surface-applied hardware, however, can be done at the installation site. Templets provided by the hardware supplier should be followed. Blocking should be provided for locks, closers, and hinges, unless the door's wood stiles and rails are wide enough to provide solid wood for attaching them.

Screws for installation should be provided with each hardware item. Generally phillips flat-head screws are used. Exposed fasteners are usually finished to match the hardware.

Hardware Installation. Hardware for swinging doors should be mounted at the locations indicated in DHI-WDHS-3, "Recommended Hardware Locations for Wood Flush Doors," published by the Door and Hardware

Institute (DHI). Hardware for sliding and folding doors should be located in the manufacturer's standard locations.

Hardware should be installed in compliance with the manufacturer's instructions and recommendations. Cutting and fitting that is required to install hardware onto or into surfaces that are not to be later painted or finished should be done carefully to avoid damaging those surfaces. When such surfaces are to be painted or finished the hardware should be removed carefully and protected for later reinstallation, or a surface protection should be applied. Surface-mounted items should not be installed until the finishes have been completed on the doors and frames.

Hardware should be installed carefully, accurately, and securely. Units should be set level, plumb, true to line and location, and be adjusted to operate properly.

5.1.8 Warranties

When a door has failed, it is important to check to see if a warranty exists and if so to examine its terms. Some warranties are valuable, others not. The better the doors, the more encompassing the warranty is likely to be— reputable manufacturers are not afraid to warrant their products. Warranties on cheap doors are often worthless, but those on good doors usually go unused, because the doors they cover seldom fail. So the best warranty is to buy good doors in the first place.

5.2 Installation of Wood Doors and Frames in New Construction

The standards listed in sections 5.1.1 and 5.1.2 apply as well to installation.

5.2.1 Delivery and Handling

Wood doors and frames should be protected during transit, storage, and handling to prevent damage, soiling, and deterioration. The "On-Site Care" recommendations in NWWDA's publication *Care and Finishing of Wood Doors* and the manufacturer's instructions should be followed. Doors should be packaged at the factory in ten-mil polyethylene film or heavy cardboard cartons.

Doors and frames should be stored in fully covered, well ventilated areas and protected from water and extreme changes in temperature and humidity. Doors should be stored flat, with their wrappings left in place until they are hung.

Items with minor damage may be repaired. Those with major damage should be discarded and new, undamaged items used.

After doors have been hung, their wrappings should be returned to them. Cutouts should be made in the wrappings only where necessary to clear hardware. Wrappings should not be finally removed until necessary either for field finishing or final approval, as applicable.

Doors should be handled with soft, clean gloves. They should not be dragged across one another or other surfaces.

Hardware should be delivered packaged in sets at the proper time and in the proper sequence to avoid delays in door installation. The sets should be stored in a secure, locked location until they are installed.

Wood frames should not be delivered to the site until concrete, masonry, and other work involving wet materials has been completed for at least ten days and spaces are ready to receive the frames.

Wood materials and prefabricated frames should be kept dry at all times. They should be protected against exposure to weather, contact with damp or wet surfaces, high humidity, and extreme changes in temperature or humidity. Such materials should be stacked to ensure proper ventilation and drainage. Air circulation within the stacks must be provided for. Humid chambers should not be permitted under the protective covers.

5.2.2 Preparation for Installation

The installer should examine the substrate and conditions under which wood door frames are to be installed and not begin the installation until conditions detrimental to proper and timely completion have been corrected.

Before doors and frames are installed, they should be examined to ensure that they are complete and have been properly fabricated, and that their frames have been back primed.

Before doors are installed, previously installed door frames should be examined to verify that they are of the correct type and have been correctly installed. Do not install doors until unsatisfactory conditions have been corrected.

5.2.3 Installation of Frames

Wood frames should be installed in accordance with the standards and controls discussed in this chapter. In addition, the installation of frames for power-operated doors and lead-lined frames must be coordinated with the installation of electrical work, the lead lining, and other associated work.

Frames should be set accurately in the correct locations and securely braced until their permanent anchors have been set. They should be left in the proper alignment, level, plumb, square, and securely fastened in place. Frames that have been installed out of square or out of plumb should be removed and reset to provide a properly operating installation of acceptable appearance. After wall construction has been completed, temporary braces and spreaders should be removed, leaving surfaces smooth and undamaged.

Frames should be fastened to anchors or blocking that are built in or directly attached to the substrates. They should be fastened to the grounds and blocking with countersunk, concealed fasteners and blind nailing as required for a complete installation. Except where removable fasteners or screws are required, fine finishing nails should be used for exposed fastening. The nails should be set for putty stopping in painted work and wood filler in transparent-finish work. Screw heads in natural-finish hardwood, should be countersunk and covered with hardwood dowels that match the hardwood in the frame. Screws in painted frames may be countersunk and filled either with putty or wood filler.

Frames components should be of the maximum practical lengths and in one piece where possible. Exposed surfaces should be sanded by machine to prepare them for finishing. Joints in exterior frames should be made to exclude water. Fire-rated frames should be installed in accordance with NFPA Standard No. 80. Labels should neither be removed nor painted over.

Wood frames should not be used as lintels to carry loads. Instead, separate structural lintels should be installed over each frame.

The perimeters of frames in concrete, stone, and unit masonry, and other frames where joints would otherwise be left open or unsatisfactory for painting should be filled with a sealant.

5.2.4 Installation of Doors

Wood doors should be installed in their frames in accordance with the recommendations of the NWWDA and the manufacturer's instructions, so that required warranties will be effective, and only after completion of other work that would raise the moisture content of the doors or damage their surfaces. Doors should be conditioned before hanging to the average prevailing humidity in the area where they will be installed. They should not be hung in rooms where the humidity is sufficiently high to cause damage to them.

Doors should be hung as required by their openings, with the clearances indicated. They should be fitted accurately in the frames, within the clearances specified in NWWDA I.S.1. Fire-rated doors should be placed with the clearances specified in NFPA Standard No. 80. Fire-rating labels must not be altered, removed, or painted over. Prefit or prefinished doors should not be cut or sized in the field. Other doors may be machined for hardware to whatever extent they were not previously worked at the factory for proper fit and uniform clearance at each edge. Cuts made in primed doors should be sealed immediately after the cutting, using the same material as the primer.

Hardware should be installed in accordance with the standards previously mentioned.

Swing doors should be installed plumb and true so that they operate freely but not loosely and are without sticking or binding after finishing.

5.2.5 Glazing and Installation of Louvers

Glazing is discussed in detail in Chapter 8.

Door louvers and their screens may be installed in either the factory or the field. They should be set accurately and securely anchored in place in accordance with the controls discussed earlier in this chapter and the instructions of the manufacturer.

5.2.6 Adjustments, Protection, and Cleaning

After doors have been hung, they should be protected from damage due to subsequent construction operations.

Damaged and broken door and frame components should be properly repaired or be removed and new, undamaged components installed. Temporary protection should be removed. Doors that do not swing freely should either be adjusted or removed and rehung.

After the installation has been completed, finish hardware should be checked, lubricated, and adjusted to ensure proper operation. Thereafter the hardware should be protected from damage.

Wood doors and frames should be left undamaged, complete, properly operating, and free from dirt, excess sealant, and glazing compounds.

5.3 Why Wood Doors and Frames Fail

The reasons wood doors and frames and their finishes fail can be grouped into three categories. The first consists of problems with the surrounding or supporting construction. It includes excess structural movement, failed steel or concrete structure, failed wood structure or wood wall or partition framing, failed metal wall or partition framing, failed concrete, stone, or masonry walls or partitions, and other building element problems. These causes are discussed in Chapter 2. While the failure causes in Chapter 2 may not be the most probable causes of the failure of doors or frames or their finishes, they are usually more serious and costly to fix than other possible causes. Consequently, the possibility that they are responsible for a failure should be investigated.

The second category has to do with the materials and finishes used in doors and frames. They include bad materials, inappropriate finish selection, improper preparation for and application of finishes, failure of the immediate substrate, failure to protect materials and finishes, failure to properly maintain applied finishes, and natural aging. These failure causes are discussed in Chapter 3.

The third category includes improper design, inappropriate door or frame type or design selection, inappropriate hardware selection, improper fabrication, improper installation, and natural aging. These failure categories and the failures they can cause are discussed in this section.

5.3.1 Improper Design

Improper design includes selecting the wrong materials for use in the door, frame, hardware, glazing, or associated accessories. Selecting hardware or fasteners that are likely to corrode for use in a corrosive atmosphere is an example.

Designing a door or frame that is too weak for the loads to be applied will lead to failure. Using 1-3/8-inch-thick doors in heavy use areas is an example. A related error is selecting fasteners or anchors that are too weak for the application, too few in number, placed too far apart, or of the wrong type, any one of which can lead to failure of the anchor or fastener.

Placing a door or frame in a location where it is likely to be damaged will often lead to failure.

5.3.2 Inappropriate Door or Frame Type or Design Selection

Selecting a door design that includes a long, narrow vision light for a door that must have a louver (see Fig. 4-19) can result in a door that is too weak.

A similar weakness occurs when vision panels or louvers are selected that are too large for the door. This leaves the portions of the door surrounding the openings too small. Even in stile-and-rail doors, the stiles and rails can be too small.

Selecting door sizes without taking finishes or operation into account can lead to misfitting doors and damaged finishes. A door that strikes carpet will surely damage the carpet, for example. Not using recommended tolerances is a similar error that can lead to problems with door or weather stripping operation.

Selecting the wrong type louver can lead to problems. A V-shaped or

Figure 5-7 It is difficult to prevent the type of failure shown in this photo when a wood door is exposed to severe weathering without protection.

Y-shaped louver may be fine for restricting vision, for example, but probably will not prevent light penetration.

Frame-design errors can also lead to failures. Designing too shallow a rabbet, for example, can result in glass loss in a storm.

Using a wood door in an exposed exterior location will often result in door failure (Fig. 5-7).

5.3.3 Inappropriate Hardware Selection

One of the more frequent problems associated with door failure is using the wrong hardware. There are many examples. Failing to coordinate hardware selection with door design is one.

Selecting hinges and butts that are too small for the door's weight can lead to sagging and even inoperative doors. Using too few hinges or butts can have a similar effect.

Selecting standard closers where entrances for the handicapped are required will result in doors that are too hard for a handicapped person to open.

Failing to require metal receivers for dead bolts can make unauthorized entry as easy as a kick.

5.3.4 Improper Fabrication

Improperly fabricated wood doors or frames are a sure precursor of failure. Perhaps the most frequent cause of improper fabrication is not following the requirements of the design, including the shop drawings and specifications and the standards of the industry.

Many types of fabrication errors involve joints. Failing to properly clean a joint that is to be glued will usually result in a failed joint.

Another common error in producing doors and frames is the joining of the components in such a way that corners or intersections are not square, plumb, and properly aligned, and so that joints are not flush, tight, and correctly fitted.

Using damaged materials in a door or frame, or damaging them during or after fabrication, must be avoided. Damaged components and materials should be repaired or removed and replaced with new ones.

5.3.5 Improper Installation

Correct installation is essential if door and frame failures are to be prevented.

Probably the number-one cause of failure due to installation errors is not following the design and recommendations of the manufacturer and recognized authorities, such as ASTM, ANSI, NWWDA, AWI, and DHI.

Failing to follow the design can mean using anchors, fasteners, and other materials that will rust where high humidity will occur or where they are set in or in direct contact with masonry or concrete. Sometimes materials that will corrode must be used for anchors. Then the error becomes failure to properly protect those metals with an applied bituminous or other appropriate coating.

Failing to follow the design and applicable standards can also lead to using anchors, fasteners, clips, and other devices that are of the wrong type or not in the proper number or location.

Failures can also occur if the area where a wood frame is to be installed is not properly prepared to receive it.

An allied error is the installation of a damaged frame or door. While the damage may or may not be the fault of the installer, installing a damaged door or frame is.

Installing a frame so that building loads that the frame is not designed to support are transmitted into it will also lead to failure. Not providing a lintel over a wood frame in a masonry wall is an example.

The installer must make sure that the various door and frame components finish in the proper plane within acceptable tolerances and that the unit is in the proper location, plumb, level, and in alignment. Improperly aligning a component may not only cause the frame to present an unpleasing appearance, but it can also force loads to be applied eccentrically, which can cause the door or frame to become damaged.

Failing to caulk around the perimeter of a wood frame will often leave an ill-appearing installation and may result in air or water leaks.

Damage to factory-applied finishes during shipment or erection is a major problem. Such damage should be repaired as soon as possible, to prevent further damage.

5.3.6 Natural Aging

Aging affects all aspects of doors and frames. Hardware may eventually simply wear out, for example. Most aging problems can be delayed for many years by proper maintenance techniques. Failure from aging, then, is mostly caused by failure to continuously protect the wood (Fig. 5-8). Most such protection is afforded by organic coatings. Refer to section 3.12.9 for a discussion of the natural aging of finishes.

5.4 Repairing Wood Doors and Frames

Existing wood doors and frames that are damaged or in an unsightly condition but still usable can often be repaired and restored. Sometimes it is necessary to remove a door or frame to effect repairs, but many repairs can be made without removing the unit.

When working with existing wood doors and frames, it is best, when possible, to use materials and methods that match the original installation. When a portion of a frame is damaged it might be possible to remove and discard only that part and provide a new one. When fasteners or attachments have failed, it will be necessary to provide new ones. It will probably be necessary to remove damaged finish hardware and install new items. Wood doors and frames that are damaged too severely to repair must of course be discarded and new units provided (Fig. 5-9).

Sometimes an apparent failure in a door or frame is actually a failure in an adjacent or supporting material or structure. Each condition must be

216   Wood Doors and Frames

Figure 5-8 The pair of doors in this photo have aged without being maintained until they are now useless. Total replacement with new doors and frames is the only solution to this problem.

examined in light of the data included in this book and the referenced sources of additional information. The condition of the damaged door or frame should then be brought as close as possible to the condition usually required when that type item is new. When damage is observed, the specific data in this book, the data in the listed sources of additional information, and the expert knowledge of someone knowledgeable about the type of problem that exists should be applied to solve it.

In addition to the requirements discussed earlier, there are additional ones where existing doors or frames are to be repaired in place or removed, repaired, and reinstalled. At least some selective demolition, dismantling, cutting, patching, and removal of existing construction is usually necessary. The comments related to such matters in sections 1.6.1 and 1.7.1 are applicable when existing wood doors or frames must be repaired.

Figure 5-9 At first glance the damage in this photo seemed repairable. Closer examination, however, revealed panel cracking and deterioration at the bottom rails to the extent that the doors were discarded and new ones installed.

5.4.1 General Requirements

This section contains some suggestions for repairing existing wood doors and frames. Because these suggestions are meant to apply in many situations, they might not apply to a specific one. In addition, there are many possible cases not specifically covered here. When a condition arises that is not addressed here, advice should be sought from the additional data sources mentioned in this book. Often, consulting with the manufacturer of the item being repaired will help. Sometimes it is necessary to obtain professional help (see section 1.4). Under no circumstances should the specific recommendations in this book be followed without careful investigation and application of professional expertise and judgment.

Before an attempt is made to repair an existing wood door or frame, the manufacturer's and finisher's recommendations for making the repairs should be obtained. When finish hardware is involved, the participation of a certified Architectural Hardware Consultant (AHC) is called for. Recommended precautions against using materials and methods that might be detrimental to the items requiring repair must be followed. Prevention of future failure requires that the repairs not be undertaken without careful, knowledgeable investigation and application of professional expertise and judgment. The repairs should be carried out by experienced workers under competent supervision.

Unless existing doors or frames must be removed to facilitate repairs to them or other portions of the building, or are scheduled to be replaced with new materials, only sufficient materials should be removed as is necessary to effect the cleaning and repairs. The more materials removed, the greater the possibility of additional damage occurring during their handling and storage. Materials removed should include those that are loose, sagging, or damaged beyond in-place repair. Other components that are sound, adequate, and suitable for reuse may be left in place or removed, cleaned, and stored for reuse.

After doors and frames have been removed, concealed damage to supporting framing or substrates may become apparent. Such damage should be repaired before repaired or new doors and frames are installed.

Existing doors and frames that are to be removed should be removed carefully and adjacent surfaces should be protected so that the process does not damage surrounding areas.

Unless the decision is made to discard them, damaged wood doors and frames and support-system components that can be satisfactorily repaired should be repaired, whether they have been removed or left in place. Failing to repair known damage may lead to additional failure later on. The methods recommended by the manufacturer of the doors and frames should be followed carefully, even when the repairs are as simple as touch-up painting. Doors, frames, and associated items that cannot be satisfactorily repaired should be discarded and new, matching items installed.

Areas where repairs will be made should be inspected carefully to verify that existing components that are to be removed have been and that the substrates and structure are as expected and undamaged. Sometimes substrate or structure materials, systems, or conditions are encountered that differ considerably from those expected, or unexpected damage is discovered. Both damage that was previously known and damage found later should be repaired before door or frame components are reinstalled. Door and frame reinstallation should not proceed until unsatisfactory conditions have been corrected.

Cleaning and repairing should be done, substrates prepared, and repairs made strictly in accordance with the material manufacturer's recommendations. Repairs should be made only by experienced workers.

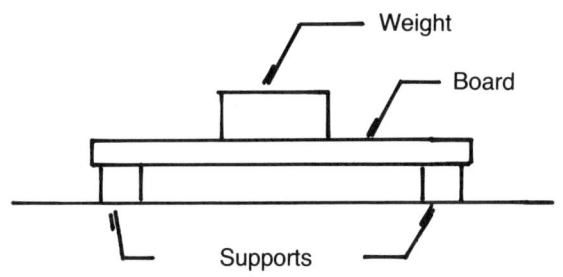

Figure 5-10 A method of straightening warped wood-frame members.

In general, the materials used to repair existing doors and frames should match existing materials in the same assembly, but should not be lower in quality than the materials recommended for new work by the appropriate industry standards.

5.4.2 Repairing Specific Types of Damage

In addition to the general requirements for repairs already discussed, the following recommendations apply to specific problems.

Warped or Twisted Doors or Frames. If the damage is not too severe it may be possible to straighten wood-frame members and warped doors. Additional nailing may suffice to straighten a bent or twisted frame member. When renailing is not sufficient, the frame member should be removed, straightened, and reinstalled. Straightening of a more severely warped member can sometimes be accomplished by wetting it, supporting it beyond the bent portion, and applying a load on the bent part to force it back into shape (Fig. 5-10). The load must be left in place for several days. When correction of a warp or twist is unsuccessful, the member must be discarded and a new one provided.

Warped wood doors can sometimes be straightened simply by installing an additional hinge. When more extensive measures are called for it may be possible to straighten a warped door using a method similar to that shown in Figure 5-10. A twisted door can sometimes be straightened using a wire-and-turnbuckle apparatus similar to that shown in Figure 5-11.

Loose, Sagging, Binding, or Out-of-Line Doors or Frames. A wood door or frame may become loose, sag, or appear to be out of line because it was poorly designed or fabricated. Sometimes a door will simply be too thin for the use to which it is being put. It is very difficult, and often impracticable, to fix a badly designed door or frame.

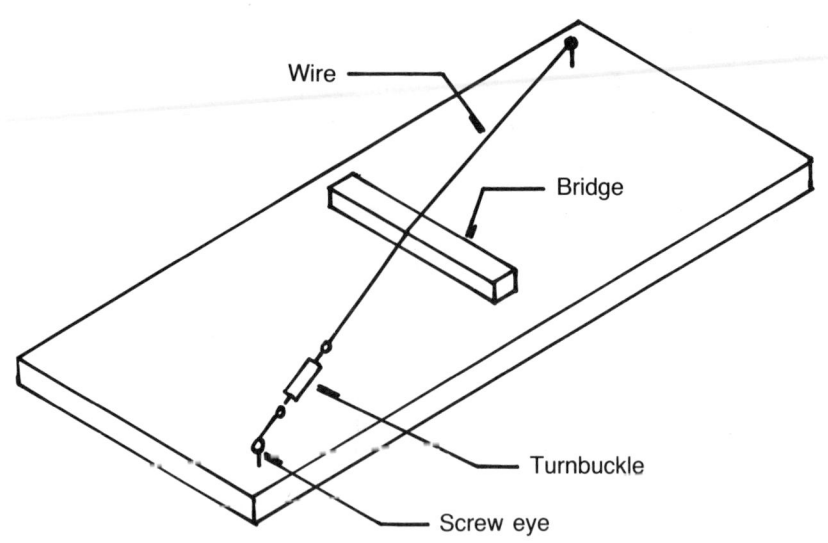

Figure 5-11 Straightening a twisted door.

Most of the time, when a wood frame is loose, sagging, out of line, or falls away from its support, the fault lies in its anchoring system. Sometimes, such problems are relatively easy to solve. For instance, additional or longer nails or screws will often satisfactorily refasten a loose frame. New anchors in the right locations will also sometimes do wonders. The appearance of the finished installation, however, is not always good. Since the existing anchors were in the wrong places or were inadequate in size or number, it is usually necessary to install new anchors where they will be visible.

Binding or loose doors may be caused by a deformation of the frame or just loose hinges. When the fault lies in frame deformation but the problem is not too severe and is not continuing to get worse, it is usually possible to correct the binding by sanding or planing the door. Loose screws, of course, should be tightened. It may be necessary to remove the screws and drive glue-coated dowels into the holes to make them tightly grip the redriven screws. When a frame has deformed slightly, it is sometimes possible to straighten a door by shimming behind the jamb side of a hinge plate with cardboard.

Binding or looseness caused by hinges being improperly installed can also often be solved by shimming or otherwise adjusting the hinges.

Shimming will also often correct misalignment between a latch bolt and its receiver. Sometimes it is necessary to relocate the strike plate.

Missing Hardware. It is usually possible to find or make replacements for missing hardware. Often the new hardware can be installed quite easily in the field.

Repairing Materials Adjacent to Wood Doors and Frames. When damage to a door or frame has also affected adjacent construction, repair of that construction is also required.

5.5 Wood Doors and Frames in Existing Construction

The requirements discussed in section 5.2 are equally applicable when a new door or frame is to be installed in an existing space. There are, however, some additional considerations that must be taken into account when they are installed into an older building. Many of those considerations are addressed in this section.

Since the number of possible wood door and frame installation conditions in existing buildings is vast, this book cannot cover them all. The following discussion is therefore generic in concept and should be viewed in that light. The suggestions here may not apply to all conditions, but the principles should be applicable to most. In the end there is no substitute for professional judgment, which is imperative to the appropriate use of the following data.

5.5.1 Door and Frame Materials, Finishes, Types, and Fabrication

There are no essential differences between wood doors and frames or associated hardware, glazing, or louvers that are installed in all-new buildings and those installed in existing construction: the same materials and finishes are also used. Refer to sections 5.1.3, 5.1.4, and 5.1.6 for discussions about wood door and frame materials, finishes, types, and fabrication and about louvers.

5.5.2 Standards, Controls, Delivery, and Handling

The standards and controls discussed in sections 1.6.1, 1.7.1, 5.1.1, and 5.1.2 apply as well to new doors and frames installed into existing construction.

While installation of new doors and frames into existing construction is under way, existing doors and frames that are not scheduled to be removed and discarded should be protected from harm. Where they become damaged they should be repaired to a condition at least equal to that before their damaging. Where acceptable repairs are not possible, the damaged door or frame should be removed and a new unit installed.

Damaged doors and frames should not be installed without being repaired. When satisfactory repairs are not possible, the damaged doors and frames should be discarded and new units provided. The requirements for installing new doors and frames also apply to reinstalling existing doors and frames.

5.5.3 Preparation for Installation

Just as preparations are necessary before new wood doors and frames can be installed in a new space, so is it necessary to make preparations before installing new doors and frames in an existing space. In the latter case, however, the preparations are likely to be more extensive.

The requirements for demolition and removal of existing items and for renovation work discussed in sections 1.6 and 1.7 are applicable to work related to wood doors and frames. The requirements in this section are in addition to those in sections 1.6 and 1.7.

As is true when doors, frames, and the installation space are all new, the site where new doors and frames will be installed in an existing building should be examined carefully by the installer. When the doors and frames will be installed in an existing space, two examinations are usually best. One should be made before demolition work begins. A report of this examination should be sent to the designer and owner so that modifications can be made in the proposed demolition, if necessary, to accommodate the door and frame design.

The second examination should be made after demolition in the space has been completed. This examination is necessary to verify that other construction to which the new frames are to be attached, and the areas and conditions where the frames will be installed, are in a proper and satisfactory condition to receive the new frames. Preparatory installation work should not be started and the frames should not be installed until unsatisfactory conditions have been corrected.

5.5.4 Installation of Doors and Frames

In general, the requirements discussed in section 5.2 apply to doors and frames installed in existing construction. This section contains modifications and additional requirements.

Should conditions be encountered that differ significantly from those expected, the person discovering the difference should immediately notify the other concerned parties, especially the owner, so that alternative methods can be developed to handle the unexpected situation.

Damaged doors and frames should not be installed. It is generally much easier to repair a damaged door or frame before it is installed. Often, damage will not be repairable after a door or frame has been installed and the item must be removed to permit repair. Damaged items that have been installed, but cannot be repaired in place should be removed and repaired or new, undamaged units installed. Installing already damaged doors or frames is one of the largest causes of later problems with them.

Doors and frames should be installed in accordance with the manufacturer's latest published instructions, the approved shop drawings, and the latest applicable industry standards and recommendations.

5.5.5 Installation of Glazing and Louvers

Glazing and louvers are installed in the same way, regardless of whether the frame is installed in new or existing construction. Neither is there a difference in installation methods whether the frame is new or existing, except that the method must be appropriate for the existing frame and will usually be the same as that originally used.

5.5.6 Patching Adjacent Materials and Surfaces

Installation of a new wood door and frame in an existing space will almost always be accompanied by a need to patch or repair adjacent surfaces. Sometimes it will have been necessary to remove adjacent materials or items, in which case they must be reinstalled. The requirements discussed in section 1.7 apply here.

5.6 Where to Get More Information

The following AIA Service Corporation's *Masterspec* Basic sections contain descriptions of the materials and installations addressed in this chapter. Unfortunately, those sections contain little that will help with troubleshooting these installations.

- The February 1989 edition of Section 06401, "Exterior Architectural Woodwork," which includes exterior wood door frames.
- The February 1989 edition of Section 06402, "Interior Architectural Woodwork," which includes interior wood door frames.
- The February 1987 edition of Section 08211, "Flush Wood Doors."
- The February 1987 edition of Section 08212, "Panel Wood Doors."

- The February 1987 edition of Section 08460, "Automatic Entrance Doors."
- The May 1984 edition of Section 08710, "Finish Hardware."

The following is a list of sections in the Architectural Woodwork Institute's *Architectural Woodwork Quality Standards, Guide Specifications and Quality Certification Program* that are applicable to wood doors:

- Section 200, "Panel Products."
- Section 900A, "Door Frames/Exterior or Interior."
- Section 900B, "Interior Door Frames."
- Section 1300, "Architectural Flush Doors."
- Section 1400, "Stile and Rail Doors."
- Section 1500, "Factory Finishing."

Every designer should have the full complement of applicable ASTM standards available for reference, of course, but anyone who needs to understand wood doors and frames should have access to a copy of the ASTM standards marked with a [5] in the Bibliography.

Janet Marinelli's 1988 article "Architectural Glass and the Evolution of the Storefront" and its accompanying sidebar, "Glass Notes," by Gordon Bock, are together an excellent overall look at their subjects. Anyone who has to deal with old wood-framed store fronts should obtain a copy.

Harold B. Olin, John L. Schmidt, and Walter H. Lewis's 1983 edition of *Construction Principles, Materials and Methods* is a good source of background data about wood doors and frames. Unfortunately, it contains little that will help specifically with troubleshooting failed metals. This book is slightly out of date, but an updated version will soon be published.

Refer also to other items marked with a [5] in the Bibliography.

CHAPTER

6

Windows and Sliding Glass Doors

This chapter discusses all kinds of steel, aluminum, and wood operating windows and associated fixed sash, and aluminum and wood sliding glass doors. It includes windows and doors with natural or painted finishes and metal- and vinyl-clad wood windows and sliding glass doors. Wood window types covered are single hung, double hung, casement, horizontal sliding, awning, hopper, and pivoting. Metal window types discussed are awning, casement, single hung, double hung, horizontal sliding, jalousie, projected, top hinged, vertical pivoted, vertical sliding, and fixed.

Metal store front sections into which metal windows might be installed are discussed in Chapter 4. Steel windows are usually used in steel store fronts, aluminum in aluminum. Wood windows are seldom installed in metal store fronts.

Aluminum curtain walls into which some of the aluminum windows discussed in this chapter might be installed are discussed in Chapter 7. Steel windows and wood windows are seldom installed in aluminum curtain walls.

Glazing for windows is discussed in Chapter 8.

6.1 General Requirements for Windows and Sliding Glass Doors

Steel and aluminum windows are more alike than different, even though different standards govern each. Window types, for example, are very similar, and installation is basically the same. The major differences are a matter of averages. Most, but not all, steel windows are face glazed, while most aluminum windows are glazed with either gaskets or aluminum stops. Few steel windows are available with thermal-break construction, while most aluminum windows are.

Many of the principles that apply to windows also apply to sliding glass doors. In many ways such doors can be thought of as just large sliding windows. There are some differences, of course, which are discussed in this chapter.

6.1.1 Controls

The general requirements related to controls in sections 1.6.1 and 1.7.1 apply also to windows and sliding glass doors installed in new construction. The controls discussed in this section are in addition to those covered in sections 1.6.1 and 1.7.1.

The window and sliding glass door manufacturer's product data, and its specifications and instructions for fabrication, factory finishing, shop painting, and installation should be submitted for the owner's approval.

Shop drawings should also be submitted to the owner for approval before the windows or sliding glass doors are fabricated. The shop drawings should include requirements for new windows and sliding glass doors in both new and existing openings; existing windows and sliding glass doors that are to be removed and discarded or turned over to the owner; existing windows and sliding glass doors that are to be removed, altered or refinished, and reinstalled; existing windows and sliding glass doors that are to be relocated; and existing windows and sliding glass doors that are to be repaired in place. The shop drawings should include all necessary information including details of the windows and doors themselves and associated cladding, weather stripping, sills, trim, casings, stools, and venetian blind inserts, and a schedule showing the location of each window and door unit. The drawings should show complete details for metal windows with thermal-break construction. For items to be repaired the shop drawings should show the methods to be used in each case. They should include a schedule showing the location of each window and door unit.

Finished samples of typical window and door sections should be submitted for acceptance by the owner. These samples should include the basic window or door and associated trim and venetian blind insert. Samples of sliding glass door frames, including jambs, head, and sill tracks, should also be submitted. The actual finishes to be provided should be submitted. Materials should be those that will be used in the installed items. The samples should show the complete range of expected colors and finishes.

Often, especially on larger projects, an actual sample of each window type proposed for use in the project is required. However, sample sliding glass doors are only required when there is a large number of similar doors. Samples should be complete with glass, venetian blind inserts, and the proposed hardware. They should be made with the proposed wood, finished with the proposed finish, clad with the proposed material, and operate in the proposed manner. The samples should be of one of the sizes that actually occurs in the project. They are usually returned to the manufacturer after the actual windows and doors have been installed, but are sometimes installed in the project.

The contractor should also submit for the owner's approval sample corner sections of the required windows and sliding glass doors showing construction details, and samples of sliding window and door tracks and each type of hardware that will be used.

Except when the owner or designer has specific experience with the windows and doors to be used or the project is too small to warrant such an expense, the manufacturer should be required to submit test reports by an acceptable independent testing laboratory attesting to compliance of the windows and doors with the project's requirements.

Before the windows and doors are fabricated, the manufacturer should be required to take measurements at the project site of each opening that is to receive a window or door, and verify the governing dimensions. Windows and doors should be fabricated to fit the actual conditions, where practicable. However, when such custom fabrication would delay a project's completion or increase the cost of the windows and doors beyond acceptable limits, they are often fabricated without taking field measurements and the tolerances are coordinated to ensure proper fit.

Windows and sliding glass doors should be installed either by the manufacturer or a firm franchised by or acceptable to the manufacturer. Windows and doors should be installed only by experienced erectors.

Finish carpentry related to windows and sliding glass doors should be done by a firm that can demonstrate successful experience in similar finish carpentry work. Only experienced personnel who are skilled in the processes and operations necessary and are working under competent supervision should be used.

Warranties beyond a normal one-year construction warranty are sometimes required for high-quality, custom-fabricated windows and sliding glass doors. The normal warranty period is three years, but some manufacturers offer a lifetime warranty. When a failure occurs, it is worth determining if there is a warranty. Extended warranties are not usually required for standard windows or doors selected from a manufacturer's catalog.

Some sources recommend that metal windows and sliding glass doors be field tested for structure performance and air and water infiltration. Such tests are seldom required, however, unless the number of windows or doors is very large or the exposure is severe. Windows in tall buildings should probably be field tested.

6.1.2 Glazing

The materials used for glazing windows may be any of those discussed in Chapter 8, and the standards discussed there apply. Thicknesses vary with window type and manufacturer. Most windows are glazed with glass, some with plastic, and a few metal windows with opaque panels such as fiber-reinforced, cementitious color-coated panels similar to Glasweld, manufactured by Eternit. Glass may be single pane, multiple pane, or insulating glass.

Many codes require that sliding glass doors have safety glazing materials that comply with the American National Standards Institute's Standard Z97.1-1984, "Safety Glazing Materials Used in Buildings."

Glazing may be done either at the factory or in the field, but the most satisfactory results are obtained when the glazing material is installed in the factory.

6.2 Specific Requirements for Steel Windows

6.2.1 Standards for Steel Windows

The only standards generally applicable to steel windows are those produced by the Steel Window Institute (SWI). Anyone responsible for designing or maintaining a large number of steel windows should probably obtain SWI's standards by contacting it at the address listed in the Appendix. Unfortunately, the SWI standards alone are often not adequate for high-quality installations. They may be supplemented with ASTM standards and, for very large projects, by requiring that the window units be tested to meet performance criteria selected by the designer. For most projects the standard used is a specific proprietary window. The window required is selected based on the designer's experience and the reputation of the selected window and its manufacturer. Other windows may be accepted when the manufacturer can prove their quality with the one used as the standard.

When steel windows are required to meet SWI standards, additional requirements are often written in by the designer to ensure a higher quality window than SWI standards alone would require. For example, a window might be required to not fail or permanently deflect when subjected to a positive (inward) test pressure of 50 psf and a negative (outward) test pressure of 25 psf when tested in accordance with the requirements of ASTM Standard E 330. Windows may also be required to meet air infiltration and water penetration requirements when tested in accordance with ASTM Standards E 283 and E 331, respectively. The test pressures used and infiltration rates should be selected by the designer to represent the worst case that might be encountered by the windows when in place. Some codes dictate minimum test pressures and performance levels.

The SWI standards include requirements for crack-opening tolerances and window grades, including window types, sizes, and thicknesses of the metals used.

Fire-resistant steel windows should comply with the National Fire Protection Association's standard *NFPA 80: Fire Doors and Windows*.

6.2.2 Materials, Products, and Finishes for Steel Windows

In general, the discussion in section 3.2 applies to the carbon steel used in windows. The most common steel products used are hot-rolled or cold-rolled carbon-steel shapes. Some formed sheet steel may be used for interior trim.

The discussions in sections 3.6, 3.7.1, 3.7.2, 3.10, and 3.11 apply to the finishes used on steel windows. After proper pretreatment and cleaning, steel windows may be shop primed and field painted or receive a shop-applied finish, which is usually an acrylic or polyester baking-finish enamel. Steel windows may also be galvanized after fabrication, in accordance with ASTM Standard A 386. Galvanized windows may be left with the galvanizing as their final finish or receive a shop primer and be field painted.

Some steel windows contain formed aluminum or stainless steel or extruded aluminum trim members, glazing beads, screen frames, weatherstripping retainers, or flashing. Such components are usually finished to match the steel windows, as discussed in section 3.11, but may be given any of the finishes discussed in Chapter 3 for such metals.

Fasteners for steel windows should be steel, galvanized steel, or stainless steel. Many sources advise using copper-alloy fasteners in steel windows, even though the Copper Development Association (CDA) advises against such use, because galvanic action between the copper alloy and the steel is likely to corrode the steel. Nevertheless, copper-alloy fasteners are often used in steel windows. In addition, the standard hardware for steel windows is often a die-cast or solid-copper alloy with steel or copper-alloy operating bars and rods.

Anchors, clips, and accessories are usually stainless steel, hot-dip galvanized steel, or copper alloy.

6.2.3 Steel Window Grades

The SWI places steel windows into eleven classifications, by weight. Several of them are also directly tied to window type (see section 6.2.4). The classifications Standard Intermediate, Heavy Intermediate, and Heavy Custom windows are of the highest quality established by the SWI. They are available in awning, casement, projected, hinged access, and hinged emergency access/egress windows. A fourth grade, Architectural Projected windows, is of a lower quality than the three just listed and includes only projected windows. A fifth grade, Commercial Projected windows, which is also lower in quality than the first three, is available in projected, reversible, and security window types. The sixth grade, Residential windows, applies only to residential-grade windows. The seventh grade, Manufacturer's Custom Grade is invoked when the window quality is established by a manufactured product. The SWI also establishes the five grades, Guard, Psychiatric, Security, Awning, and Curtain Wall. These last five are not well defined, however.

6.2.4 Steel Window Types

Windows are identified by types that are related to the operating arrangement of the window's sashes. Steel windows are generally available in the types addressed in this section. The types of windows referred to in this section can be seen in Figure 6-1. Other types of steel windows may also be available in some locations. Special types can also be made to match existing windows, but it may be difficult to find a fabricator willing to make them.

Awning windows contain one or more top-hinged out-swinging operating sashes in a single frame. All of the ventilators are operated simultaneously by a single device.

Casement windows have a pair of out-swinging, side-hinged operating sashes.

Double-hung windows have two sashes that slide vertically in separate tracks, bypassing each other and making an opening either at the top or bottom or both. Single-hung windows are similar, except that only one sash moves, but they may be difficult to find in steel windows.

Reversible steel windows rotate about a central axis. The center of rotation may be either vertical as shown or horizontal. Reversible windows can be latched both when they are 180 degrees from the plane of the window and when they are fully reversed.

Projected windows are probably the most often used window type.

Awning   Casement   Double hung

Vertical reversible or pivoted   Projected   Top hinged

Hinged access   Hinged emergency access/egress   Sliding

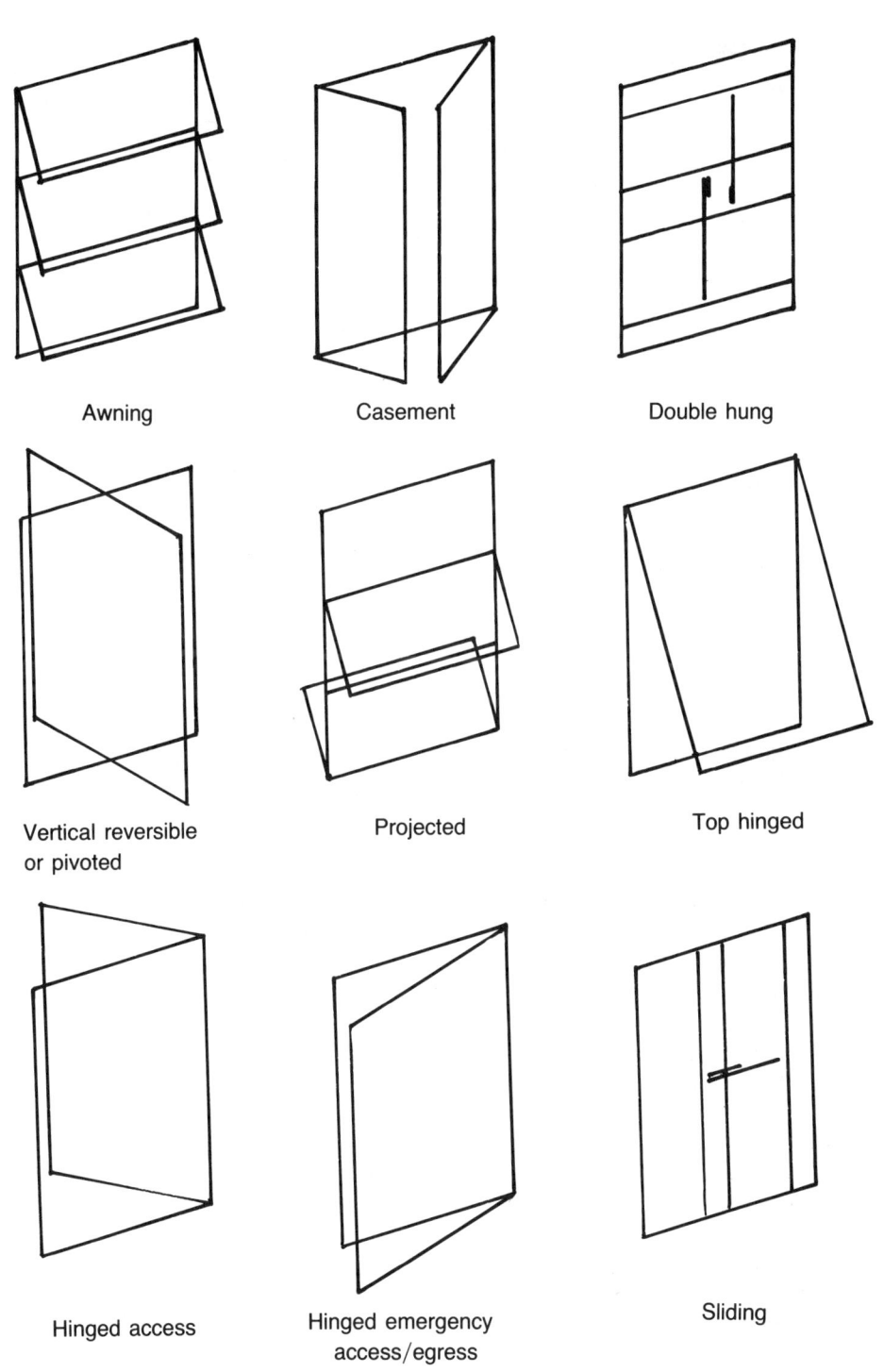

Figure 6-1 Window types.

They have in- or our-swinging top- or bottom-hinged sashes, and usually both. They may also have fixed sashes. The configuration shown in Figure 6-1 is a common one with a fixed sash, an in- or out-swinging middle sash and an in- or out-swinging bottom sash, and many other configurations are possible.

Top-hinged out-swinging windows are often used in continuous strips and operated by mechanical devices.

Hinged-access windows are either side- or top-hinged and swing in for cleaning.

Hinged emergency access/egress windows are out-swinging, side-hinged windows that can be opened in emergency for use as a fire exit.

Fixed windows can be made from any window type. They usually consist of a frame element and a fixed sash.

Combination windows are made up of two or more window types. Projected sash windows, for example, often contain fixed portions and casement windows are often installed over hopper vents.

Any window type can be used in retrofit situations. See section 6.10.2 for a discussion of retrofit casings and trim.

6.2.5 Steel Window Hardware

Suitable hardware should be provided to properly operate and securely lock window access sashes. The operating hardware may be manually locking or key-lock type, depending on the owner's requirements and the type of window.

The operating hardware for steel windows may be individual hardware for each unit or a bank of devices that operates more than one window at a time.

Each type of window has its own specific operating hardware. For some window types, there are several options. Most awning windows, for example, are supported on pivots and balance-support arms and operate by worm gears and a crank handle, but can also be operated by pole operators or remote devices. Casement windows hang on hinges. Some have gear-type rotary operators; others open manually with a finger pull. Projected windows may either hang on hinges or ride on pivots. Pole operators are available for use with some types of hardware to provide access to high windows.

Group operating systems may be used with awning and top-hinged, out-swinging windows. They may be manually operated by means of cranks or chains or electrically by motors.

The operating hardware for double-hung windows is a balancing mechanism that may consist of springs, pulleys, and sash weights, or friction mechanisms or other devices.

6.3 Specific Requirements for Aluminum Windows and Sliding Glass Doors

6.3.1 Standards for Aluminum Windows and Sliding Glass Doors

In general, aluminum windows and sliding glass doors should comply with the applicable standards of the National Association of Architectural Metal Manufacturers (NAAMM) and the American Architectural Manufacturers Association (AAMA).

Windows and sliding glass doors fitting into the AAMA's residential, commercial, and heavy-commercial-grade designations should conform with the AAMA's ANSI/AAMA 101-1985 standards "Voluntary Specifications for Aluminum Prime Windows and Sliding Glass Doors." Heavier-duty architectural-grade windows, which used to be called monumental windows, should comply with the AAMA standard GS-001, "Voluntary Guide Specifications for Aluminum Architectural Windows." Aluminum windows and sliding glass doors should be certified by the AAMA as complying with a particular AAMA grade and performance class. Where such a certification is not possible, the manufacturer should be required to submit test reports certifying that the windows and doors have been tested for conformance with the AAMA's certification label requirements.

The tests required for window and sliding glass door performance by ANSI/AAMA 101-1985 and AAMA GS-001 include the following:

Structural performance: ASTM Standard E 330.

Air infiltration: ASTM Standard E 283.

Water infiltration: ASTM standards E 331 and E 547.

Condensation Resistance Factor (CRF): Windows with thermal breaks are required to be tested in accordance with the requirements in AAMA publication 1502.7-1981.

Sound Transmission Class (STC): ASTM Standard E 413.

Deglazing: Sliding glass doors are required to pass a deglazing test in which the glazing is not disengaged when tested according to ASTM Standard E 987.

Forced-entry resistance: ASTM Standard F 842.

6.3.2 Materials, Products, and Finishes for Aluminum Windows and Sliding Glass Doors

In general, the discussion in section 3.3 applies to the aluminum used in windows and sliding glass doors. The most common aluminum products used are extruded shapes that should conform with ASTM Standard B 221 and usually are made from alloy 6063-T5 or T6. Sheet aluminum and bent-plate products are also used. They usually are required to have the properties recommended in ASTM Standard B 209 and are made either from alloy 5005, 5086, or 6061, as is suitable for the required finish.

The discussions in sections 3.6, 3.8, 3.10, and 3.11 apply to the finishes used on aluminum windows and sliding glass doors. While field painting and other factory-applied finishes are also used, the exposed surfaces of most aluminum windows and sliding glass doors are either clear or color anodized or finished with a fluorocarbon polymer-based coating.

Exposed fasteners for aluminum windows and sliding glass doors should be countersunk aluminum or stainless steel screws finished to match the windows or doors. Concealed fasteners may be aluminum, stainless steel, cadmium- or zinc-plated steel, or epoxy adhesive.

Anchors, clips, reinforcements, and accessories are usually either aluminum or stainless steel. Iron, carbon steel, and hot-dip galvanized steel may also be used if they are properly separated from the aluminum.

6.3.3 Aluminum Window and Sliding Glass Door Grades and Performance Classes

Although they differ somewhat in details, the AAMA's publications ANSI/AAMA 101-1985 and AAMA GS-001 both classify aluminum windows by type, grade, and class. The AAMA's designation system for aluminum windows includes all three of these elements. Thus, the AAMA designation C-HC50 is for a casement (C) window in the heavy commercial (HC) grade with a class of performance of 50.

The AAMA's ANSI/AAMA 101-1985 also classifies aluminum sliding glass doors by grade and class, but there are no separate types. Actually, "sliding glass doors" is the type.

The AAMA publication ANSI/AAMA 101-1985 lists three window and sliding glass door grades: residential (R), commercial (C), and heavy commercial (HC). Architectural windows, which some consider a fourth grade, are covered in AAMA GS-001. Architectural windows are not further subdivided into grades but rather are listed in three separate categories: Standard Architectural Windows, Modified Standard Architectural Windows, and Custom Architectural windows.

The performance class of windows and sliding glass doors in the AAMA system is denoted by a factor known as design pressure. A minimum design pressure, called its Standard Performance Class (PC), is established for each window and sliding glass door grade. The residential grade PC is 15 pounds per square foot (psf), the commercial grade 20 psf, and the heavy commercial grade 40 psf. The AAMA recognizes that the actual design pressure required is often larger than the minimum in the AAMA system and thus establishes additional design pressures, called Optional Performance Classes, for each grade, starting at 5 psf above the Standard PC and increasing in 5-psf increments up to 50 psf.

Based on the PC, the AAMA also establishes the actual pressures to be used in testing windows and doors. The structural test pressure is 150 percent of the design pressure. The water-infiltration test pressure is 15 percent of the design pressure, with a 2.86 psf minimum. At that rate no water is permitted to pass through the window.

Air-infiltration test pressures vary with window type. Heavy commercial casement, fixed, top hinged, vertically pivoted, and projected windows and sliding glass doors are tested at 6.24 psf, which is the pressure induced by a fifty-mile-per-hour wind. Other heavy commercial window types, and residential and commercial windows and sliding glass doors, are tested at 1.57 psf, the pressure induced by a twenty-five-mile-per-hour wind. The allowable air-infiltration rate also varies with window type and grade and door grade.

The design pressures given in the AAMA GS-001 for windows in the architectural window category are established for each test rather than for each window grade. They are usually higher than the design pressures listed in ANSI/AAMA 101-1985, often beginning where its requirements stop. For example, the AAMA GS-001 air-infiltration test rate is 6.24 psf, the water-infiltration test rate 8 psf. The AAMA GS-001 air-infiltration amounts are also more stringent than those for the windows covered by ANSI/AAMA 101-1985.

6.3.4 Aluminum Window Types

A window's type identifies its method of operation. There are at least twenty aluminum window types on the market. Here we will discuss only those most commonly used in buildings other than houses. They include awning, casement, single hung, double hung, horizontal sliding, jalousie, projected, top hinged, vertical pivoted, and fixed windows. Many of the window types mentioned can be seen in Figure 6-1. Omitted are vertical sliding, solar shading, greenhouse, sound insulating, security, and other highly specialized window types.

Awning windows contain one or more top-hinged, out-swinging operating sashes in a single frame. All the ventilators are operated simultaneously by a single device.

Casement windows have a pair of out-swinging, side-hinged operating sashes.

Single- and double-hung windows have sashes that slide vertically in tracks. Their operating hardware includes a balancing mechanism, which usually consists of springs and pulleys or counterweights.

Horizontal sliding windows contain one or more sash units that slide in tracks. Sliding sashes should be removable.

Jalousie windows consist of a series of overlapping horizontal glass louvers in the same frame. The louvers are operated in banks by a single device. A frame may contain more than one bank of louvers and therefore more than one operating device. Because of their design, jalousie windows are necessarily narrow.

Projected windows are probably the most used window type. They have in- or out-swinging, top- or bottom-hinged sashes, usually both. They also often have fixed sashes.

Top-hinged, in-swinging windows have a sash that is hinged at the top to swing into the building to permit cleaning.

Vertically pivoted window sashes are pivoted at the center of their head and sill so that they rotate in a complete circle. They are usually designed to latch both when fully closed and when partly open.

Fixed windows can be made from any window type.

Combination windows contain more than one type window. Projected-sash windows, for example, often contain fixed portions and casement windows are often installed over hopper vents.

Any window type can be used in retrofit situations. See section 6.10.2 for a discussion of panning trim used in retrofit cases.

6.3.5 Aluminum Window and Sliding Glass Door Hardware

Suitable hardware should be provided to operate properly and securely lock access sash and to operate venetian blind inserts, where they are required. Operating hardware may be manually locking or key-lock type, depending on the owner's requirements.

The operating hardware for aluminum windows may be individual hardware for each unit or a bank of devices that operate more than one window at a time. Each type of window has its own specific type of operating hardware. For some there are several options. For example, awning windows are supported on pivots and balance-support arms. Most are operated by worm gears and a crank handle, but they can also be operated by a combination lever and handle and cam-type lock. Worm-gear operators may be cranked manually or by means of a pole operator or remote device.

Group operating systems may be used with window types in which all the sash in a single window are operated by a single device. They may be manually operated by cranks or chains, or electrically by motors.

Sliding glass door hardware includes roller assemblies, door pulls, and locks.

6.4 Specific Requirements for Wood Windows and Sliding Glass Doors

6.4.1 Standards for Wood Windows and Sliding Glass Doors

In general, wood windows manufactured today should be expected to conform with the applicable standards of the National Wood Window and Door Association's (NWWDA) I.S. 2-1988, "Industry Standard for Wood Window Units." Wood sliding glass doors should be expected to conform with the NWWDA's I.S. 3-1988, "Industry Standard for Sliding Patio Doors." Wood window and casings and trim for wood windows and sliding glass doors should also conform to the requirements of sections 200, "Standing and Running Trim," and 1000, "Exterior Windows," of *Architectural Woodwork Quality Standards, Guide Specifications and Quality Certification Program*, published by the Architectural Woodwork Institute (AWI).

Water-repellant treatment of wood windows and sliding glass doors and their trim should comply with the requirements of NWWDA I.S. 4-1981, "Industry Standard for Water-Repellent Preservative Non-pressure Treatment for Millwork."

NWWDA has a certification program to ensure that wood windows and sliding glass doors comply with its standards. Unfortunately, the standard is relatively new (1987), and few manufacturers have submitted their products for certification. Consequently, most wood windows and sliding glass doors found in existing buildings today will not have an NWWDA, or any other, certification. The lack of such certification does not necessarily mean that a window or door is not a good one or does not meet NWWDA standards. It may mean simply that the window or door is older than the standard or that the manufacturer has not had its units tested.

Even when not labeled by the NWWDA, new wood windows and sliding glass doors should be required to be tested by an independent testing laboratory, in accordance with the test methods recommended by the NWWDA and such other tests as the owner or designer may deem appropriate. The following test standards may be required:

Structural performance: ASTM Standard E 330.

Air infiltration: ASTM Standard E 283.

Water infiltration: ASTM standards E 331 and E 547.

Sound Transmission Class (STC): ASTM Standard E 413.

Thermal performance: ASTM Standard C 236 or C 976.

Forced-entry resistance for windows: ASTM Standard F 588.

Forced-entry resistance for sliding glass doors: ASTM Standard F 842.

6.4.2 Materials, Products, and Finishes

In general, the discussions in section 3.1 apply to the wood in windows and sliding glass doors. Specific wood materials and aluminum and vinyl cladding are discussed in section 3.1.2. Materials for use in trim that is not a part of the window unit are discussed in section 3.1.3. Preservative treatment is discussed in section 3.1.4.

The discussion in section 3.11 applies to the finishes used on wood windows and sliding glass doors. The exposed wood portions of windows and sliding glass doors are either given a shop coat of paint for field finishing or are left unpainted. Wood surfaces to receive a transparent finish may be finished either in the factory or the field. Aluminum cladding is usually finished with a baked-on enamel. The finish of vinyl cladding is integral, usually white, but is paintable.

Exposed fasteners for wood windows and sliding glass doors should either be zinc coated or nonferrous nails and screws. Hardware-mounting screws are often brass. Exposed fasteners should not be used in either aluminum or vinyl cladding. Exposed fasteners in wood should be set and filled.

Hardware items for wood windows and sliding glass doors are manufactured from many materials and given many different finishes. Materials used include, but are not limited to, cast iron, steel, stainless steel, aluminum, copper and its alloys brass and bronze, zinc, chromium-plated steel, and chromium-plated aluminum. Metal materials are discussed in sections 3.2, 3.3, 3.4, and 3.5. Finishes for metal hardware items include those discussed in sections 3.6 and 3.11.

Anchors, clips, and accessories for wood windows and sliding glass doors are usually either aluminum, zinc-coated steel, or stainless steel.

Sealants are used between windows and doors and adjacent construction to seal joints for finishing and fill unsightly cracks. Elastomeric sealants are usually used on exteriors. Interior sealants may be either elastomeric types or acrylic-latex type.

6.4.3 Performance Requirements for Wood Windows and Sliding Glass Doors

The NWWDA I.S. 2 and I.S. 3 classify wood windows and doors in three categories called performance grades. They are grades 20 (Residential), 40 (Light Commercial), and 60 (Heavy Commercial). The numbers refer to the air pressure used to test units for structural performance. A grade-20 window or sliding glass door, for example, must be tested for structural performance under a wind load of 20 pounds per square foot (psf).

Unlike aluminum windows, where test pressures and other criteria vary with window type, all wood window types are tested using the same pressures, and the same results are required. There are some differences in the results demanded of doors, however, when compared with windows. Grade-60 wood sliding glass doors, for example, are permitted only a 0.1 percent deflection under load, while grade-60 windows are permitted to deflect by 0.4 percent.

Wood windows and sliding glass doors are usually required to comply with a Grade 10 rating for forced-entry resistance. They also may be required to have a certain sound-transmission class and a specific thermal-performance level.

6.4.4 Wood Window Types

A window's type identifies its method of operation. There are many possible types, but this text discusses only those that are widely available, including double hung, casement, horizontal sliding, awning, bow, bay, and fixed windows. Some other types that are sometimes used but are not specifically discussed here are hopper, single hung, tilt-turn, and pivoting windows. The principles here, however, apply to them as well.

Double-hung windows (see Fig. 6-1) have two sashes that slide vertically, bypassing each other so that they open from both the top and the bottom.

Casement windows (see Fig. 6-1) have single, side-hinged ventilators that open by swinging outward.

Horizontal sliding windows (see Fig. 6-1) contain one or more sash units that slide in tracks. The sliding sashes should be removable.

Awning windows (see Fig. 6-1) contain a single, top-hinged, out-swinging operating sash in a single frame. All the ventilators are operated simultaneously by a single device.

Bow and bay windows are assemblies with either operating or fixed windows or both. Bow windows curve outward. Bay windows have angled ends and a straight front that parallels the building wall (Fig. 6-2). Angled mullion posts and interior and exterior trim for the special shapes involved are usually part of bow and bay windows. Interior seats and overhead soffits are also often furnished as part of these windows.

Fixed windows may have special frames (Fig. 6-3) or may be made using components from an operating window type. Many fixed windows are available today with decorative shapes such as oval, arched, elliptical, octagonal, or circular ones.

Combination windows are made up of two or more other window types assembled into a single window.

Any window type may be used for retrofit. See section 6.10.2 for a discussion of panning trim for retrofitting.

Figure 6-2 A bay window.

6.4.5 Hardware for Wood Windows and Sliding Glass Doors

Suitable hardware should be provided to properly operate and securely lock access sashes and operate venetian blind inserts where they are required. The operating hardware may either be a manually locking or a key-lock type, depending on the owner's requirements.

Each type of window has its own specific type operating hardware. For some window types there are several options. For example, awning windows are supported on pivots and balance-support arms. Most are operated by worm gears and a crank handle, but they may also be operated by a combination lever handle and cam-type lock. Worm-gear operators may either be cranked manually or by a pole operator or remote device.

Older windows may have hardware quite different from that used today. New wood double-hung windows are likely to have channel-spring balances,

Figure 6-3 A fixed window with a special frame.

for example, some form of tube or block-and-tackle-type balances, or simply a friction mechanism. Older double-hung windows are more likely to have a weight and pulley or clock-spring-tape balance system.

Sliding glass door hardware includes roller assemblies, door pulls, and locks.

6.5 Miscellaneous Window and Sliding Glass Door Components and Accessories

6.5.1 Thermal Barriers

A thermal barrier is a nonconductive material interposed between the outdoor and indoor portions of a metal window or sliding glass door to separate metal that contacts outdoor air from metal that is in contact with indoor air, thus preventing heat from traveling from one side to the other.

Some hollow steel windows are available with a thermal barrier. They are also usually designed to receive insulating glass. Most steel windows are not so designed.

Most aluminum window types and sliding glass doors are available today either with or without thermal barriers. Each window and sliding glass door manufacturer has its own thermal-barrier material. Chemically curing, high-density polyurethane is a commonly used one.

6.5.2 Weather Stripping

Some steel windows have no weather stripping, but aluminum and wood windows and sliding glass doors are usually weather stripped. Different window types have different weather stripping. There are two types of weather stripping used on windows and sliding glass doors: compression, and sliding. Weather-stripped single-hung, double-hung, and sliding windows and sliding glass doors require both types.

Compression weather stripping in metal windows and sliding glass doors is made from several materials, including molded EPDM or neoprene gaskets that comply with ASTM Standard D 2000, Designation 2BC415 to 3BC620 or, for aluminum windows only, with AAMA SG-1; expanded EPDM or neoprene gaskets complying with ASTM Standard C 509, Grade 4; molded PVC gaskets complying with ASTM Standard D 2287; urethane foam; thermoplastic olefins; and others.

Sliding-type weather stripping for metal windows and sliding glass doors should comply with the American Architectural Manufacturers Association's AAMA 701.2-1974. Most of it today is polypropylene.

Several different types of compression-type weather stripping are used in wood windows. The oldest is made from nonferrous spring-metal strips. More-modern types include molded PVC gaskets complying with ASTM Standard D 2287, and expanded EPDM or neoprene gaskets complying with ASTM Standard C 509, Grade 4. Urethane foam and thermoplastic olefins are also used.

Compression-type weather stripping in wood sliding glass doors is made from one of several materials. The most usual are expanded EPDM or neoprene gaskets complying with ASTM Standard C 509, Grade 4, or molded PVC gaskets complying with ASTM Standard D 2287. Urethane foam, thermoplastic olefins, molded EPDM or neoprene gaskets, and other materials are also used.

Sliding-type weather stripping should comply with the American Architectural Manufacturers Association's AAMA 701.2-1974. Materials used include polypropylene or wool or nylon pile backed with resin-impregnated fabric and aluminum. Some proprietary semirigid polypropylene products are also used.

6.5.3 Frames and Trim

Most window and sliding glass door frames are furnished as part of the window or door. For example, the frames and trim for arched, circular, and other decorative windows are often part of the window. The frames and trim for bow and bay windows are usually furnished with the windows. Some windows, however, are designed to fit within a separate frame. Sometimes the separate frame is a different construction or material altogether. Standard windows are often installed in curtain wall construction, for example.

Even when trim is not furnished with the window, it is often needed to make a window fill the opening provided for it, especially when the opening is in an existing building. Trim is usually of the same material as the window.

The design of frames and trim must permit unrestricted expansion and contraction.

6.5.4 Window Sills and Stools

Steel windows usually have sills of another material, such as brick, stone, or cast stone. Aluminum and wood windows may also have brick, stone, cast stone, or other such sills, but they often have extruded- or formed-aluminum sills. Even when other sills are present, aluminum and wood windows often have secondary extruded- or formed-aluminum sills. Some wood windows may also have wood sills.

Stools for steel, aluminum, and wood windows may be wood, slate, tile, or another material different from that of the windows. Stools are also made from sheet steel or aluminum.

6.5.5 Sliding Window and Door Tracks

Sliding tracks for steel windows are usually steel. Tracks for sliding aluminum windows are usually made from extruded or formed aluminum. Tracks for sliding wood windows are either extruded aluminum or plastic. Many plastic tracks are rigid polyvinyl chloride, but they may be another weather-resistant plastic.

Sliding glass door tracks are usually extruded aluminum regardless of the material of the doors.

6.5.6 Window Mullions

Strips of windows require mullions between the individual units regardless of the window type. Mullions are usually made from the same material as the window. They must permit unrestricted thermal movement in both the windows and the mullions.

Mullions in bow and bay windows are a part of the window.

6.5.7 Muntins

Traditionally, steel windows have been divided into small panes by muntins, and many modern steel windows also have them. Aluminum windows usually do not have muntins, but some may. Snap-in aluminum false muntins are available for some aluminum windows. Aluminum windows with false muntins are usually found where it is necessary to simulate the appearance of existing windows.

Before the days of large glass panels and insulating glass, wood windows had to be divided into small panels by wood muntins, so that large openings could be made using small glass panes. Some people still like the look of small panes, and some manufacturers still make windows with true wood muntins and small glass panes. The use of true muntins, however, is not always practicable because of today's manufacturing processes and energy-efficient window construction, and the high cost of glazing small panes with insulating glass or multiple panes of glass. To create the look of small panes while holding prices down, wood window manufacturers provide what they call grilles, which are false muntins that snap into a window's frame over large glass panes (Fig. 6-4). False muntins are available in many designs and are made of either wood or extruded polyvinyl chloride.

6.5.8 Storm Windows

Removable aluminum-framed storm windows are available for use with many wood window types. They are used on windows that are not glazed with insulating glass.

6.5.9 Venetian Blinds

Some modern double- or triple-glazed aluminum and wood windows have built in, remotely operated, horizontal venetian blinds in a space between the glass panes. Such blinds usually have approximately one-inch-wide aluminum slats. They should have long-lasting polyester fiber cords and should be equipped for tilting, raising, and lowering by standard operating hardware located on the interior face of the sash.

6.5.10 Insect Screens

Insect screens are usually provided for each operating window sash and sliding glass door. The exception is sashes that are not openable by occupants without using special tools, as is often the case in high-rise air-conditioned buildings.

Screens are usually mounted on the interiors of awning, out-swinging

Figure 6-4 The muntins in the window shown in this photo are false. The window is aluminum, as is the storm window. This is a poor—albeit energy efficient—attempt to match the look of wood windows in an historic building.

casement, jalousie, and projected windows; on the exteriors of single-, double-, and triple-hung, horizontal sliding, in-swinging casement, and vertical sliding windows; and on the interiors of sliding glass doors. Interior-mounted screens must permit access to a window's operating hardware. Hinged access panels in the screens are sometimes used, but they are usually unsightly and may not be acceptable in many applications.

Galvanized-steel, stainless steel, aluminum, copper-alloy, and glass-fiber mesh are all used as insect screens, but the standard used by most window and sliding glass door manufacturers is plastic-coated glass-fiber mesh. Copper-alloy screens should not be used on galvanized steel or aluminum windows, because corrosion of the window material may result. Wire-fabric insect screens should comply with Federal Specification RR-W-365. Glass-fiber-mesh insect screens should comply with Federal Specification L-S-125.

Screen wire for windows should be mounted in a frame that is usually of steel, stainless steel, or aluminum for steel windows and of aluminum for aluminum and wood windows. The wire should be held in place by a removable metal or plastic insert. Screens for sliding glass doors should be mounted in an aluminum frame. The screen-frame corners should be mitered and reinforced. Screen frames for windows should come complete with mounting hardware and associated fasteners and should be securely fastened to the window. Screens for sliding doors should slide in a slot in the top and bottom tracks and should have their own latching hardware. Screens for sliding glass doors should comply with the Screen Manufacturers Association's ANSI/SMA 2005-1976, "Specifications for Aluminum Sliding Screen Doors." Screens for both sliding glass doors and windows should also comply with ANSI/SMA 1004-1976, "Specifications for Aluminum Tubular Frame Screens for Windows."

6.5.11 Accessories

Clips, anchors, fasteners, closures, flashings, reinforcement, water sheds, and other items, materials, and devices necessary to make a complete installation should be delivered to the site along with associated windows and sliding glass doors.

Reinforcement may be necessary to make standard frames and mullions able to meet the proper performance requirements.

The design of the needed accessories must permit unrestricted expansion and contraction.

6.6 Fabrication of Windows and Sliding Glass Doors

Windows and sliding glass doors should be fabricated in accordance with the applicable industry standards and the recommendations of the association representing the producers of the particular type of windows or doors at hand. Where a particular proprietary window or door has been selected to be the standard, the windows or doors used should conform to the design and construction of and have the same performance characteristics as the selected window or door.

Windows and sliding glass doors should be designed, manufactured, and installed so that it is possible to reglaze them without dismantling window or door components. Glazing stops may be screw-applied or a snap-in type, as is standard for the window type or sliding glass door being used. Glazing gaskets should be removable.

When they are required, thermal breaks must be complete, without metal bridges of any kind.

6.6 Fabrication of Windows and Sliding Glass Doors

Metal frame and sash corners can be lapped, butted, or mitered. They can be mechanically joined using fasteners and clips or be welded, or be both mechanically fastened and welded, depending on the selected window or door design. Mitered and welded corners that are also secured mechanically are usually considered the better construction, but if a window or door conforms with the established performance standards its corner construction is probably immaterial.

Metal windows and sliding glass doors should have weep holes to bleed infiltrated water and condensation to the exterior. The bottom tracks in sliding wood windows and wood sliding glass doors should also have weep holes.

Side-hinged ventilators should be capped with water-shed devices to prevent water running down the sash from entering the building.

Windows and sliding glass doors should have subframes where necessary. They should be of the same material and finish as the windows and doors.

Wood windows and sliding glass doors and associated trim should be fabricated at the factory or mill to be rigid, neat in appearance, free from defects and warp, and have their wood surfaces free of machine and tool marks. Hardwood dowels to cover screw heads should be provided where applicable.

Wood windows and window and door trim should be fabricated in accordance with the project's requirements, the approved shop drawings, and applicable Architectural Woodwork Institute (AWI) and National Wood Window and Door Association (NWWDA) standards, to the dimensions, profiles, and details required. Joints should be made tight and should be constructed to conceal shrinkage. Trim, stops, and moldings should be mitered at exterior angles and coped at interior angles and returns. The materials used should show no excessive warp.

Except for members with their ends exposed, the backs of flat wood-trim members should be routed or grooved, and wide members should be kerfed. The edges and corners of solid-wood members that are less than one inch thick should be eased to a 1/16-inch radius, thicker members to a 1/8-inch radius.

When possible, fabrication, assembly, finishing, hardware application, and other work should be completed before frames are shipped to the project site. Windows and sliding glass doors should be delivered complete with fasteners, anchors, clips, and accessories. Fasteners should be concealed whenever possible.

Windows and sliding glass doors are not designed to support loads other than themselves and therefore require separate lintels to support walls or partitions above the frame.

Necessary mullions, closures, cover plates, and other accessories should be provided.

6.7 Installation of Windows and Sliding Glass Doors in New Construction

6.7.1 Delivery and Handling

Windows and sliding glass doors should be protected during transit, storage, and handling to prevent damage, soiling, and deterioration. Items with minor damage may be repaired and installed, but those with major damage should be discarded and new, undamaged items used.

Wood windows and sliding glass doors should not be delivered to the construction site until the project is ready to receive them. Concrete, stone, brick, concrete unit masonry, and other wet work should have been completed and permitted to dry for at least ten days before wood doors or windows are delivered.

Immediately after windows and doors have been delivered to the construction site, their protective coverings and crating should be removed. They should then be stored in a dry place and protected from damage. Their coverings should not be a non-breathing material or be applied so that they create a humid chamber.

Wood materials for trim should be kept dry and protected against exposure to weather, high humidity, extreme changes in temperature or humidity, and contact with damp or wet surfaces. Materials should be stacked to ensure proper ventilation and drainage. Air circulation must be provided within the stacks. Humid chambers should not be permitted under the protective covers.

6.7.2 Preparation for Installation

Before windows or sliding glass doors are installed, their locations should be inspected and unsatisfactory conditions corrected. Head and sill flashings must be in place, properly designed and installed, and undamaged. Other adjacent work must be complete.

Openings must be square, true, plumb, of the correct size, and properly prepared to receive windows and sliding glass doors. Sills must be level and straight. Surrounding masonry and concrete must be dry and free of excess mortar and other contaminants. Wood framing must be well nailed and nails that would interfere with window installation must be driven flush with the wood. Joints should be smooth and without offsets. Metals in and near the openings should be clean, dry, and free of grease, oil, dirt, rust, and other corrosion. The metal itself should have no sharp edges or joint offsets.

Before wood windows or sliding glass doors are installed, they should be examined to ensure that they are complete and properly fabricated and have been back primed.

6.7.3 Installation

Windows and sliding glass doors and associated hardware, operators, mullions, sills, and other components should be installed in accordance with the shop drawings, the recommendations of the window or door manufacturer, and applicable industry standards. Installed windows and doors should be rigidly and securely anchored in place, square, plumb, level, true to line, at proper elevation, and in alignment with other work. Anchors should be concealed.

The proper sealants should be used within frame assemblies and in concealed locations between the window and door assemblies and the substrates.

Metal window sills, sliding door sills, and window and door frames should be set in a bed of sealant. Joints in sills, and between sills and windows, should be made watertight using sealants or gaskets.

When mullions and closures are installed, the shop drawings and manufacturer's recommendations should be followed exactly. They should be fastened in place using nails or screws through predrilled holes. The entire installation should be back sealed and thoroughly calked.

Exposed wood components should be finished as soon as possible after installation, to help prevent water intrusion into the wood.

Windows and sliding glass doors that have not been glazed in the shop should be glazed in the field. After windows and sliding glass doors have been installed and glazed, their operating sashes and venetian blind inserts should be adjusted so that they operate properly. Hardware and operating parts should be lubricated. Screens should be properly installed where they are required. The perimeter of each window and door unit should be calked, and the entire installation left watertight.

6.7.4 Adjustments, Cleaning, and Protection

Where metals are placed in contact with dissimilar metals, masonry, mortar, concrete, plaster, or pressure-preservative-treated wood, the concealed contact surfaces should be covered with a heavy coat of alkali-resistant bituminous paint, or contact should be prevented, using plastic sheet materials or another appropriate method.

Operating sashes, venetian blind inserts, screens, and associated hardware should be protected by keeping the operating sash and screens tightly closed and locked.

Cementitious materials, tape, and paint should be immediately removed from installed units before they harden or damage the finish. Excess sealing compound should be removed from exposed metal surfaces before it cures.

Wood, metal, and vinyl surfaces should be thoroughly and carefully cleaned as soon as the windows and doors have been installed. Mild soap and water or solvents that will not damage the finish should be used. Thereafter, the installation should be protected from damage, soiling, and staining.

6.8 Why Windows and Sliding Glass Doors Fail

Windows and sliding glass doors and their finishes fail for many reasons, most of which can be grouped into three categories. The first consists of problems with the surrounding or supporting construction. It includes excess structural movement, failed steel or concrete structure, failed wood structure or wood wall or partition framing, failed metal wall or partition framing, failed concrete, stone, or masonry walls or partitions, and other building element failures that are discussed in Chapter 2. While the failure causes in this first category may not be the most probable ones for the failure of metal windows and sliding glass doors or their finishes, they are usually more serious and costly to fix than the other possible causes. Consequently, the possibility that they are responsible for a failure should be investigated.

The second category concerns the materials and finishes used in metal windows and sliding glass doors. It includes bad materials, inappropriate finish selection, improper preparation for and application of finishes, failure of the immediate substrate, failure to protect materials and finishes, failure to properly maintain applied finishes, and natural aging. These failure causes are discussed in Chapter 3.

The third category includes improper design, inappropriate window or sliding glass door selection, inappropriate hardware selection, improper fabrication, improper installation, and natural aging. These failure categories are discussed in this section.

6.8.1 Improper Design

Selecting fasteners or anchors that are too weak for the application, too few in number, placed too far apart, or of the wrong type can lead to failure of the anchor or fastener. Requiring that incompatible fasteners be used can also lead to failure. Steel fastened with copper-alloy bolts, for example, will corrode. Requiring that incompatible anchors and attachments be used is a similar problem.

Requiring that the wrong type joining method be used will lead to failure. Insisting that aluminum be welded or brazed in the field is one example. Requiring that thin metal be welded to a thick one will almost always cause problems. Designing open joints for use where a unit must exclude water will not necessarily lead to actual failure of the joint, but the unit will not perform its intended function.

Requiring that butt joints be used in anodized aluminum or copper-alloy panels will often produce an unsatisfactory appearance. No two anodized aluminum or finished copper-alloy panels will ever exactly match in color and shade. This type of problem can be solved by introducing contrasting mullions or other dividers between large panels of either anodized aluminum or finished copper alloys.

Failing to allow for enough expansion and contraction can lead to several kinds of failures, including glass loss, open joints, and leaks. A related error is failing to require enough rebate for glazing, which can result in glass loss.

Using windows and sliding glass doors that are inefficient thermally may result in cold spots and a poorly functioning heating and ventilating system. It is almost impossible to properly balance a mechanical system when the windows and doors permit excessive heat transmission. In wood windows and sliding glass doors this error is more likely to relate to the glazing selected than the window itself. Similar errors can result, however, from permitting too much air infiltration.

A related problem is using metal windows and sliding glass doors without thermal breaks where the indoor humidity is high. The combination of high humidity and cold metal will sometimes lead to heavy layers of condensation.

Using face glazing with compounds without stops on the glazing side in large sash openings will sometimes lead to glass loss. Using glazing beads will help prevent such loss.

Interior glazing is often selected for use in multistory buildings where reglazing would be difficult from the exterior would require using scaffolds or lifts, or would otherwise be difficult to accomplish from the exterior. Face glazing with compounds or gaskets in such cases can result in the glazing being blown into the building, because wind pressure is resisted only by the glazing points or stops and putty or gaskets. When the panes are outside-glazed, wind pressures are resisted by the fixed glazing stop. Even though inside glazing using loose stops installed with screws may prevent panes from being blown out of the window, windows with inside glazing are much more likely to leak than those with outside glazing.

Exterior glazing can have its own problems, of course. Panes may be sucked out of the window if outward-acting pressures become large, which sometimes happens on the downwind side of tall buildings. Then the only solution is to use fixed glazing beads. Rebates, of course, must be properly designed in any case, and the glass thickness must take into account the conditions that will be encountered. Many steel windows are designed for exterior face glazing with compounds, which precludes their use in situations where glazing stops are required.

6.8.2 Inappropriate Window or Sliding Glass Door Grade or Window Type Selection

Selecting a window or sliding glass door that is too weak for the wind loads to be applied will lead to its failure. This error includes selecting windows or doors with the wrong performance grade. These items may fail because they are treated more roughly than their design permits. A residential-class sliding glass door may be too light for heavy commercial use, for example.

A window or sliding glass door may fail from wind and weather conditions that exceed its design limitations, even when the proper grade was apparently selected. Some failures occur because the designer fails to determine whether the windows or doors to be used meet the performance-class requirements needed. Sometimes a designer will select a particular window or door from a manufacturer's catalog without requiring submission of test results proving that the unit conforms with the project's performance requirements and then does not require field tests to prove conformance.

Selecting windows that must be operated manually for use where they cannot be reached will not necessarily lead to their failure, but it can make the windows difficult to clean and impossible to use for ventilation.

Windows that open out can be dangerous, especially when used near the floor in multistory buildings, and most especially when they will be accessible to small children or handicapped people.

6.8.3 Inappropriate Hardware Selection

Failing to coordinate hardware selection with window type may be a problem during the design stage of a project, but is seldom a problem in existing buildings. A more often found problem there results from failing to select remote hardware or pole operation for windows that cannot be easily reached, or failure to use a group operating system where the windows are hard to reach and must usually be opened as a group. While neither of the latter two problems is likely to cause actual failure of the hardware or window, either may be responsible for difficulty in operation and a failure to properly maintain the windows.

Selecting the wrong type of operator can sometimes make a window a dangerous hazard. For example, using a combination lever handle and cam lock on an awning window where the windows must be kept closed except for cleaning will defeat the design requirement. In such a case rotary operators with removable cranks are a better idea. Again, this type of error will not necessarily lead to window failure, but it will result in failure of the window to carry out its intended function.

Requiring or permitting inferior or the wrong weather stripping can sometimes spoil an otherwise well-designed window. Air infiltration through ill-fitting or deteriorated weather stripping can create more heat loss than failing to use insulating glass. Steel windows, some of which do not have weather stripping, may be precluded from some applications for this reason.

6.8.4 Improper Fabrication

Improperly fabricated windows and sliding glass doors will usually fail. Perhaps the most frequent cause of improper fabrication is failure to follow the requirements of the design, including the shop drawings and specifications and the standards of the industry.

Many fabrication errors involve joints. A common error, for example, is joining the components in such a way that corners or intersections are not square, plumb, and properly aligned, and so that joints are not flush, tight, and correctly fitted, or do not exclude water.

Joining errors in metal windows and sliding glass doors include improperly making welded joints. Welding without proper preheating can lead to stress-corrosion cracking, for example. Using the wrong weld metal can result in cracking in the weld itself or in adjacent metal. Welding or brazing metals, especially thin metals, from the visible side will usually lead to unsightly joints. Failing to properly clean a joint that is to be welded, brazed, or soldered will almost always result in a failed joint. Failing to use flux before brazing or soldering a joint will probably result in failure of the brazing metal or solder to adhere to the metal, and the joint will fail.

Using damaged materials in a window or door, or damaging them during or after fabrication, must be avoided. Damaged components and materials should be repaired or removed and replaced with new ones.

Damage to preapplied finishes or cladding during fabrication, shipping, or handling is a problem. Damaged finishes should be touched up and damaged cladding repaired before the loss of its protection permits the window or door to suffer damage.

Failing to provide weep holes in the bottom frames of metal windows and in the tracks of wood and metal sliding windows and sliding glass doors—and keep them open—will lead to leaks or collection of condensation. Weep holes may become clogged either during construction or later.

Failure to make thermal breaks continuous will defeat their purpose and affect the heating and cooling of a building. If the mechanical design is done with the assumption that thermal-break construction will be used and it does not function properly, the heating and air-conditioning systems may not work as they should.

Failure to provide watersheds over out-swinging window ventilators may result in rain water pouring into a building if its windows are left open even slightly.

Windows and associated mullions and trim, and sliding glass doors that are installed in a way that does not permit sufficient expansion and contraction to occur, may leak, twist, buckle, or pull away from the opening.

6.8.5 Improper Installation

Correct preparation and installation are essential if failure in windows and sliding glass doors or their finishes is to be prevented.

The primary cause of failure due to installation error is not following the design and recommendations of the manufacturer and recognized authorities, such as ASTM, ANSI, AAMA, NAAMM, AWI, or NWWDA.

Failing to follow the design can mean using anchors, fasteners, and other materials that will rust or otherwise corrode where high humidity will occur or where they will be set in or in direct contact with masonry or concrete. Sometimes materials that will corrode must be used for anchors. Then the error becomes one of failing to properly protect those metals with an applied bituminous or other appropriate coating.

Not following the design and standards can also lead to using anchors, fasteners, clips, and other devices of the wrong type for the installation or that are too small or weak to perform the necessary function or are not in the proper number or location.

Omitted or improperly installed flashings are a major cause of water leaks in the vicinity of windows and sliding glass doors.

Installing windows or sliding glass doors into improperly prepared openings or into openings in wet materials can cause a poor fit, improper anchorage, and damage to wood or metal materials or window or door anchors.

An allied error is installing a damaged window or sliding glass door. While the damage itself may or may not be the fault of the installer, installing a damaged unit is.

Installing a window or sliding glass door so that loads are transmitted into it will surely cause the unit to fail.

The installer must be careful to install windows and sliding glass doors so that their various components finish in the proper plane within acceptable tolerances and so that the units are in the proper location, plumb, level, and in alignment. Improperly aligning a component may not only cause the unit to present an unpleasing appearance, it can force loads to be applied eccentrically, which can cause the window or sliding glass door to become damaged. Windows and sliding glass doors that are not set plumb, square, and in correct alignment may not operate properly. Ventilators may bind and refuse to close tightly. Weather stripping may become twisted and fail to properly seal.

Windows and sliding glass doors that are not securely anchored in place may become loose over time, break their perimeter calking, and permit water or air penetration. Loose units may also vibrate and rattle in a wind.

Failing to seal around the perimeter of a wood window or sliding glass door will leave an ill-appearing installation and may result in air or water leaks. Sills, mullions, and window and sliding glass door perimeters that are not properly calked may leak air or water into the building.

A common error with anodized aluminum units is that of failing to remove anodizing acids from the aluminum. If the acid used to anodize the aluminum is not properly removed from the final product, it will eventually wash off and damage the aluminum and adjacent surfaces. Acids trapped in hollow sections are a particular problem, as are ones trapped in joints between assembled sections. Weep holes are usually needed in hollow sections and assemblies to ensure proper removal of acids.

Another error in aluminum windows and sliding glass doors is failing to remove strippable coatings and tape soon after installation. These materials may become difficult or even impossible to remove after prolonged exposure to sunlight.

6.8.6 Natural Aging

Aging affects all aspects of windows and sliding glass doors. Hardware may eventually simply wear out, for example. Most aging problems can be delayed for many years by performing proper maintenance. Refer to section 3.12.9 for a discussion about the effects of aging on materials and finishes.

In addition, the natural deterioration of glazing compounds in face-glazed windows presents a particular problem that must be dealt with regularly, by removing old glazing compounds and reglazing the windows. Similar problems may occur with plastic glazing gaskets and stops. Unfortunately, because replacement plastic stops may not be available when needed, it may be necessary to remove existing windows and install new ones.

6.8.7 Improper Modifications

Building owners create all sorts of problems for themselves when they upgrade or modify their buildings. This is especially true when they decide to air condition an older building or upgrade its heating system. As Figures 6-5 through 6-8 illustrate, there is no end to the kinds of changes that are made in windows to accommodate changes to mechanical systems. Sometimes such intrusions cause leaks and their vibration is responsible for broken glass and even damage to the windows. Two broken panes can be seen in Figure 6-6, for example.

Other sorts of modifications can also cause problems. Placing a sidewalk above the bottom of a window, as at the left side of Figure 6-9, is sure to

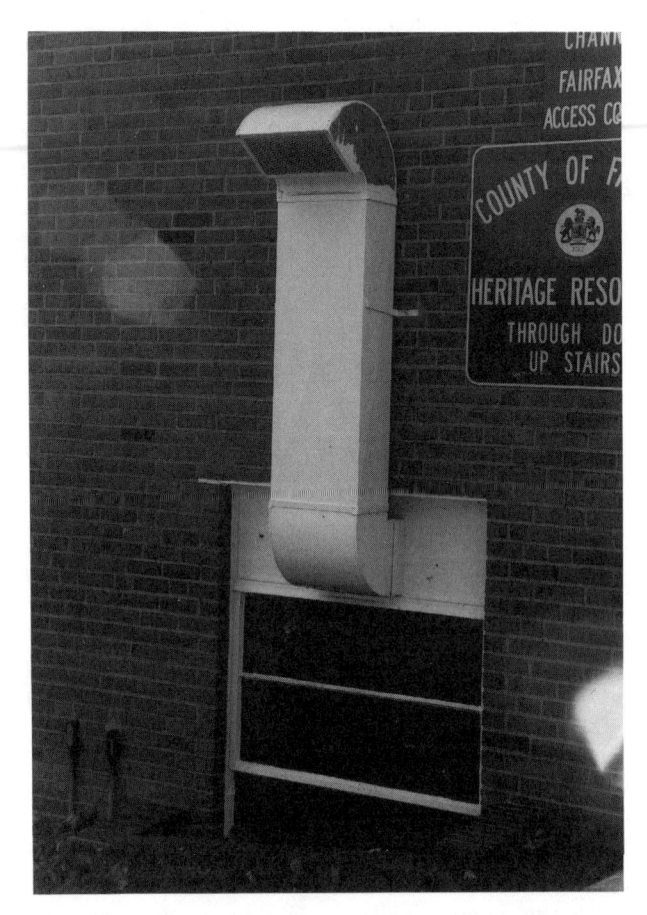

Figure 6-5
Sometimes a window is the easiest place to penetrate an existing wall.

lead eventually to leaks. Imagine a few inches of snow lying against the window, then consider what a snow shovel could do to that glass.

6.9 Repairing Windows and Sliding Glass Doors

Existing windows and sliding glass doors that are damaged or unsightly but still usable can often be repaired and restored to presentable condition. Sometimes it is necessary to remove a window or door to effect the repairs, but many repairs can be made without removing the unit. Often, damage to a window or sliding glass door, or repair of such damage, will damage the unit's finish. Refer to section 3.13 for a discussion about cleaning finishes and repairing wood and metal materials and their finishes.

6.9 Repairing Windows and Sliding Glass Doors 257

Figure 6-6 The air-conditioning unit in this photo is braced against the wall below, but that did not prevent its vibrations from cracking the glass above it.

When working with existing windows and sliding glass doors it is best to use materials and methods that match the original installation. When fasteners or attachments have failed, it will be necessary to provide new ones. It will probably be necessary to remove damaged hardware and install new items. Windows and sliding glass doors that are damaged too severely to repair, of course, must be discarded and new units provided.

Sometimes an apparent failure in a window or sliding glass door is actually a failure in an adjacent or supporting material or structure. Each condition must be examined in light of the data included in this book and in the referenced sources of additional information. The condition of the damaged unit should then be brought as close as possible to the condition usually required for that type of item when new. When damage is observed, the specific data in this book and in the sources listed for additional information, and the expert knowledge of someone knowledgeable about the type of problem that exists should be applied to solve the problem.

When existing windows or sliding glass doors are to be repaired in place or removed, repaired, and reinstalled, at least some selective demolition, dismantling, cutting, patching, and removal of existing construction

Figure 6-7 The air conditioner shown in this photo is actually hung from the window, which is sure to cause problems.

is usually necessary. The comments related to such matters in sections 1.6 and 1.7 are applicable.

6.9.1 General Requirements

This section contains some suggestions for repairing existing windows and sliding glass doors. Because these suggestions are meant to apply to many situations, they might not apply to a specific one. In addition, there are many possible cases that are not covered here. When a condition arises that is not addressed here, advice should be sought from the additional data sources mentioned in this book. Often, consultation with the manufacturer of the item being repaired will help. Sometimes it is necessary to obtain professional help (see section 1.4). Under no circumstances should

Figure 6-8 The air-conditioning unit in this photo prohibits the window from closing, but there it stands nevertheless, waving in the breeze.

the specific recommendations in this book be followed without careful investigation and the application of professional expertise and judgment.

Before an attempt is made to repair an existing window or sliding glass door, the manufacturer's recommendations should be examined. It is necessary to follow recommended precautions against using materials and methods that might be detrimental to the items requiring repair. Prevention of future failure requires that repairs not be undertaken without careful, knowledgeable investigation and the application of professional expertise and judgment. Repairs should be carried out by experienced workers under competent supervision.

Unless existing windows or sliding glass doors must be removed to facilitate repairs to them or to other portions of the building, or are scheduled to be replaced with new materials, only sufficient materials should be removed as is necessary to effect the cleaning and repairs. The more materials removed, the greater the possibility of additional damage occurring during their handling and storage. Materials that are removed should include those that are loose, sagging, or damaged beyond in-place repair. Com-

Figure 6-9 There was no concrete sidewalk when this fixed-glass window was installed; the problem was added later.

ponents that are sound, adequate, and suitable for reuse can either be left in place or be removed, cleaned, and stored for reuse.

After windows or sliding glass doors have been removed, concealed damage to their framing or substrates may become apparent. Such damage should be repaired before repaired or new units are installed.

Existing windows and sliding glass doors and supports that are to be removed should be removed carefully, and adjacent surfaces should be protected so that the process does not damage the surrounding area.

Unless a decision is made to discard them, damaged windows and sliding glass doors and support-system components that can be satisfactorily repaired should be repaired, whether they have been removed or left in place. Failing to repair known damage may lead to additional failure later. The methods recommended by the manufacturer of the units should be followed carefully when making repairs. Windows, doors, and associated items that cannot be satisfactorily repaired should be discarded and new, matching items installed.

Areas where repairs will be made should be inspected carefully to verify that existing components that should be removed have been and that the substrates and structure are as expected and not damaged. Sometimes substrate or structural materials, systems, or conditions are encountered that differ considerably from those expected, or unexpected damage is discovered. Both the damage that was previously known and damage found later should be repaired before windows or sliding glass doors are reinstalled. Reinstallation should not proceed until unsatisfactory conditions have been corrected.

6.9.2 Repairing Specific Types of Damage

In addition to the general requirements for repairs already discussed, the following recommendations apply to specific problems.

Bent, Twisted, or Deformed Metal Windows. When the affected metal window component is not too severely damaged, repairing it may be possible. Sometimes ventilating sashes become twisted by careless or abusive handling. Often, however, when a window or sliding glass door member becomes bent, twisted, or deformed, the cause is related to the adjacent construction, not the window or door itself. Such external damage must be corrected before window or door repair is undertaken.

It may be possible to straighten a twisted window frame or sash, but removing it will probably be necessary. Even when a window component can be straightened in place, removing the glazing is usually necessary. If the original damaging did not break the glass, straightening often will do so.

Broken or Deteriorated Wood Components. When part of an operating component has become cracked or broken, it is usually possible to remove the operator from the window or door, to facilitate repairs. When the damage is not too severe, broken frame components can often be repaired in place. Where the break is clean and the broken component is not part of an operator, it may be possible to repair the break, using screws and adhesives. Even when the wood is crushed or part of it is missing, fixed wood components can be repaired using epoxy fillers, plastic wood, and adhesives. Parts of operators can occasionally be similarly repaired, but most of the time such parts will encounter too much stress for such repairs to hold. Then it is necessary to remove the damaged parts and rebuild the

Figure 6-10 When the muntin at the left side of the window shown in this photo was broken, the owner took the easy way out and removed it entirely. The repair works, but looks somewhat strange.

operator, using new matching wood members. In severe cases it may be necessary to replace the operator with a new one.

When wood window and sliding glass door components or sills begin to deteriorate, usually from rot, it is sometimes possible to scrape away the rotted wood and repair the member with a mixture of sawdust and resorcinol glue, wood putty, or epoxy. Holes and open joints can be similarly repaired. It may be difficult to find a contractor experienced at making such repairs, however, especially one proficient in using epoxy. When these methods are not satisfactory, replacing the component becomes necessary. Sometimes, especially when the damaged member is not load bearing, part of the member can be removed and a new piece spliced in. As Figure 6-10 demonstrates, other solutions are also possible.

Windows and sliding glass doors that are damaged beyond the point where they can be repaired must be removed and new units installed.

Loose, Sagging, Sticking, or Out-of-Line Windows or Sliding Glass Doors. A window or sliding glass door may become loose, sag, stick, or appear to be out of line because it was poorly designed or fabricated. It is very difficult to fix a poorly designed window or sliding glass door.

Usually, when a window or sliding glass door becomes loose, sags, or is out of line, the fault lies either with undue movement of the supporting wall, partition, or structure or with a failure in the window or door's anchoring system. Repairing such supports and structures, which can be complicated, is beyond the scope of this book. Sometimes, when an anchoring system is at fault, the problem is relatively easy to solve. For example, using additional or longer nails or screws in lieu of nails will often satisfactorily refasten a loose wood window or sliding glass door frame. New anchors in the right locations will sometimes do wonders. The appearance of the finished installation, however, is not always good. Because the existing anchors were in the wrong places or inadequate in size or number, it is usually necessary to install new ones where they will be visible.

When the deformation being repaired has broken welds or damaged mechanical joints, it will probably be necessary to remove the unit to make the repairs.

Most sticking of sliding windows, particularly of double-hung ones, is caused by paint or dirt accumulating in the grooves or on the edges of the stops, by rust or other corrosion in the grooves or on the sashes or frames, or weather stripping that fits too snugly. In addition, sticking wood windows may be due to expansion of the wood, caused by humidity or other wetting of the sash or frame. An accumulation of dirt or minor corrosion are easy to solve; just remove the paint, dirt, or corrosion and lubricate the track or sash. But faulty weather stripping on metal doors and frames will probably need to be replaced. Weather stripping that fits too snugly and swollen wood windows can often be made to work properly either by tapping a block of wood along the groove to expand the frame or simply by lubricating the contact surfaces. If neither method works, it is necessary to remove the wood sash and plane it to fit.

Sticking crank-operated casement or awning windows are often a hardware problem. The concealed gear mechanism may be worn out, corroded, or caked with solidified grease. Badly worn gears will need replacement. Corrosion and caked-on grease must be removed and the gears relubricated.

The sticking of windows may also be caused by corrosion on the contact surfaces of hardware or by bent hardware components.

Missing or Damaged Hardware. Detached hardware can usually be reinstalled. Unless the windows or sliding glass doors have some special or unusual hardware, finding replacements for missing items is not usually difficult. Replacement hardware for some older windows, however, particularly the types of casement windows popular in the 1940s and 1950s may not be easily found today. In such cases, missing hardware may be cause for removing existing windows and installing a different type.

It may also be difficult to obtain replacement hardware parts for old double-hung windows. Sash weights, for example, are hard to find, especially for large sashes, and suitable pulleys may not be available. When existing components are missing or damaged so that they are not reusable, it may be necessary to remove old weight-and-pulley balancing systems and replace them with modern spring-operated systems. Doing so offers the opportunity to fill the old window pocket with insulation, but the new system may not be as effective or last as long as the original one.

Cladding Failure. Depending on the type of failure, it may or may not be possible to repair cladding. Torn vinyl cladding may be repairable, for example, while torn aluminum may not be. Patches in aluminum will almost certainly be visible, especially if the damaged cladding is not completely refinished.

Air or Water Leaks. Probably the largest complaint that owners have about windows and sliding glass doors concerns leaks. Water leaks are an immediate problem, of course. They must be stopped as soon as possible, to prevent damage to interior materials. Water leaks may be caused by missing or deteriorated calking or by damaged or missing weather stripping. In sliding units, leaks may be caused by clogged weep holes in the track. Leaking can be exacerbated by free water cascading over a window or door. Repairs require first finding the reason for the leak, then recalking or repairing weather stripping, clearing weep holes, and providing a means of diverting water away from the window, as appropriate.

Air leaks, while not as urgent, can be even more costly than water leaks. The energy lost through cracks around a building's windows and doors can be its single largest source of heat loss. The source of many air leaks is deteriorated or missing calking. The repair, of course, is simply to recalk around the windows or sliding glass doors.

When failed or old weather stripping is responsible for air leaks, the failed materials should be removed and new weather stripping of the proper kind installed.

6.10 Windows and Sliding Glass Doors in Existing Construction

The requirements discussed in section 6.7 also apply when those types of items will be installed in existing construction. There are, however, some additional considerations that must be taken into account when they are installed into an older building. Many of those considerations are addressed in this section.

Since the number of possible window and sliding glass door conditions in existing buildings is large, it is impossible to cover every conceivable one in this book. Therefore, the discussion here is generic in concept and should be viewed in that light. These suggestions may not apply to all conditions, but their principles should be applicable to most. In the end, there is no substitute for professional judgment.

6.10.1 Standards, Controls, Delivery, and Handling

The standards and controls in sections 1.6.1, 1.7.1, 6.1.1, 6.2.1, 6.3.1, and 6.4.1 also apply to windows and sliding glass doors installed in existing construction.

In addition, existing windows and sliding glass doors should be removed and new units installed by the manufacturer of the new items or by a firm franchised or approved by the manufacturer. Only experienced erectors should be permitted to do this work.

While the installation of new windows and sliding glass doors into existing construction is under way, existing windows and sliding glass doors that are not scheduled to be removed and discarded should be protected from harm. When existing units that are to remain become damaged, they should be repaired to a condition at least equal to that existing before the damage occurred. Where acceptable repairs are not possible, the damaged window or door should be removed and a new unit installed.

Damaged windows and doors should not be installed without repair. When satisfactory repairs are not possible, damaged windows and doors should be discarded and new units provided.

6.10.2 Windows and Sliding Glass Doors and Associated Hardware, Glazing, and Miscellaneous Components

With few exceptions, windows and sliding glass doors in existing buildings and those in new construction are essentially the same. Therefore, most of the requirements discussed in sections 6.1, 6.2, 6.3, 6.4, 6.5, and 6.6 are applicable also to windows and sliding glass doors that are to be installed in an existing building. The following requirements are in addition to those in the referenced sections.

Retrofit casings (also called panning) and associated closure members and trim are available for use in trimming new windows in existing openings. These are special units designed to fit over an existing casing, so that the complete removal of the existing material is not necessary when a new window is installed where an existing window sash has been removed.

Many designs are available, depending on the manufacturer. Retrofit casings are usually of metal. They should fit around the existing frame and completely conceal the remaining parts of the existing windows. Unfortunately, most retrofit casings do not preserve the appearance of the original window and are therefore not satisfactory when that is a requirement. An additional negative aspect of panning is that it makes it impossible to inspect the underlying wood. Thus, should such wood become wet or have insects invade it, the resulting damage is concealed and goes unchecked until considerable damage has been done.

The design of frames, trim, and retrofit casings must permit unrestricted expansion and contraction.

6.10.3 Preparation for Installation

Just as preparations are necessary before new windows and sliding glass doors are installed in a new space, so is it necessary to make preparations before installing such items in an existing space. In the latter case, however, the preparations are likely to be more extensive.

The requirements for demolition and removal of existing items and for renovation work discussed in sections 1.6 and 1.7 are applicable to work related to wood windows and sliding glass doors. The requirements in this section are in addition to those in these two sections.

As is true when the windows, sliding glass doors, frames, and space are all new, the site where new windows and sliding glass doors will be installed in an existing building should be examined carefully by the installer. Making two examinations is usually best, one before demolition work begins. A report of this examination should be sent to the designer and the owner so that modifications in the proposed demolition can be made, if necessary, to accommodate the window and sliding glass door design.

The second site examination should be made after demolition in the space has been completed. This examination is necessary to verify that other construction to which the new windows and sliding glass doors are to be attached, and the areas and conditions where they will be installed, are in a proper and satisfactory condition to receive them. Preparatory installation work should not be started, and the windows and doors should not be installed, until unsatisfactory conditions have been corrected.

Should conditions be encountered that differ significantly from those expected, the person discovering the difference should immediately notify the other concerned parties, especially the owner, so that alternative methods can be developed for handling the unexpected situation.

When new windows will be installed where there are existing ones, the existing window's sashes, hardware, and glazing material must be completely removed. The frame may also be removed, but it is usually less expensive to leave the existing frame in place and cover it with a new retrofit casing (panning and trim). When the frame will be left in place, the mullions should be cut off where they join the frame. The new windows will probably require new mullions.

Before new windows or sliding glass doors are installed, it is a good idea to verify that their head and sill flashings are in place, properly designed and installed, and undamaged. Problems should be corrected before new items are installed. Improperly installed or damaged flashings are a major cause of leaks associated with windows and sliding glass doors.

6.10.4 Installation of Windows and Sliding Glass Doors, and Their Glazing

The recommendations in section 6.7 also apply to windows and sliding glass doors installed in an existing building. The suggestions in this section are in addition to those in that one.

Damaged windows and sliding glass doors should not be installed; it is generally much easier to repair such items before they are installed. Often, damage will not be repairable after a window or sliding glass door has been installed, and the item must be removed to permit repair. Damaged items that have been installed but cannot be repaired in place should be removed and repaired or new, undamaged units installed. Installing already damaged windows or sliding glass doors is one of the largest causes of later problems with them.

Doors and frames should be installed in accordance with the project's drawings and specifications, their manufacturer's latest published installation instructions, the approved shop drawings, and the latest issues of applicable industry standards and recommendations.

When installing retrofit casings (panning) and associated closures, trim, and other components, the manufacturer's recommendations should be followed exactly. Such items should be fastened in place using screws through predrilled holes. The entire installation should be back sealed and thoroughly calked.

Voids in window frames, mullions, and between panning trim and closures and existing construction should be filled using the window manufacturer's standard glass-fiber insulation.

Glazing is installed in the same way, regardless of whether the windows or sliding glass doors are installed in new or existing construction. Neither is there a difference in the installation methods used whether the window or sliding glass door is new or existing, except that the method must be appropriate for the existing unit and will usually be the same as that originally used.

6.11 Where to Get More Information

The following AIA Service Corporation *Masterspec* Basic sections contain excellent descriptions of the materials and installations addressed in this chapter. Unfortunately, they contain little that will help with troubleshooting these installations.

- The February 1989 edition of Section 06401, "Exterior Architectural Woodwork," includes exterior wood window trim.
- The February 1989 edition of Section 06402, "Interior Architectural Woodwork," includes interior wood window trim.
- The August 1988 edition of Section 08311, "Aluminum Sliding Glass Doors," contains a discussion about current standards applicable to aluminum sliding glass doors.
- The August 1988 edition of Section 08312, "Wood Sliding Glass Doors," contains a discussion about current standards applicable to wood sliding glass doors.
- The August 1986 edition of Section 08510, "Steel Windows," contains a detailed discussion of the Steel Window Association's standards and their use in specifying steel windows. Anyone designing or selecting steel windows or responsible for caring for large numbers of them should obtain a copy and study it for general information.
- The August 1986 edition of Section 08520, "Aluminum Windows," and the November 1986 edition of Section 08525, "Aluminum Architectural Windows," contain discussions of current standards applicable to aluminum windows.
- The August 1986 edition of Section 08610, "Wood Windows," contains a discussion of the Wood Window and Door Association's standards for wood windows.

The following is a list of sections in the Architectural Woodwork Institute's *Architectural Woodwork Quality Standards, Guide Specifications and Quality Certification Program* that are applicable to wood windows:

- Section 300, "Standing and Running Trim & Rails (Interior and Exterior)."
- Section 1000, "Exterior Windows."
- Section 1700, "Installation of Architectural Woodwork (Interior)."

Every designer should have the full complement of applicable ASTM standards available for reference, of course, but anyone who needs to understand wood doors and frames should have access to a copy of the ASTM standards marked with a [6] in the Bibliography.

Lyon D. Evans's 1985 article "Aluminum Windows and Sliding Glass Doors: The New Specs" is a good discussion about the origins, effects, and uses of current American Architectural Manufacturers Association (AAMA) standards for aluminum windows.

John H. Myers's 1981 article "Preservation Briefs: 9, The Repair of Historic Wooden Windows" is equally applicable to the repair of windows that are not of historic significance. It divides window repair into three groups. Repair Class I: Routine Maintenance includes the removal of excess paint, reputtying, replacing sash cords, and the refinishing of wood windows. Repair Class II: Stabilization discusses what to do with a window that shows signs of physical deterioration, including partial decay. It describes how to repair cracks and build up decayed wood-window portions using wood putty, mixtures of sawdust and glue, and epoxy. Repair Class III: Splices and Parts Replacement describes the methods necessary to make repairs when portions of a window must be replaced. This publication also includes information on weatherizing windows using storm windows and weather stripping, and on window replacement using new windows.

Harold B. Olin, John L. Schmidt, and Walter H. Lewis's 1983 edition of *Construction Principles, Materials, and Methods* is a good source of background data about wood windows and sliding glass doors. Unfortunately, it contains little that will help specifically with troubleshooting failed windows or doors. This book is slightly out of date, but an updated version will soon be published.

Morgan Phillips and Judith Selwyn's 1978 article "Epoxies for Wood Repairs in Historic Buildings" is also applicable to buildings that are not of historic significance.

Richard Rush's 1987 article "Refining Window Energy Performance" contains a review of two computer programs that permit design analysis of the energy performance of windows.

Heinz Trechsel's 1988 article "Specifying an Energy Efficient Thermal Window" is a good capsule look at the subject.

Refer also to other Bibliography entries marked with a [6].

CHAPTER

7

Glazed Curtain Walls

This chapter discusses the five kinds of aluminum glazed curtain wall systems classified by the American Architectural Manufacturers Association: stick, unit, unit-and-mullion, panel, and column-cover-and-spandrel systems. It also includes sloped glazing. Curtain walls made primarily from wood, glass, or other metals are not discussed here. The principles in this chapter, however, are applicable to glazed curtain walls made from other metals because, even though they are made from formed sections rather than extrusions, the problems that other metal systems have are similar to those that aluminum-glazed curtain walls have.

Metal doors, frames, and store fronts are discussed in Chapter 4. Refer to the definitions in the Glossary for a description of the differences between curtain walls and store fronts.

Wood doors, frames, and store fronts are included in Chapter 5, wood and metal windows and sliding glass doors in Chapter 6.

Suspended glass, glass mullion, and flush (structural silicone) glazing-type curtain wall systems are discussed in Chapter 8, as is glazing for the curtain walls addressed in this chapter.

7.1 General Requirements for Glazed Curtain Walls

Every exterior wall that supports only its own weight is a curtain wall. Therefore, non–load bearing masonry, precast concrete, and stone are curtain walls, as are exterior walls made up of horizontal or vertical masonry strips separated by strips of windows. The AAMA's *Aluminum Curtain Wall Design Guide Manual* says that a curtain wall made mostly of metal or a combination of metal, glass, and other materials supported by a metal framework is a metal curtain wall.

There are many metal curtain wall systems in use, but the AAMA classifies only the five previously mentioned: stick, unit, unit-and-mullion, panel, and column-cover-and-spandrel. Their names recognize their installation methods. Other systems include sloped glazing (a special application of the stick system), suspended glass, glass mullion, and flush glazing, also called structural silicone.

The components of a glazed curtain wall system may include any or all of the items in the following list:

- Metal framing system.
- Column covers, soffits, sills, and similar border and filler items.
- Entrance doors and frames, although these are ordinarily considered entrance and store front work (see Chapter 4).
- Glass and other glazing materials, including glass spandrel materials.
- Aluminum, ceramic, stone, and other non-glass panels glazed into the curtain wall.
- Glazing gaskets.
- Interior curtain wall components.
- Louvers associated with the curtain wall.
- Curtain wall copings and associated flashings.
- Cap flashings associated with the curtain wall.
- Operable windows, including sills and stools (see Chapter 6).
- Blind tracks.
- Incidental items including reinforcements, anchorages, shims, fasteners, accessories, and support brackets for components of the curtain wall system.
- Insulation and firestopping directly related to the curtain wall.
- Joint-sealing work associated with the curtain wall.

7.1.1 Systems

Stick systems, in their several variations, are the more common glazed curtain wall systems. They consist of vertical mullions, usually of extruded aluminum that form the main support for the other components, insulated horizontal non-glass panels, and glass panels. Stick systems may also include horizontal rails, operating windows, sills, fascias, copings, and interior and exterior trim. Most stick systems are composed of unmodified stock components. When using just stock components is not desirable, however, stick components can be especially designed to meet almost any design requirements. It is not unusual to find a glazed curtain wall system composed of a mixture of stock and custom components.

The vision panels in stick systems are usually glazed from the interior using gaskets or a glazing compound. Spandrel panels are glazed from the exterior. A special form of stick system features exterior glazing with lockstrip gaskets. These gaskets are slipped into the frame and the glazing material installed using a zip-in strip gasket.

Stick systems must be glazed in the field, because they are field assembled.

Unit curtain wall systems are preassembled in the factory as modular panels and shipped to the construction site in one piece. Such panels are often glazed before shipment. Some such systems require the use of field-applied joint covers.

Unit-and mullion systems combine the stick and unit systems. The main framing mullions are installed in the field and the unit panels are then lowered into place between them.

Panel systems are similar to unit systems, except that panel systems have homogeneous sheet or cast panels with few joints, while unit systems are assembled from components.

As their name implies, column-cover-and-spandrel systems consist of column covers and spandrel panels that span between them.

Sloped glazing systems are a form of stick system in which portions are either sloped or curved.

7.1.2 Standards

The materials and finishes used in aluminum-glazed curtain walls should comply with the applicable standards of the National Association of Architectural Metal Manufacturers (NAAMM), the American Architectural Manufacturers Association (AAMA), the Aluminum Association (AA), ASTM, and the American National Standards Institute (ANSI). Standards for materials and finishes are addressed in Chapter 3. Glazed curtain wall systems should comply with the AAMA standards mentioned in the body of the text, in section 7.6, in the Bibliography and followed by a [7], or in more than one of those places.

The standards for glazed curtain wall performance are addressed in section 7.1.5.

Tests should be performed by an independent testing laboratory rather than the manufacturer. The standard for qualifying such independent laboratories is ASTM Standard E 699.

The wind loads to be used for testing glazed curtain wall systems for performance should be taken from the map in ANSI A58.1, but should in no case be less than those required by applicable building codes.

The specific materials and items used in glazed curtain wall systems should comply with the standards in the following list:

- Aluminum extrusions: ASTM Standard B 221. Alloys 6063-T5 or -T6 are frequently used, but extrusions may be made from any other alloys and tempers, depending on the design.
- Sheet aluminum and bent plate: These should have the properties recommended in ASTM Standard B 209. They are usually made from alloys 5005, 5086, or 6061, as suitable for the required finish.
- Hot-dip galvanized assembled steel products: ASTM Standard A 386.
- Hot-dip galvanized steel inserts, bolts, and fasteners: ASTM Standard A 153, Class C or D, as applicable.
- Window-cleaning devices: ANSI A39.1.
- Hat-shaped furring channels: ASTM Standard C 645.
- Extruded polyvinyl chloride glazing gaskets: ASTM Standard D 2287.
- Extruded or molded neoprene glazing gaskets: ASTM Standard D 2000.

Firestopping should comply with AAMA TIR-A3-1975.

Structural rubber lockstrip gasket glazing should comply with ASTM Standard C 964. The materials used should comply with ASTM standards C 542 and C 716.

7.1.3 Controls

The general requirements related to controls in sections 1.6.1 and 1.7.1 apply also to glazed curtain wall systems. The controls discussed in this section are in addition to those in sections 1.6.1 and 1.7.1.

The first step in effective control of the quality of a glazed curtain wall system is the preparation by a competent architect of drawings and specifications for the work. The drawings and specifications should indicate the spacing of members, the profiles and sizes of components, other dimensional requirements, the materials, fabrication methods, performance requirements, and similar design requirements. Depending on the product, there may be minor differences between the product the architect chooses to illustrate the design requirements and the actual product installed in the project, without compromising the design. Manufactured systems of the same quality vary slightly from each other. General appearance, finish, and compliance with the performance requirements are usually the governing factors.

Glazed curtain wall components and systems should have been initially designed by a structural engineer employed by the manufacturer. The design criteria for establishing a curtain wall system's performance requirements should be selected by a structural engineer hired by the owner and familiar with glazed curtain wall design. Stock system components, custom-designed vertical wall systems, and every sloped glazing system should be designed by a structural engineer. Such systems should be designed specifically for the particular project. For custom systems and sloped glazing the manufacturer should be required to submit design calculations prepared by a licensed professional engineer.

Unless there is a good reason for not doing so, all the components of a curtain wall system should be products of the same manufacturer or should at least be provided by the manufacturer of the major components.

The curtain wall system manufacturer's product data and specifications and instructions for fabrication, factory finishing, shop painting, installation, glazing, and maintenance should be submitted for the owner's approval. These data should include certified test reports prepared by an independent testing laboratory attesting to compliance with the project's requirements for system performance. When the system will be the manufacturer's standard, preconstruction tests up to three years old may be acceptable. When the system will be custom designed or require extensive adaptation of a standard system, the preconstruction tests should be new.

Preconstruction testing should be done using a mock-up of the system to be used set up in the testing laboratory's facilities. The mock-up should be complete with all components that will appear in the final installation, and should be constructed in the same way as will the actual work. It should including glazing, operating windows, non-glass panels, fascias, spandrels, corners, splices, sealants, anchors, and other necessary components.

Shop drawings should be submitted to the owner for approval. They should include all necessary information, including details showing adaptations of the manufacturer's standard system components for the particular project, typical system elevations, dimensions, member profiles, anchorage systems, non-glass panels, windows, glazing, drainage gutter and weep systems, and interfaces with building construction. When existing construction is involved, the shop drawings should show it and its interface with the new glazed curtain wall. Design calculations should also be shown, including the section moduli of wind-load-bearing members and calculations of stresses and deflections for performance under design loading. The shop drawings should show clearly where and how the manufacturer's standard system and components deviate from the drawings and specifications. For curtain walls with thermal-break construction they should show complete details of that construction. For items to be repaired, the shop drawings should show the methods to be used in each case.

Shop drawings should be submitted for louvers and grilles that will be installed in the glazed curtain wall. They should show louver sizes and locations, methods of installation and support, blade shapes and sizes, metal thicknesses, relationships to other curtain wall components, and methods of anchoring them in place. The drawings should also indicate the free area of each louver.

The curtain wall frame manufacturer should furnish written certification attesting that the metal surfaces have been finished as required.

Samples of the metals that will be used with the required finishes applied should be submitted for approval. They should show the range of colors and textures to be expected in the curtain wall.

At least twelve-inch-long samples of each full-sized extrusion should be submitted for approval. These samples should be finished as they will be in the completed installation. Unless the owner and architect are very familiar with the product to be used, it is usually a good idea to require the submission of fabrication samples showing prime members, non-glass panels, joinery, anchorage and expansion provisions, glazing, profiles, intersections, and similar details. Samples of louvers and operating windows should also be provided. Refer to section 6.1.1 for sample requirements for windows.

The curtain walls' installer should be a firm with not less than five years of successful experience in erecting curtain wall systems similar to those for the project at hand, and it should be acceptable to the system's manufacturer. When possible, work closely associated with curtain walls should be assigned to the installer. Such associated work includes, but is not necessarily limited to, curtain wall metal component erection, and work related to glass and glazing, non-glass panels, louvers, venetian blind pockets, window sills, copings, thermal insulation, firestopping, flashing, and joint sealing.

Unless a standard manufactured system with which the owner and architect are very familiar will be used, a field mock-up should be built to ensure that the system components to be installed actually match the selected components and the system presents the appearance desired. The mock-up should accurately represent the project's conditions, including joints, sealants, vision and spandrel glass, non-glass panels, glazing, anchors, thermal and safing insulation, and finishes. The mock-up should be tested using a garden hose and following the procedures recommended in AAMA 501-1983. The approved mock-up should remain in place throughout the construction period to serve as the standard for the glazed curtain wall throughout the project. The mock-up is often, but not always, left in place as part of the actual building wall.

The water-hose test conducted on the mock-up should be repeated on the remaining curtain wall whenever a question arises about construction quality.

7.1.4 Warranties

Warranties beyond the normal one-year construction warranty are usually required for large projects and when custom-designed systems are used or a manufacturer's standard system is drastically altered for the building. The normal warranty period is five years, but may be only two years for smaller projects or where alterations to a standard system are not too extensive. When a failure occurs, it is worth determining if there is a warranty. Extended warranties are not usually required for stock systems used on small projects.

Extended warranties should be signed by the general contractor, the installer, and the manufacturer. They should agree to repair or replace defective materials and workmanship of glazed curtain wall systems during the warranty period. *Defective* is defined to include abnormal deterioration, aging, or weathering, glass breakage, the failure of operating parts to function normally, deterioration or discoloration of finishes, and failure of the system to meet its performance requirements. It is normal to stipulate that repairs or replacements required because of acts of God that cause conditions exceeding the design criteria used to establish the system's performance requirements, vandalism, inadequate maintenance, alterations, failure of the structure supporting the curtain wall system, or other causes beyond the control of the manufacturer, fabricator, installer, or contractor be completed by them and paid for by the owner at reasonable, prevailing rates that are agreed upon at the time of the repair or replacement work. An extended warranty should be carefully constructed so that it does not deprive the owner of other available actions, rights, or remedies.

7.1.5 Performance Requirements

A curtain wall system's performance requirements are applicable to every component, including framing, glass and non-glass panels, and louvers. Such requirements should be based on the system's location and use. When a failure is discovered, it is necessary to compare the proper requirements for the particular project with the actual installation. Failure to meet necessary performance requirements is a major cause of failure in glazed curtain walls.

ANSI A58.1 contains a map that suggests the wind velocity that glazed curtain wall systems should be tested for. The minimum is seventy miles per hour, but higher velocities may be necessary. For structural performance it is often advantageous to establish different velocities at different heights above the ground.

Air infiltration should be measured in accordance with ASTM Standard E 283. AAMA 501-1983 suggests the minimum air leakage and static-air-pressure differential to use in the tests, but a particular project may require higher values.

Water penetration should be measured in accordance with ASTM Standard E 331. Water leakage is defined in, and minimum test pressures suggested by, AAMA 501-1983. A particular project may require higher test pressures.

Condensation resistance is tested in accordance with AAMA 1502.7-1981, which contains the test procedures that should be used and recommends the Condensation Resistance Factor (CRF) that should be obtained. The CRF is less important in warm climates where condensation is not as much a problem.

It is normal to require that a glazed curtain wall system be able to withstand the amount of thermal movement that will be generated by the extreme range of temperatures expected at the project's location, without damage of any kind. For most of the United States the required ambient temperature differential is 120 degrees Fahrenheit, which results in a metal surface temperature differential of 180 degrees Fahrenheit. Some extreme northern or southerly locations may have greater differential temperatures.

Structural strength is measured in terms of deflection resistance in accordance with the tests in ASTM Standard E 330. ASTM Standard E 1233, which was new at the time of this writing, may also be used, but to date the AAMA has not suggested minimum test criteria for it. ASTM Standard E 1233, which is actually a supplement to ASTM Standard E 330, contains the recommended tests for cyclic loading. Vertical wall systems should be tested for deflection both parallel and perpendicular to the plane of the wall and in both the inward and outward directions. Sloped glazing systems should be tested for both upward- and downward-acting loads, as well as lateral wind load. Sloped glazing systems should also be required to carry a concentrated load applied in any direction on the framing members. The concentrated load is usually two hundred pounds, but may be higher if the structural engineer thinks that higher concentrated loads will be applied to the installed system. Under the test pressure no system component or associated glazing, louvers, grilles, or sealants should experience either permanent deflection or failure of any kind. The test pressures to be used for perpendicular-to-the-wall tests should be 150 percent of the structural engineer's selected design wind pressures. Design wind pressures may vary with height above the ground and may be higher for curtain wall elements that are within fifteen feet or so of corners, especially on portions that are fifty feet or more above the ground. Test loads for parallel-to-the-wall tests should be the total applied dead load of the wall and anything it carries. The loads for sloped glazing should be established by the owner's structural engineer, depending on the actual loads to be expected.

The maximum allowable deflection of framing members perpendicular to the face of the wall is recommended by the AAMA to be 1/175th of the span. Some manufacturers and other industry sources recommend that the maximum deflection be only 1/240th or even 1/1,000th of the span of any member. The absolute maximum is 3/4 inch over the total length of any single member, regardless of the fraction used. The applicable building code may have other requirements. Under no circumstances should the allowable deflection permit glazing panels to be dislodged. This requirement is important in any curtain wall, but is especially critical in sloped glazing systems, where the glass is more likely to fall.

The maximum deflection of non-glass panels is recommended to be 1/120th of the span, with a maximum of 3/4 inch.

The allowable deflection parallel to the face of the wall is dictated by the glass and non-glass panels' edge conditions. The usual requirement is that when the system is fully deflected the glass bite cannot be reduced to less than 75 percent of the designed depth and the space between glass or non-glass panel edges and metal or between the framing members and other fixed members cannot be less than 1/8 inch. Even greater restrictions are sometimes imposed by assembly requirements.

Deflection should not reduce the width of sealant joints surrounding the curtain wall by less than one-half of the designed width of the joints. The reduction should be even less when so recommended by the sealant's manufacturer.

In no case should the stresses that will be imposed on glazed curtain wall components exceed the allowable values established by the standards specified in section 7.1.2 or the yield point of the metal.

Under no circumstances should the temporary deflection of an anchor exceed 1/8 inch or its permanent deflection exceed 1/16 inch, in any direction.

The Sound Transmission Classification (STC) of a glazed curtain wall system is determined by testing it in accordance with ASTM Standard E 90. The required STC rating is determined by the requirements and location of the project. Buildings in ordinary locations may not have an STC requirement. Buildings in the vicinity of airports, freeways, and other high-noise-level environments often do.

7.1.6 Materials

Glazed curtain walls can be made from wood, steel, stainless steel, copper alloys, or aluminum. Extruded aluminum, however, is the only material from which standard systems are currently being manufactured. Therefore, the majority of existing glazed curtain wall systems are made from extruded aluminum. This chapter addresses in detail only extruded aluminum systems. The principles here, however, also apply to systems made from other materials.

Metals. In general, the discussion in section 3.3 applies to the aluminum used in glazed curtain walls. The most common aluminum products used are extruded shapes. Sheet-aluminum and bent-plate products are also used, especially in non-glass panel construction.

Exposed fasteners for glazed curtain walls should be aluminum or non-magnetic stainless steel screws finished to match the material being fastened. Concealed fasteners may be aluminum, stainless steel, cadmium- or zinc-plated steel, or an epoxy adhesive.

Anchors, clips, reinforcements, and accessories are usually either aluminum, stainless steel, or cadmium-plated steel. Iron, carbon steel, hot-dip galvanized steel, and zinc-painted mild steel can also be used if they are properly separated from the aluminum. Where moisture will be present, non-magnetic stainless steel is a good choice.

Brackets and reinforcements should, where feasible, be the manufacturer's standard high-strength aluminum units. Where greater strength is required, they may be non-magnetic stainless steel. Brackets not exposed to weather or abrasion may be hot-dip galvanized steel. The use of aluminum castings is questionable.

Shims for installation and alignment of glazed curtain walls should be non-staining and non-ferrous.

Window cleaner's bolts should be non-magnetic stainless steel.

Concealed flashing is usually dead-soft 26-gage stainless steel of a type selected by the curtain wall manufacturer for compatibility.

Fasteners and accessories should be the curtain wall system manufacturer's standard non-corrosive items that are compatible with the materials used in the framing system. Their exposed portions should match the finish of the curtain wall system. Where fasteners anchor into aluminum less than 0.125 inch thick, non-corrosive pressed-in splined grommet nuts or other reinforcement should be provided to receive the fastener threads.

Bolts for attachment of aluminum framing members to the supporting structure should be cadmium-plated steel, non-magnetic stainless steel, or zinc-coated steel. Fasteners for frame assemblies should be non-magnetic stainless steel or aluminum. Inserts for use in concrete or masonry should be made from cast iron, malleable iron, or hot-dip galvanized steel.

Copings, trim, and soffits are usually extruded aluminum, but may also be formed from aluminum sheet. Venetian blind tracks and window stools are usually formed from aluminum or zinc-coated, commercial-quality steel sheet.

Steel and aluminum for porcelain-enamel panels should be as discussed in section 3.10.2.

Miscellaneous Materials. Many other materials are necessary to produce a complete glazed curtain wall system. Glass, glazing sealants, and fillers are discussed in Chapter 8.

Framing-system gaskets and joint fillers are usually the curtain wall manufacturer's standard permanent-type components. Their design should take the expected joint movement and sealing requirements into account.

Sealants and joint fillers should be elastomeric types suitable for the amount of joint movement expected. Their types will probably vary with use. A different type sealant will be used, for example, in non-moving metal-to-metal joints than between metal components and adjacent construction. The metal-to-metal sealant is often a gun grade, polyisobutylene, or other elastomeric material that is formulated to remain permanently plastic. Sealants for concealed locations between metals and porous surfaces are usually also gun-grade elastomeric sealants that have been formulated specifically for that purpose.

Where movement within the curtain wall system is expected, slip-joint linings should be provided. They should be made from sheets, pads, shims, or washers of fluorocarbon resin, or a similar material recommended by the curtain wall manufacturer.

7.1.7 Finishes

The discussions in sections 3.6, 3.8, 3.10, and 3.11 apply to the finishes used on aluminum glazed curtain walls.

A few glazed curtain wall components are field painted. Framing members are sometimes finished with baked enamel, siliconized polyester, and other factory-applied finishes. Metal panels are finished with any one of those or with porcelain enamel or even ceramic tile. The exposed surfaces of most aluminum glazed curtain wall components, however, are clear or color anodized, or finished with a fluorocarbon polymer-based coating.

Anodized finishes are usually architectural Class I (see section 3.8.3). Concealed aluminum is usually left with its mill (as fabricated) finish.

Steel window stools, venetian blind tracks, and similar components are usually finished with baked-enamel.

7.1.8 Design and Fabrication

Glazed curtain walls should be designed and fabricated in accordance with applicable industry standards. When a proprietary system is the standard, the curtain wall system should conform to the design and construction of that system.

Glazed curtain wall systems should be designed so that their vision panels are reglazable without dismantling the framing.

Components should be fabricated rigid, neat in appearance, and free from defects, warp, and buckle. When possible they should be fitted and assembled in the manufacturer's plant before they are given their finish. Systems that cannot be permanently factory assembled before shipment should be clearly identified and marked to assure proper assembly at the project site.

A condensation-removal system should be provided. Removal of condensation and minor amounts of water that may have intruded through the wall is important in any curtain wall system, but is an especially acute problem in sloped glazing systems. There it may be necessary to design gutters and channels to remove such water and provide weep holes to lead it to the exterior.

Effective, quick removal of storm water from exterior surfaces of sloped glazing systems is also important. Storm water passing over the exterior of the system without direction will build up against horizontal framing and glazing members, which will then tend to leak. Gutters are often included in the design of such systems.

When they are required, thermal breaks must be complete, without metal bridges of any kind.

Metal frame and sash corners may be lapped, butted, or mitered. They can be mechanically joined using fasteners and clips, be welded, or be both mechanically fastened and welded, depending on the selected design.

Some glazed curtain walls are designed to be watertight. They have sealants at the exterior of the frame and panels to exclude water. Unfortunately, because the interior pressure usually is lower than the exterior pressure, water will be sucked through every fault in such a system faster than it can run out through the weep holes. Even weep holes will serve as conduits for water entry, unless they are located where no rain water can reach them. These systems usually leak.

Most experts think that such total exclusion of water from the system is not only impossible to achieve but unnecessary. They advise that minor amounts of water be permitted to infiltrate, then be led back to the exterior by a system of water channels and weep holes. This advice is based on the assumptions that all curtain wall sealant joints are differential seals, that water will penetrate almost any seal under those conditions, and that the seals will function properly only when water is prevented from reaching the sealant. The sealant is thus concealed within the system. This arrangement is dependent on controlling the amount of water that can enter the system. The control mechanism is called the rain-screen principle. A rain screen is a device arranged to prevent rainwater from entering a building cavity. A cavity is created within the curtain wall and protected by the exterior portion of the framing, which acts as a rain screen. Another aspect of these designs is called pressure equalization. Because the cavity is open to the outside air and separated from the interior air by the sealant, it is at the same pressure as the exterior air, which negates the tendency of water to flow through the wall because of the lower interior pressure. These concealed sealant seals work better when they are installed in the shop. Raymond Ting's 1988 article, "Designing a Leak-Free Curtainwall System" and L. J. Heitmann's 1987 article "The Rain Screen Principle Can Work for You" offer excellent and more detailed discussions of the subject.

Joints should be made tight and should be constructed to conceal shrinkage.

Incidental aluminum tubing, structural angles, channels, flats, and other supporting members, reinforcements, fasteners, trim, sleeves, drips, closer angles, and other components necessary to complete installation should be furnished with the system. Aluminum or steel reinforcement should be provided as necessary to meet the performance requirements of the framing.

Mechanical joints should be accurately milled and fit to make hairline joints. Joints in exterior frames should be watertight. Concealed reinforcements should be used to make rigid connections. Metal-to-metal joints should be filled with sealant.

Welds should be deep-penetration type, on the insides of shapes, and should be pickled and washed. No weld discolorations should appear on finished surfaces.

Trim, glazing beads, and gaskets should provide the minimum grip on the glazing recommended in the industry standards mentioned in Chapter 8 for the type of glass and size of opening and should comply with the performance requirements of the system. Glazed curtain walls are usually glazed with 1/4-inch-thick single-pane glass or one-inch-thick insulating glass. Other thicknesses of single pane or insulating glass may be used, however.

Framing anchorages should be designed to permit both horizontal and vertical expansion of the system.

Many types of panels are used in glazed curtain walls. Spandrel glass panels are addressed in Chapter 8. Other types include insulated metal faced, ceramic tile, stone, brick, and opaque fiber-reinforced cementitious color-coated panels similar to Glasweld, manufactured by Eternit. Most stock systems come with aluminum-faced insulated panels.

Many types of metal-faced panels are available for use in curtain wall systems. A common type of construction consists of a backing sheet, a core, and a facing sheet. The backing sheet is usually aluminum or galvanized steel. The core is rigid closed-cell polyisocyanurate or polystyrene, rigid glass fiberboard, polyethylene, or a kraft-paper honeycomb. The facing sheet is aluminum. Sometimes a stabilizer sheet of hardboard or gypsum board is used to keep the facing flat. A gypsum board stabilizer sheet is also used when a fire rating is required.

Panels should be flat within a tolerance of 1/16 inch in each two feet and 1/8 inch over the entire panel. They should be factory fabricated and ready for field installation in one piece. Their edges should be factory fabricated for glazing into the framing and to receive glazing members, windows, and other components.

Components should be delivered to the installation location complete with fasteners, anchors, reinforcements, and accessories. Fasteners should be concealed whenever possible. Necessary exposed fasteners should be countersunk, flat phillips-head screws and bolts.

7.1.9 Glazing in Curtain Walls

Glazing includes glass and plastic glazing materials, spandrel glass, and operating windows. Operating windows are discussed in Chapter 6. Glazing materials and glazing sealants and fillers used for vision glazing and spandrel glass are discussed in Chapter 8, and the standards, installation methods, and other subjects discussed there apply. Glazing types and thicknesses vary with a project's requirements. The vision panels in most glazed curtain walls are glass, but some are plastic. The glass may be single pane, multiple pane, or insulating glass. Many codes require that glazed panels near the floor be of safety glazing materials.

Depending on the curtain wall system, glazing may be done either at the factory or in the field, but the most satisfactory results are obtained when the glazing material is installed in the factory. Unfortunately, stick systems must be field glazed.

Glazing gaskets are generally furnished by the curtain wall system's manufacturer. Common gasket materials include extruded polyvinyl chloride, extruded or molded neoprene, extruded or molded EPDM, and structural rubber lockstrip gaskets.

7.1.10 Miscellaneous Components and Accessories

Thermal Barriers. A thermal barrier is a non-conductive material interposed between the outdoor and indoor portions of a metal curtain wall framing member to separate metal that contacts outdoor air from metal that is in contact with indoor air. They prevent heat from traveling from one side of the frame to the other.

Louvers and Grilles. Louvers and grilles are routinely used in glazed curtain wall systems for air intake and exhaust. They are often directly associated with unit-type heating and air-conditioning devices.

Louvers and grilles may be part of the system's opaque panels, glazed into its framing members using beads or gaskets, or mounted mechanically into the framing.

Louvers in aluminum glazed curtain walls should be aluminum. Their finish usually matches the adjacent materials, but may be different, depending on the design.

Many proprietary louver blade shapes are available, but all should be of a weatherproof design such as the Z-shaped one shown in Figure 4-15.

Louvers that do not connect to ducts require insect screens. Those that do connect to ducts should have bird screens. The screens should be mounted on the interior face of the louver, in a removable frame. Most louvers in glazed curtain walls are fixed, but some may be backed up by operating louvers. Louver frames should be designed so that water that enters the louver is drained away. Wide louvers may require bracing, the most satisfactory type being concealed mullions fastened to the insides of the louvers.

Louver blades may be welded in place or mechanically attached, using crimped lugs or bent legs fastened to the frame.

The amount of free air required in louvers and grilles must be sufficient to accommodate the required air flow. The minimum is usually 40 percent free air before they are screened, but some installations may require more.

Where ducts connected to louvers are smaller than the louvers, unused portions of the louver should be blanked out on the inside face, using 1/2-inch-thick glass-fiber insulation or its equivalent between two sheets of aluminum.

Accessories. Clips, anchors, fasteners, closures, concealed flashings, reinforcement and other items, materials, and devices necessary to make a complete installation should be provided with glazed curtain wall systems.

Reinforcement may be necessary to make standard glazed curtain-wall framing components able to meet the proper performance requirements. The design of the needed accessories must permit unrestricted expansion and contraction. Exposed fasteners should be countersunk.

Inserts, bolts, and fasteners should be the glazed curtain wall manufacturer's standard units. Steel units to be built into exterior walls should be hot-dip galvanized. Anchors should be of the shapes and sizes required by the adjoining type of wall construction. They should be located as necessary to ensure compliance with the system's performance requirements. In every case, countersunk holes should be provided in the system's components for exposed anchorage.

7.2 Installation of Glazed Curtain Walls in New Construction

The manufacturer's instructions for protecting, handling, and installing fabricated curtain wall components should be followed. Particular care and attention should be given to preservation of applied finishes. Components should be protected during transit, storage, and handling to prevent damage, soiling, and deterioration. Items with minor damage may be repaired and installed, but those with major problems should be discarded and new, undamaged items used instead.

Components should be installed as soon as possible after they reach the site, to avoid prolonged storage and the accompanying potential for damage. Where storage at the site is unavoidable, components should be stored under cover in a manner that will prevent corrosion and damage. Unvented plastic or canvas shelters that could create humidity chambers should be avoided.

7.2.1 Preparation for Installation

Before glazed curtain walls are installed, their locations should be inspected and unsatisfactory conditions corrected.

Inserts that are to receive curtain wall anchors should be furnished at the proper times for setting in concrete formwork or masonry as it is erected.

Areas to receive curtain wall components must be properly prepared. The substrate and structure must be sufficiently square, true, and plumb so that the curtain wall framing can be properly attached and kept in the proper location, plumb and true. Adjacent masonry and concrete must be dry and free of excess mortar and other contaminants. Metals near or in contact with the glazed curtain wall materials should be clean, dry, and free of grease, oil, dirt, rust, and other corrosion. These metals should have no sharp edges or joint offsets. Other metals must be painted or otherwise separated from the aluminum in the curtain wall.

7.2.2 Installation

Glazed curtain wall systems should be installed in accordance with the shop drawings, the recommendations of the manufacturer, and applicable industry standards.

The system's framing components should be anchored securely in place and non-glass panels installed in the manner indicated on the approved shop drawings. Framing members should be shimmed as necessary and they and non-glass panels should be left square, plumb, level, true to line, at the proper elevation, and in alignment with other work within acceptable tolerances. Permissible tolerances vary from project to project and system to system. Usually, systems are required to be plumb within 1/8 inch in any ten-foot-long segment with a maximum of 1/4-inch variation in any forty feet; level within 1/8 inch in any twenty feet, with a maximum of 1/4 inch in any forty-foot segment. Surfaces should be in alignment within 1/8 inch, except that the maximum offset should be limited to 1/16 inch when the elements are supposed to be flush with each other. The surfaces may be offset by as much as two inches when separated by a protruding element or a reveal. Systems should also be within 3/8 inch of the planned location.

Curtain wall components that have been installed out of square or plumb, or are misaligned or out of location beyond acceptable tolerances should be removed and reset, to provide a proper installation of acceptable appearance.

The curtain wall installation should allow for movement resulting from changes in thermal conditions. Separators and isolators should be installed to prevent "freeze-up" of moving joints. Anchors should be concealed.

Curtain wall framing members should be securely anchored to the building frame or construction. Necessary anchors and inserts should be properly set in concrete and masonry. Units should be anchored securely, with concealed fasteners where practicable.

Exposed connections should fit together accurately to form tight, hairline joints. Connections that are not to be left as open joints or be screwed, bolted, or riveted together, should be welded together in the shop.

Framing members should be reinforced using structural sections of the same material used in the rest of the member when necessary to comply with the system's performance requirements. Members used as reinforcement should be anchored securely to the structure.

Accessories and trim should be installed rigidly, plumb, square, level, and true, in the proper location, in alignment with other work, and properly scribed to fit when necessary.

Exposed fasteners should be used sparingly. Those that are necessary should be countersunk phillips-head screws finished to match the metal being fastened.

Firestopping should be provided in glazed curtain walls when required by the building code or other governing regulations. Temporary closures should be provided behind curtain walls to prevent the accumulation of debris in concealed locations. Debris that does accumulate should be removed during erection, and firestopping should be installed. Firestopping should be securely anchored through metal flanges or equivalent provisions, to prevent dislocation.

The proper sealants should be used within the system, in concealed locations between the curtain wall and the substrates, and at the perimeter, to make the system weathertight.

Louvers and their screens and blank-off panels are often installed in non-glass panels at the factory, but can be installed in the field. Louvers in the framing of stick-type systems must be installed in the field. In any case, they should be set accurately and securely anchored in place.

7.2.3 Adjustments, Cleaning, and Protection

Where metals are placed in contact with dissimilar metals, masonry, mortar, concrete, plaster, or pressure-preservative-treated wood, the concealed contact surfaces should be given a heavy coat of alkali-resistant bituminous or zinc-chromate paint, or contact should be prevented using plastic sheet materials or by another appropriate method.

Cementitious materials, tape, and paint should be immediately removed from installed components before they harden or damage the finish. Excess sealing compound should be removed from exposed metal surfaces before it cures.

The entire installation should be protected from damage, soiling, and staining, both during erection and after initial cleaning. The complete system should be cleaned after completion, without damaging the finish or sealants. Strippable coatings and other temporary protection should be removed. Damaged and broken components should either be properly repaired or be removed and new, undamaged components installed.

7.3 Why Glazed Curtain Walls Fail

Glazed curtain walls and their finishes fail for many reasons, most of which can be grouped into three categories. The first consists of problems with the surrounding or supporting construction. It includes excess structure movement, failed steel or concrete structure, failed wood structure or wood wall or partition framing, failed metal wall or partition framing, failed concrete, stone, or masonry walls or partitions, and failed other building elements. The first categories' causes are discussed in sections 2.1 through 2.6. While the failure causes in the first category may not be the most probable cause for the failure of glazed curtain walls or their finishes, they are usually more serious and costly to fix than the other possible causes. Consequently, the possibility that they are responsible for a failure should be investigated.

The second category has to do with the materials and finishes used in glazed curtain walls. It includes bad materials, selecting inappropriate materials, selecting inappropriate finishes, improper preparation for application of finishes, improper finish application, failure of the immediate substrate, failure to protect materials and finishes, failure to properly maintain applied finishes, and natural aging. These failure causes are discussed in sections 3.12.1 through 3.12.9.

The third category includes improper design, improper fabrication, improper installation, and natural aging. These failure causes are discussed in this section.

7.3.1 Improper Design

The most serious possible design error is selecting or designing a glazed curtain wall system that is too weak for the loads to be applied, which will lead to its failure due to excess deflection. This error includes using metals that are too thin to support the loads or framing members that have inadequate cross-sections for the loads and spans they will encounter. Excess deflection can cause permanent deformation of members, stress failure in the metal used, water and air leaks, and glass loss.

A related error is selecting fasteners or anchors that are too weak for the application, too few in number, placed too far apart, or of the wrong type, any one of which can lead to failure of the anchor or fastener. Requiring that incompatible fasteners be used can also lead to failure. Requiring that incompatible anchors and attachments be used is a similar problem.

Frame configuration design errors can also lead to failures. Designing too shallow a glazing channel, for example, can result in glass loss from deflection of either the glass or the system's framing components under a load.

A common error with anodized aluminum units is failing to remove anodizing acids from the aluminum. If the acid used to produce anodized aluminum is not properly removed, it will eventually leak out and damage both the aluminum and adjacent surfaces. Acids trapped in hollow sections are a particular problem, as are ones trapped in joints between assembled sections. Weep holes are usually needed in hollow sections and assemblies to ensure proper removal of acids.

Requiring that the wrong type joining method be used can lead to failure. Insisting that aluminum be welded or brazed in the field is one example. Requiring that a thin metal be welded to a thick one will almost always cause problems. Designing open joints for use where water must be excluded will lead to leaks.

Requiring that butt joints be used in anodized aluminum panels will produce an unsatisfactory appearance. No two anodized panels will ever exactly match in color and shade.

Failing to allow for enough expansion and contraction can lead to several kinds of failures, including glass loss, open joints, and leaks. A related error is failing to require sufficient rabbet for glazing, which can result in glass loss.

Using glazed curtain wall systems that are inefficient thermally may result in cold spots and a poorly functioning heating and ventilating system. It is almost impossible to properly balance a mechanical system when glazed curtain walls permit excessive heat transmission. In some cases this error may relate to the glazing selected more than to the other glazed curtain wall components. Similar errors can result, however, from permitting too much air infiltration.

A related problem is using glazed curtain wall framing that does not have thermal breaks where indoor humidity is high. The combination of high humidity and cold metal will sometimes lead to heavy layers of condensation.

Failing to include appropriate condensate drainage channels and weeps will lead to damage of adjacent surfaces by condensate, and possible corrosion of metal components of the system.

Failing to include an effective exterior draining system to lead water away from a sloped glazing system's horizontal glazing and framing members can lead to leaks.

Placing a glazed curtain wall in a location where it is likely to be damaged will often lead to failure.

Selecting poor or incompatible shapes for components can produce sections that do not fit together well. There are limits on sizes and configurations that would prevent some designs from being made at all, but the kinds of problems that find their way into existing construction are more subtle, such as misfitting joints and surfaces that do not align.

Failure to provide sufficient, appropriate means for fastening the components of a frame together can lead to joint separation and both air and water leaks.

The improper design of spandrel cavities is a continuing source of problems in many curtain wall installations. Glass, concrete, metal, and many other materials are used to close the spandrel cavity. Insulation is also necessary, either loose or applied to the closing material or interior wall material. In any case, a vapor retarder is needed to help prevent condensation from forming in the cavity. A gutter should be incorporated, to carry condensation that does form and water from minor leaks to the exterior. The space in the cavity between the insulation and the spandrel panel material should be vented, to prevent heat and moisture buildup. Contact between the spandrel panel and other materials in the cavity should

Figure 7-1 A combination of a faulty panel and water caused the surface deposits seen in this photo.

be prevented. Improper design or construction of the cavity can cause failure of coatings and opacifiers on the inside of the spandrel panel, collection of foreign deposits on the interior of the spandrel panel, moisture showing through the spandrel panel or on the interior of the building, deterioration of the insulation, and destruction of the spandrel panel through corrosion or other damage.

Not providing a positive pressure differential between the space behind a glazed curtain wall and the exterior can lead to failure. If the pressure behind the space is lower that that on the exterior, water will easily pass through small defects in the system, migrating toward the lower pressure. Pressure equalization is critical.

Poor design can lead to unsightly stains and discoloration in spandrel panels. Refer to Chapter 3 for a general discussion of discolorations on metal finishes and what to do about them. Design problems may also cause the discoloration of other materials used as spandrel panels. Masonry and ceramic tile, for example, may be subject to efflorescence and other deposits when water from the cavity penetrates the spandrel and evaporates (Fig. 7-1).

Sealants that are inadequate for the job may pick up dirt and become discolored.

7.3.2 Improper Fabrication

Improperly fabricated glazed curtain wall components will often fail. Perhaps the most frequent cause of improper fabrication is failure to follow the requirements of the design, including the shop drawings and specifications and the standards of the industry.

Many fabrication errors involve joints. A common error, for example, is joining components in such a way that their corners or intersections are not square, plumb, and properly aligned and so that the joints are not flush, tight, and correctly fitted or do not exclude water.

Joining errors in glazed curtain walls include improperly making welded joints. Welding without proper preheating can lead to stress-corrosion cracking, for example. Using the wrong weld metal can result in cracking either in the weld itself or in the adjacent metal. Welding or brazing metals, especially thin ones, from the visible side will usually lead to unsightly joints. Failing to properly clean a joint that is to be welded or brazed will almost always result in a failed joint. Failing to use flux before brazing or soldering a joint will probably result in failure of the brazing metal or solder to adhere to the metal and the joint will fail.

Using damaged materials or damaging them during or after fabrication must be avoided. Damaged components should be repaired or removed and replaced with new ones.

Damage to preapplied finishes during fabrication, shipping, or handling is a problem. Such damage should be repaired before the loss of protection permits the metal to suffer damage.

Failure to keep drainage channels and gutters and associated weep holes clean and open will cause water to back up, with subsequent damage. Channels, gutters, and weep holes may become clogged either during construction or later.

Failure to make thermal breaks continuous will defeat their purpose and affect the heating and cooling of a building. If the mechanical design is done with the assumption that thermal-break construction will be used and it does not function properly, the heating and air-conditioning systems may not work as they should.

Glazed curtain wall systems that are installed in a way that does not permit sufficient expansion and contraction to occur may leak, twist, buckle, or pull away from the substrates or adjacent construction.

Gaskets and seals that are improperly made so that they do not fall within design tolerances will not seal tightly and will leak. Sometimes seals will extrude and produce a condition like that shown in Figure 7-2.

Figure 7-2 The sealer in the curtain wall frame member shown in this photo has extruded, leaving an unsightly mess.

7.3.3 Improper Installation

Correct preparation and installation are essential if failure in glazed curtain walls or their finishes is to be prevented.

The primary cause of failure due to installation error is not following the design and recommendations of the manufacturer and recognized authorities, such as ASTM, ANSI, AAMA, or NAAMM.

Failing to follow the design can mean using anchors, fasteners, and other materials that will rust or otherwise corrode where high humidity will occur or where these items will be set in or in direct contact with masonry or concrete. Sometimes materials that will corrode must be used for anchors. Then the error becomes one of failing to properly protect those metals with an applied bituminous or other appropriate coating.

Failing to follow the design and industry standards can also lead to using anchors, fasteners, clips, and other devices that are the wrong type for the installation, too small or weak to perform the necessary function, or not in the proper number or location.

Installing glazed curtain wall systems onto improperly prepared substrates and adjacent construction or onto wet materials can cause a poor fit, an improper anchorage, and damage to the curtain wall's components or anchors.

An allied error is the installation of a damaged curtain wall component. While the damage itself may or may not be the fault of the installer, installing a damaged component is.

Installing a glazed curtain wall in such a way that building loads are transmitted into it will surely cause it to fail.

The installer must be careful to install glazed curtain wall components so that they finish plumb, level, properly aligned, and in the proper plane, within acceptable tolerances. Improperly aligning a component may not only cause the system to present an unpleasing appearance; it can force loads to be applied eccentrically, which can damage the system. For example, glazing gaskets may become twisted and fail to seal properly, operating ventilators may not work properly, and joints may leak.

Glazed curtain wall framing members that are not securely anchored in place may become loose over time, break their perimeter calking, permit water or air penetration, or even fail structurally. Anchors may become dislodged, fasteners may loosen or break, and framing members may separate from their supports, with disastrous results. Even when loose components do not fail, they may vibrate and rattle in a wind.

Using old or the wrong types of gaskets and seals and installing them improperly are major causes of leaks. Gaskets must be butted tightly and be of the proper length to completely fill the spaces provided, without gaps of any kind. Seals must be properly formed, using appropriate sealants— not all sealants are usable in every joint. Sealants with a low range of movement should not be used in moving joints, for example. Old gaskets and sealants will not function properly and will probably contribute to leaks.

Failing to seal around the perimeter of a glazed curtain wall system will leave an ill-appearing installation and almost certainly result in air or water leaks. A related error is failure to place sealants in metal-to-metal joints in the system when they are required by the design, as is usually the case.

Another error in aluminum glazed curtain walls is failing to remove strippable coatings and tape soon after installation. Such materials may become difficult if not impossible to remove after prolonged exposure to sunlight.

7.3.4 Improper Maintenance

Refer to section 3.12.8 for a discussion on the effects of failing to maintain finished metals. Other maintenance is also necessary if glazed curtain walls are not to fail. Weather seal maintenance, for example, is critical. Failing to maintain them can permit moisture and even free water to reach corrodible anchors and attachments, resulting in their corrosion and eventual failure. Failure to maintain and reapply glazing sealants and gaskets when necessary can result in leaks and eventual glass loss.

Some panel materials require weatherproof coatings to maintain watertightness. Failure to renew such coatings when required can result in damage to the panels.

7.3.5 Natural Aging

Aging affects all aspects of glazed curtain walls. Most aging problems can be delayed for many years by performing proper maintenance. Refer to section 3.12.9 for a discussion about the effects of aging on materials and finishes.

In addition, the natural deterioration of liquid glazing compounds must be dealt with regularly by removing old glazing compounds and reglazing the curtain walls. Similar problems may occur with rubber and plastic glazing gaskets and stops, even though they last longer than liquid glazing compounds.

7.4 Repairing Glazed Curtain Walls

Even though existing glazed curtain-wall components may be damaged and unsightly, they can often be repaired and restored to a presentable condition. Sometimes it is necessary to remove a component, but many repairs can be made without doing so. Often, damage to a glazed curtain wall component or repairing of such damage will harm the component's finish. Refer to section 3.13 for a discussion.

When working with existing glazed curtain wall components, it is best to use materials and methods that match the original installation. When fasteners or attachments have failed, it will be necessary to provide new ones. Curtain wall components that are damaged too severely to repair, of course, must be discarded and new components provided.

Sometimes an apparent failure in a curtain wall is actually a failure in an adjacent or supporting material or structure. Each condition must be examined in light of the data included in this book and in the referenced sources of additional information. The condition of the damaged system should then be brought as close as possible to the condition usually required for that type of system when it is new. When damage is observed, the specific data in this book and in the listed sources of additional information, and the expert opinion of someone knowledgeable about the type of problem should be applied to solve it.

When existing glazed curtain wall components are to be repaired in place or be removed, repaired, and reinstalled, at least some selective demolition, dismantling, cutting, patching, and removal of existing construction may be necessary. The comments related to such matters in sections 1.6 and 1.7 are applicable.

7.4.1 General Requirements

This section contains some suggestions for repairing existing glazed curtain wall components. Because the suggestions are meant to apply to many situations, they might not apply to a specific one. In addition, there are many possible cases that are not specifically covered here. When a condition arises that is not addressed here, advice should be sought from the sources of additional data mentioned in this book. Often, consulting with the manufacturer of the item being repaired will help. Sometimes it is necessary to obtain professional help (see section 1.4). Under no circumstances should the specific recommendations in this book be followed without careful investigation and the application of professional judgment.

Before an attempt is made to repair an existing glazed curtain wall component, the manufacturer's recommendations should be obtained. A knowledgeable structural engineer should also be hired by the owner to ensure that unrecognized structural damage is not present and that the contemplated repairs will not cause structural failure. It is necessary to be sure that the manufacturer's recommended precautions against using materials and methods that may be detrimental to the items requiring repair are followed. Preventing future failure requires that repairs not be undertaken without careful investigation and the application of professional judgment. The repairs should be carried out by experienced workers under competent supervision.

Unless existing glazed curtain wall components must be removed to facilitate repairs to them or other portions of the building, or they are scheduled to be replaced with new materials, only sufficient material should be removed as is necessary to do the cleaning and repairs. The more materials removed, the greater the possibility of additional damage during their handling and storage. Materials that are loose, sagging, or damaged beyond in-place repair should be removed. Components that are sound, adequate, and suitable for reuse may be left in place or removed, cleaned, and stored for reuse.

After glazed curtain wall components have been removed, concealed damage to supporting elements, such as framing or substrates, may become apparent. Such damage should be repaired before repaired or new curtain wall components are installed.

Existing curtain wall components and supports that are to be removed should be removed carefully and adjacent surfaces should be protected so that the process does not damage the surrounding area.

Unless the decision is made to discard them, damage curtain wall components and support system elements that can be satisfactorily repaired should be, whether they have been removed or left in place. Failing to repair known damage may lead to additional failure later on. The methods recommended by the manufacturer of the system and the structural engineer should be followed carefully when making repairs. Components that cannot be satisfactorily repaired should be discarded and new, matching items installed.

Areas where repairs will be made should be inspected carefully to verify that existing components that should be removed have been and that the substrates and structure are as expected and not damaged. Sometimes, substrate or structural materials, systems, or conditions are encountered that differ considerably from those expected, or unexpected damage is discovered. Both damage that was previously known and that which is found later should be repaired before glazed curtain wall components are reinstalled. Reinstallation should not proceed until unsatisfactory conditions have been corrected.

7.4.2 Repairing Specific Types of Damage

In addition to the general requirements for repairs already discussed, the specific recommendations that follow apply to specific problems. The recommendations in section 6.9.2 related to repair of windows also apply to windows in curtain walls.

Bent, Twisted, or Deformed Components. When the affected metal component is not too severely damaged, repairing it may be possible. Often when a glazed curtain wall component becomes bent, twisted, or deformed, the cause is related to the adjacent construction and not the curtain wall. Such external damage must be corrected before curtain wall repair is attempted.

It may be possible to straighten a twisted glazed curtain wall component, but removing it will probably be necessary. Even when a component can be straightened in place, removal of its glazing is usually necessary. If the original damage did not break the glass, straightening often will do so. Often it is necessary to send the component to a shop for the repairs to be made. Rebending metal may set up stresses in it that can only be relieved by heat treatment of some sort, which is usually impracticable to do in the field.

Often, straightening a badly damaged curtain wall component is impracticable. Then removing the damaged component and providing a new one may be the only alternative. Stock components may still be available.

Custom-designed replacement components may have to be made, which can be expensive. Unfortunately, even when the component can be replaced, the finish on the new one is not likely to match the adjacent older materials.

Loose, Sagging, or Out-of-Line Curtain Wall Components. A curtain wall component may become loose, sag, or be out of line because it was poorly designed or fabricated. It is very difficult to "fix" a poorly designed component.

Usually, when a curtain wall component becomes loose or sags or is out of line, the fault lies either with undue movement of the supporting construction or with a failure in the system's anchors. Failure of the supporting construction should be investigated and repaired if it is at fault. Such repairs are beyond the scope of this book.

Failed anchors are a serious problem for curtain wall systems, because they are the only thing holding the system to the building. In most systems, the failure of a single anchor can result in sagging or loose members and possible glass loss. The failure of a single anchor in a sloped glazing system can result in collapse of the entire system.

In some types of curtain wall systems, the solution to failed anchors may be as simple as installing a new anchor. When installing too few anchors has caused the problem, additional ones can be provided. If the wrong material was used to make the anchor, it can be replaced. Unless they are prohibited from doing so, some sloped glazing manufacturers will use cast aluminum anchors and structural connection components, which sometimes fail under the repetitive application and removal of loads. If the new anchors are exposed, the appearance of the finished installation will be less than desirable. Installing new concealed anchors requires removal of at least some components of the system or part of the adjacent construction. It may also make it necessary to reinforce the supported curtain wall component at the anchor locations. In aluminum systems, field welding of reinforcements will not be practical. Therefore, unless the reinforcements will be installed using exposed fasteners, the component will have to be removed and sent to a shop.

Sometimes a failed anchor can be uncovered and removed and a new anchor installed. Care must be taken, however, to prevent repeating the problem that caused the original failure. If it occurred because the original anchor was too small or inadequately fastened in place or to the curtain wall component, a different design or a larger anchor and fasteners must be used. The important thing is to discover the underlying reason for the failure and make sure it does not happen again.

When deformation has broken welds or damaged mechanical joints, it will probably be necessary to remove the component to make repairs.

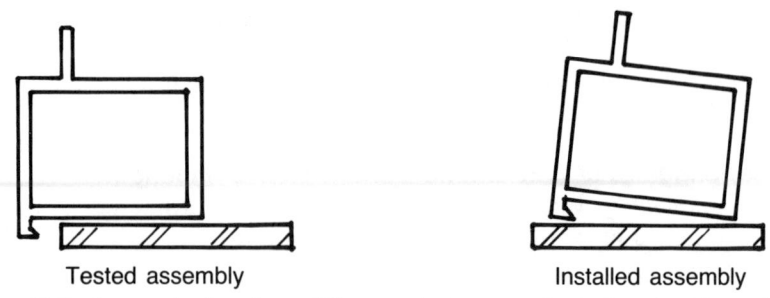

Figure 7-3 A seemingly minor difference between a tested assembly and the one actually installed can cause untold problems.

Air or Water Leaks. Probably the largest problem with glazed curtain walls is their propensity to leak. Water leaks are an immediate problem, of course, that must be stopped as soon as possible to prevent damage to interior materials. Apparent water leaks may be condensation, which can be quite profuse under some temperature and humidity conditions. True water leaks may be caused by missing or deteriorated calking, damaged or missing glazing sealants or gaskets, or open joints in the system. Both condensation and true water leaks may be exacerbated by clogged weep holes or blocked drainage channels. Leaking can be contributed to by free water cascading over a sloped glazing system. Repairs require first finding the reason for the leak then recalking or repairing glazing sealants or gaskets, clearing weep holes and drainage channels, repairing open joints, and providing a drainage system to divert water that falls on a sloped glazing system and direct it to the ground.

Although air leaks do not require as urgent a response as do water leaks, they can be even more costly. The energy lost by infiltration through and around glazed curtain walls can be a significant contributor to heat loss in a building. The source of many air leaks is deteriorated or missing calking or glazing gaskets or compounds. When failed or old glazing sealants or gaskets are responsible for air leaks, the failed materials should be removed and new materials of the proper kind should be installed.

Sometimes leaks happen because the curtain wall is not installed properly. A typical example is shown in Figure 7-3, where the system passed all the required tests with the drip hanging over the structural support, but leaked when the member was installed with the lip on top of the support. Failure to install or properly install sealants in joints within the system will also lead to leaks.

Some leaks happen because improper installation of the system makes proper sealant application at the perimeter impossible. Others occur because the sealant was improperly installed, even when the joint was properly constructed. Figure 7-4 shows two of many possible errors related to sealing around the perimeter and one acceptable application. The designer must keep in mind that sealants around curtain walls must act as differential seals, because the pressure on the inside is less than the pressure outside. Under these conditions, only a perfect seal will prevent water migration to the interior.

Lack of Flatness in Panels. Panels may not be flat for several reasons. The undulating effect called oil-canning is the most normal manifestation of a lack of flatness. Lack of flatness in panels can be reduced by using heavier metal skins on the panels, roll forming the metal face sheets rather than brake forming them, avoiding foamed-in-place panels in which foam shrinkage will sometimes make depressions in the surface, and by more carefully controlling the manufacturing process. Once a panel has an uneven surface that has occurred because of its design or manufacture, correction may be impossible.

Lack of panel flatness may also be caused by improper installation. The panel may have been twisted to fit misaligned framing, for example, or the fasteners or anchors holding it in place may have been located improperly. Unless the error has caused permanent deformation, such errors can sometimes be corrected by removing the incorrectly installed component and properly reinstalling it.

Panel Skin Undulation, Blistering, and Rippling. Thermal conditions can cause undulating, blistered, or rippled panel skins on foam-core panels.

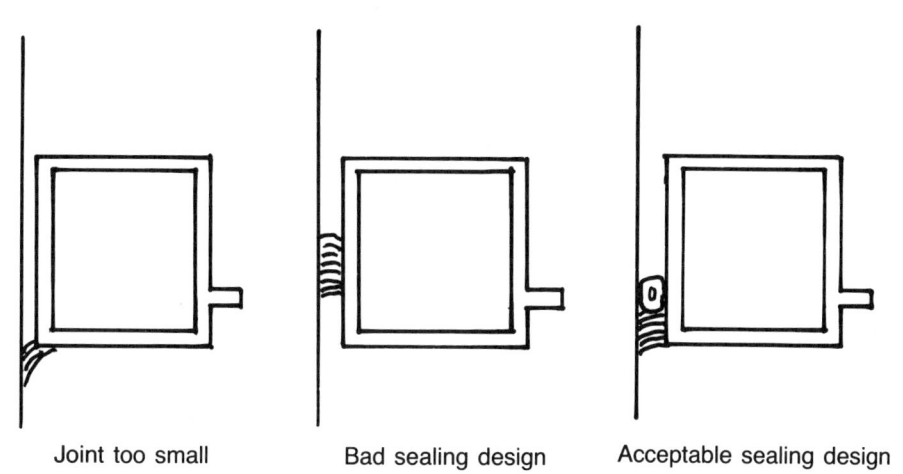

Figure 7-4 Curtain wall perimeter sealing conditions.

Under the influence of the sun, insulated panels become hotter in the center than along the edges, which can result in undulations of the skin. This condition can be avoided only by reducing the thermal value of the insulation: when the value is high, undulation will occur.

Blistering happens because it is not possible for a panel's skin to bond completely to its core, and the pressure of expanding gas escaping from the insulation will create bubbles where the skin does not firmly adhere. Small blisters can be relieved simply by drilling a hole in the skin and releasing the gas, large ones by doing the same thing and then injecting an adhesive into the pocket.

Metal panels can bow or ripple after installation when exposed to hot sunlight. The exterior face panel expands, but the thermal barrier created by the panel's insulation keeps the interior skin cooler so that it expands less than the exterior skin. This effect is strongest when the panels have been erected in cold weather and the interior face is particularly cold. After the building has been heated, the differential is less and rippling is less likely to occur. In short panels this bowing or rippling is not usually noticeable when sufficient space has been provided for thermal movement. In long panels with intermediate supports, however, the stresses generated may exceed the buckling resistance of the skin and create permanent transverse ridges in it. The way to avoid this type of bowing and rippling is to properly design the panels to resist the stresses that will be generated. Once rippling has occurred, it is often not correctable.

7.5 Glazed Curtain Walls in Existing Construction

There are two basic ways to install new glazed curtain walls on an existing building. First, whether the existing exterior wall material is masonry or an existing metal curtain wall, it can be removed and the new glazed curtain wall installed as it would be on a new building. Second, when the cost of removing the existing wall material is too high or the effect of such removal on building use is unacceptable, the new glazed curtain wall may be installed directly over the existing wall and fastened through it into the building's structure.

The requirements discussed in sections 7.1 and 7.2 also apply when the curtain walls will be installed on an older building. There are, however, some additional considerations that must be taken into account. Many of them are addressed in this section.

Since the number of possible glazed curtain wall installation conditions in existing buildings is large, it is not possible to cover them all in this book. The discussion here is therefore generic in concept and should be viewed in that light. These suggestions may not apply to all conditions, but the principles should be applicable to most. In the end, there is no substitute for professional judgment.

7.5.1 Standards, Controls, Delivery, and Handling

The standards and controls in sections 1.6.1, 1.7.1, 7.1.2, and 7.1.3 apply to glazed curtain walls installed in existing construction.

In addition, existing glazed curtain wall components should be removed and the new system installed by the manufacturer of the new system or by a firm franchised or approved by the manufacturer. Other contractors may be used to remove masonry and other types of construction from buildings that are to receive new glazed curtain walls. Only experienced personnel should be permitted to do any of this work.

While the work related to the removal of existing materials and the installation of a new glazed curtain wall system is under way, existing materials and items that are not to be removed and discarded should be protected from harm. If they become damaged, they should be repaired to a condition at least equal to that existing before the damaging. Where acceptable repairs are not possible, the damaged materials and items should be removed and new materials or items installed.

Damaged glazed curtain wall components should not be installed without repair. When satisfactory repairs are not possible, the damaged components should be discarded and new ones provided.

7.5.2 Glazed Curtain Wall Materials, Finishes, Types, and Fabrication

Glazed curtain wall systems used in existing buildings and those for new construction are essentially the same. Therefore, most of the requirements discussed in section 7.1 are applicable also to glazed curtain walls that will be installed on an existing building. Special anchoring devices will, of course, be necessary when an existing curtain wall will remain in place.

7.5.3 Preparation for Installation

Just as preparations are necessary before a new glazed curtain wall system can be installed on a new building, so is it necessary to make preparations before installing them on an existing structure. In the latter case, however, the preparations are likely to be more extensive.

The requirements for demolition and removal of existing items and for renovation work discussed in sections 1.6 and 1.7 are applicable to work related to glazed curtain walls. The requirements in this section are thus in addition to the ones in those two sections.

As is true when the glazed curtain wall and the building's structure are both new, the location where a new glazed curtain wall system will be installed in an existing building should be examined carefully by the installer. Making two examinations is usually best, the first before demolition work begins. A report of this examination should be sent to the designer and the owner so that modifications in the proposed demolition can be made, if necessary, to accommodate the curtain wall design.

The second examination should be made after demolition has been completed to verify that other construction to which the new curtain walls are to be attached, and the areas and conditions where they will be installed, are in a proper and satisfactory condition to receive them. Preparatory installation work should not be started and the curtain wall should not be installed until unsatisfactory conditions have been corrected. Local repair of the structure may be necessary, for example, before new anchors can be installed.

Should conditions be encountered that differ significantly from those expected, the person discovering them should immediately notify the other concerned parties, especially the owner, so that alternative methods can be developed for handling the unexpected situation.

When a new curtain wall system will be installed where there are existing windows, the existing windows are rarely left in place. Sometimes the openings are even filled with masonry or other materials to reduce heat loss or gain in the building.

When there is an existing glazed curtain wall system, its framing members and non-glass panels may either be removed or left in place when they do not interfere with the new system. Sometimes the existing glazing and operating windows are also left in place.

7.5.4 Installation of Glazed Curtain Walls

The recommendations in section 7.2 apply also to curtain walls installed in an existing building. The suggestions in this section are in addition to those in section 7.2.

Damaged glazed curtain wall components should not be installed, because it is generally much easier to repair a damaged component before it is installed. Often damage will not be repairable afterward, and the component must then be removed to permit repair. Damaged components that have been installed and cannot be repaired in place should be removed and repaired or new undamaged components installed. Installing already damaged components is one cause of later problems with them.

Glazed curtain wall systems should be installed in accordance with the manufacturer's latest published instructions, the approved shop drawings, and the current applicable industry standards and recommendations. A higher than normal safety factor should be used for the anchoring system. A subgirt system is usually applied over existing masonry walls to support the new curtain wall.

Non-glass panels, louvers, and glazing are installed in the same way, regardless of whether the curtain wall is installed in new or existing construction.

Spandrel areas require special consideration. Unless insulated spandrel panels will be used, insulation must be installed continuously in the spandrel space. A vapor retarder should be provided. If there is an existing retarder, it must be removed unless its location is such that it will not cause condensation instead of preventing it. Air spaces between the spandrel panel and the insulation must be vented to the outdoors to prevent excess heat buildup.

When a new system is installed over an existing one, the new system must be fastened through the existing one to the underlying structure. Fastening to the old system is not satisfactory.

The space between a new insulated curtain wall and an existing wall should be sealed and particular care exercised to select panel components that will not be subject to thermal rippling due to the wintertime temperature differential between the exterior and the space between the walls. When the new system is not insulated, the space must be ventilated. Sometimes exhaust air from the building is directed into the space to pressurize it. In this case the old wall must be well sealed and the new one designed to permit the exhaust air to escape to the exterior.

7.6 Where to Get More Information

The AIA Service Corporation's *Masterspec* Basic Version Section 08920, "Glazed Aluminum Curtain Walls," contains helpful information about glazed curtain walls. Unfortunately, it contains little that will help with troubleshooting such installations.

The following American Architectural Manufacturers Association (AAMA) publications contain information valuable to anyone interested in glazed curtain walls:

- AAMA TIR-A1-1975, "Sound Control for Aluminum Curtain Walls and Windows."
- AAMA TIR-A3-1975, "Fire-Resistive Design Guidelines for Curtain Wall Assemblies."
- *Aluminum Curtain Wall Design Manual* (CW-I-9). A nine-volume manual that should be on the shelf of every designer of glazed curtain walls and those who must deal with existing glazed curtain walls.

304 Glazed Curtain Walls

- AAMA 501-1983, "Methods of Test for Metal Curtain Walls."
- AAMA 1502.7-1981, "Voluntary Test Methods for Condensation Resistance of Windows, Doors, and Glazed Wall Sections."
- AAMA 1503.1-1988, "Voluntary Test Method for Thermal Transmittance and Condensation Resistance of Windows, Doors and Glazed Wall Sections."
- AAMA 1504-1988, "Voluntary Standard for Thermal Transmittance and Condensation Resistance of Windows, Doors and Glazed Wall Sections."
- *Curtain Wall Manual*, vol. 10, *Care and Handling of Architectural Aluminum from Shop to Site.*
- *Curtain Wall Manual*, vol. 11, *Design Windloads for Buildings and Barrier Layer Wind Tunnel Testing.*
- *Curtain Wall Manual* (CWM-1).

Every designer should have the full complement of applicable ASTM standards available for reference, of course, but anyone who needs to deal with problems with glazed curtain walls should obtain and understand the standards referenced in this chapter.

Refer also to other items marked with a [7] in the Bibliography.

CHAPTER 8

Glazing

This chapter discusses glazing materials and their installation in doors, windows, store fronts, and curtain walls. Included are glass and plastic materials and the glazing of those materials as well as non-glass panels other than plastic. The glass types covered are primary, heat treated, coated, laminated, sealed insulating, and mirror glass. Plastic glazing materials are either acrylic or polycarbonate sheets. Other non-glass panels include metal, concrete, stone, ceramic tile, brick, and other types of panels.

Metal swinging and revolving doors into which the materials in this chapter are installed are discussed in Chapter 4; wood swinging, solid and paneled sliding and folding doors in Chapter 5; wood and metal windows and sliding glass doors in Chapter 6; and aluminum curtain walls in Chapter 7.

8.1 General Requirements for Glazing

The term *glazing* means the installation of glass, transparent plastic sheets, and the types of non-glass panels discussed in Chapter 7. *Glazing materials* are the glass and other panels to be installed and the sealants, compounds, and gaskets used to set them. *Glazing accessories* are the various clips, points, and the like necessary for proper installation of glazing materials.

8.1.1 Standards

Glazing materials and accessories and glazing should comply with the recommendations of the manufacturers of the glazing materials, but should not be of a lesser quality than that recommended by the Flat Glass Marketing Association's (FGMA) *Glazing Manual* and *Sealant Manual*. In addition, insulating glass units should be set in compliance with the recommendations of the Sealed Insulating Glass Manufacturers Association (SIGMA), except as otherwise specifically recommended by the glass and sealant manufacturers.

Primary glass materials should comply with the requirements of ASTM Standard C 1036, which classifies glass according to type, class, quality, and style. Patterned and wire glasses are further classified according to form, finish, mesh type, and pattern. This standard replaces the old standards found in Federal Specification DD-G-451.

Heat-treated glass materials should comply with the requirements of ASTM Standard C 1048, which replaces the old standards found in Federal Specification DD-G-1403. ASTM Standard C 1048 classifies glass according to kind, condition, type, quality, class, and form.

Fire-resistance-rated wire glass should also be able to meet the tests described in ASTM Standard E 163. It should be labeled and listed by the UL or another nationally known testing and inspecting agency acceptable to the government authorities that have jurisdiction.

Wire glass and safety glass should comply with the requirements of ANSI Z97.1-1984. Safety glass, including tempered and laminated glass, must be able to meet the applicable testing requirements of the Consumer Product Safety Commission's 16 CFR Part 1201. Wire glass does not comply with the latter standard and is inappropriate for locations where compliance with it is required. Wire glass was used as a safety-glazing material in buildings constructed before 1980, however, because such usage was legally acceptable until that time.

Tempered glass should comply with the requirements of ANSI Standard Z97.1-1984.

Sealed insulating glass units should comply with ASTM Standard E 774.

Silicon glazing sealants should comply with ASTM Standard C 920.

8.1.2 Controls

Copies should be submitted to the owner for approval of the manufacturer's specifications and the installation and maintenance instructions for each needed type of: primary glass; laminated, insulating, and other fabricated glass products; plastic glazing materials; glazing sealants, compounds, and gaskets; and associated miscellaneous materials. These submittals should also include, for each type of glazing material, either the manufacturer's published technical data or a letter of certification or certified testing laboratory's report attesting that each material complies with the project's requirements, is intended for the particular application, is the proper thickness, and that the system has the proper edge retainage for the glazing material that is to be installed.

To permit verification that the materials that will be installed are the same as those that were selected, the manufacturer or installer should be required to submit twelve-inch-square samples of each type of primary glass, fabricated glass product, and plastic glazing material, and twelve-inch-long samples of each color required for each type of glazing sealant, compound, and gasket that will be exposed to view. The sealant, compound, and gasket samples should be contained between two strips of material that represent the color and finish of the material that will adjoin the glazing material.

When there is the slightest doubt about the products to be installed or their installer, samples should be required of each type of tape, shim, setting block, and other accessories that will be used.

Glazing materials should arrive at the installation site bearing their manufacturers' labels indicating all data necessary to permit their identification. When practicable, the labels should be permanent and located in the lower corner of each light. When the glass is not cut to size by the manufacturer but is rather furnished as unlabeled material taken from local stock, an affidavit should be required stating the quality, thickness, type, and manufacturer of the material furnished.

For each glazing material a certificate prepared by the manufacturer should be submitted attesting that the material complies with the project's requirements. There is no need, however, to require separate certification for materials bearing the manufacturer's permanent labels designating the material's type and thickness, and a label establishing compliance with a quality-control program requiring tests conducted by a recognized certification agency or independent testing laboratory acceptable to the authorities having jurisdiction.

It would help an owner to exercise control over the product being installed if new insulating-glass units were required to bear a certification label from either the Insulating Glass Certification Council (IGCC) or the Associated Laboratories, Inc. (ALI). All manufacturers do not participate in such programs, however, so existing insulating-glass units may not bear such labels. When labels are required, they should be permanently marked either on the unit's spacers or on at least one component pane. A listing of the participating manufacturers in each program is contained in the IGCC's *Certified Products Directory—Sealed Insulating Glass* and the ALI's *Certified Products Directory—Fenestration Products.*

Safety glass bearing the label of the Safety Glazing Certification Council (SGCC) or another nationally known certifying agency will probably be of a high quality. But some manufacturers who produce perfectly acceptable insulating glass units do not participate in such programs, so an existing insulating glass unit can be perfectly satisfactory without bearing a label. The SGCC's *Certified Products Directory—Safety Glazing Materials Used in Buildings* contains a list of manufacturers whose products are available with labels.

The manufacturers of glass, plastic glazing materials, non-glass panels, gaskets, glazing accessories, and framing members that will be used in contact with glazing compounds or sealants should be required to submit samples of those items to the glazing compound and sealant manufacturers. The samples should then be tested in accordance with those manufacturers' standard testing methods for compatibility and adhesion capability.

Sealant, glazing compound, and gasket manufacturers should be required to submit statements attesting that the glass, plastic and non-glass panels that will be used have been tested to ensure their compatibility with the applicable sealants, compounds, and gaskets. These documents should also state that the sealants and compounds will adhere properly to the glazing materials. These statements should interpret the test results relative to the materials' performance and include recommendations for the primers and substrate preparations needed to ensure adhesion.

Except on very small projects, when glazing will be done in the field it is a good idea to hold a meeting at the project site before work related to glazing is begun, to review the procedures and time schedule for glazing and discuss coordination of glazing with other work. Present at this meeting should be the owner, the designer, the glass framing erector, and the glazier; technical representatives of the manufacturers of sealants, glazing compounds, and gaskets; the persons responsible for the work of other trades that will affect the glazing; and representatives of other trades that will be affected by the glass installation.

On large projects, field-constructed mock-ups are sometimes required. When the quantity of such items warrants it, full-sized glazed sample windows, doors, store fronts, and curtain wall sections should be erected at the site. Such mock-ups should be built to demonstrate the glazing systems required, including typical light size, framing system, and glazing materials and methods. The mock-ups should be retained, undisturbed, during construction as a standard for judging the completed work. Mock-ups for curtain walls should include each type of vision and spandrel panel that will appear in the completed building.

For a structural silicone sealant glazing system, the mock-up should be tested using the "hand pull" test for adhesion in the AAMA's *Curtain Wall Manual*, volume 13, *Structural Sealant Glazing Systems*. Additional adhesion tests should be conducted on the actual system as it is installed.

To ensure consistent quality in appearance and performance, all the materials used in each glazing application should have been produced by a single manufacturer. Each primary glass used in a single laminated or insulating-glass unit should also be the product of the same manufacturer.

Work related to art glass, which is also called stained glass, should be done by a firm with successful experience in such work, and by experienced personnel skilled in the processes and operations necessary, working under competent supervision. A competent representative of this firm should examine existing art glass. When the art glass is in finished wood framing, the art glass repair firm's representative should be accompanied on this examination by the finish carpenter. Together they should identify damaged and missing components of the art-glass work and establish procedures for coordinating the leaded-glass work with the finish carpentry. As will be discussed later, art glass must frequently be removed to a shop to facilitate repairs. At the shop the representative of the art-glass repairer should further examine in detail each panel of the art glass to determine the extent of needed repairs and recaming.

8.1.3 Warranties

Warranties beyond the customary one-year construction warranty are often required for coated, insulating, and laminated-glass products and for mirrors. Such warranties should be in addition to—not a limitation of—other rights the owner may have under the law or the owner–contractor agreement. Warranties should be written documents signed by both the manufacturer and the installer of the warranted item.

The signers of warranties for coated glass should agree that within the warranty period they will remove coated-glass units that develop manufacturing defects and provide new, approved units at no cost. Manufacturing defects are usually defined as peeling, cracking, or deterioration in the glass's coating due to normal conditions and not from handling or installation or cleaning practices contrary to the glass manufacturer's published instructions. The warranty period should be the manufacturer's standard but should be not less than five years.

The signers of a warranty on insulating glass should agree that within the warranty period they will remove insulating glass units that develop manufacturing defects and provide new, approved units at no additional cost. Manufacturing defects are defined as failure of the hermetic seal of the air space, unless the failure is due to glass breakage. Failure is evidenced by intrusion into the air space of dirt or moisture, internal condensation or fogging, deterioration of a protected internal-glass coating, and other visual indications of seal failure or performance. The warranty will usually exclude from coverage damage resulting from violations of the manufacturer's instructions for handling, installing, protecting, and maintaining the insulating glass. The warranty period may be either five or ten years.

Laminated safety glass should be warranted against failure of every kind for a period of five years. The warranty should provide that if failure should occur during the warranty period the failed units will be removed and new, acceptable laminated-glass units provided at no cost. The replacement of failed laminated safety glass included in insulating-glass units should be included in the warranty covering the laminated glass and should also be included as part of the warranty covering the insulating-glass units.

The silvering on mirrors should be warranted for at least five years against failure of any kind.

In every case, units installed in place of units removed under the warranty should be similarly warranted during the remainder of the original warranty period.

Excepted from warranties are damage caused by deliberate, malicious vandalism and damage due to failure of the supporting structure.

8.1.4 Performance Requirements

Installed glazing materials should be required to comply with the same requirements applicable to the glazed door, window, store front or curtain wall into which they are glazed. Refer to Chapters 4 through 7 for those requirements.

Installed glazing materials should have been produced, fabricated, and installed so they will withstand with no failure normal temperature changes and wind loading, and doors and windows to withstand applicable impact loading. Failure includes loss or breaking of glass, failure of compounds, sealants, or gaskets to remain watertight and airtight, deterioration of glazing materials, and other defects in the work. Normal thermal movement is usually defined as that resulting from an ambient temperature range of 120 degrees Fahrenheit and from a consequent temperature range within the glazing material and framing members of 180 degrees Fahrenheit. Deterioration of insulating glass is defined as failure of the hermetic seal due to causes other than breakage that result in the intrusion of dirt or moisture, internal condensation, or fogging, the deterioration of a protected glass coating resulting from seal failure, and other visual evidences of failure. Deterioration of coated glass is defined as the development of manufacturing defects, including peeling, cracking, or other deterioration in the coating during normal use.

Tinted glass should be required to meet specific levels of visible-light transmittance and have the shading coefficient selected by the designer. These values vary greatly from product to product.

Coated glass should be required to meet specific requirements for U-value and visible-light transmittance as selected by the designer from the characteristics of the desired material. These characteristics vary greatly from product to product.

The maximum tensile and shear stresses in structural silicon sealant joints must be the minimum permissible for the loads expected.

8.2 Materials

Windows and doors are usually glazed with either glass or plastic. Store fronts and curtain walls may be glazed with glass, plastic, or other non-glass panels. Refer to Chapter 7 for information about non-glass panels other than plastic ones.

8.2.1 Glass Types and Quality

The transparent or translucent material, glass, is a peculiar substance that behaves like a solid but has its atoms arranged in the random, disordered fashion characteristic of a liquid. Because it has the characteristics of both a liquid and a solid, glass is sometimes called a supercooled liquid. Two of glass's most important characteristics are its fatigue resistance and its transparency. The first is important because it permits glass to bend up to its breaking point without undergoing permanent deformation. When the load is removed, the glass returns to its original shape. Thus, glass can be severely deflected by wind without permanent damage, which permits the use of thinner sheets than would otherwise be necessary. The advantages of glass's transparency are obvious, of course, but also important is its ability to have its degree of transparency altered readily in a number of ways, which we will discuss as we go along.

Another important characteristic of glass is its optical fidelity, or lack of distortion, in sheets of a constant thickness. Glass is also highly resistant to corrosion and is non-absorptive, which are characteristics that help it to remain relatively unchanged after years of exposure to sun, wind, and precipitation.

About 90 percent of the glass produced in this country, and almost all the glass used in glazing buildings, is soda-lime glass. It is a fused mixture of about 72 percent silica, 15 percent sodium oxide (soda), 9 percent calcium oxide (lime), and 4 percent various other ingredients. Other types of glass are available, but they are seldom used to glaze doors, windows, store fronts, or curtain walls.

The chief ingredient of soda-lime glass is a type of sand that contains about 99 percent silica. Glass can be made by simply melting such sand, but this requires very high temperatures. Lower temperatures are needed and the glass is more workable when soda and lime are added. In the early days of modern glass manufacturing, most glass was made using potash, a form of potassium, instead of soda. Nickel, cobalt, silver, gold, copper, and selenium oxides and arsenic and other additives are now used to impart color (tint) to glass.

The glasses used to glaze windows, doors, store fronts, and curtain walls can be placed in five categories: primary, heat treated, coated or reflective, laminated, and insulating. The last four kinds all start as a type of primary glass. The number of primary glass types made in this country is small, but each of them can be processed further to make a large assortment of glass products with a wide range of characteristics. Additional treatment can be done to affect a glass's appearance, light and energy transmission, and structural performance. Heat-treated and coated or reflective glass are made by altering primary glass. Laminated and insulating glass are built up composites made from panes of primary, heat-treated, or coated or reflective glass. Often more than one type is used in a single laminated or insulating-glass unit.

Primary Glass Products. The main types of primary glass used today are clear float glass, tinted float glass, patterned glass, and wire glass. Early glass used in windows and doors was made by blowing cylinders of glass, then cutting, unrolling, and reheating the cylinders to flatten them into small, flat panes. They were irregular in thickness and not very flat on the surface, so they tended to distort the images seen through them. Their color and opacity were also difficult to control. Blown-glass panes can be seen in many historic preservation projects. Unfortunately, when such a pane becomes broken, finding a replacement is difficult, if not impossible.

Sheet glass was a vast improvement over blown glass. It was made by drawing molten glass out of a tank using a steel rod called a bait and a series of rollers. When the bait was inserted into the molten glass, which has a very high viscosity, the glass adhered to it. When the bait was removed, the glass was pulled upward with it. Rollers then caught the molten glass, advanced it, and flattened it. As the glass moved along, flame jets fire polished it, making it transparent. Sheet glass was difficult to keep consistent in width and had an inherent wave or distortion across the sheet. Though not too objectionable in small pieces, the imperfections were very apparent in large sheets. Sheet glass had the advantage of being cheap to produce.

8.2 Materials

The large-sheet distortion problem was virtually eliminated when a process was developed to cast glass into the familiar plate glass, also called polished plate glass. At first, plate glass was made by pouring molten glass on a casting table, then rolling it flat. After cooling, it was ground and polished. Later, a continuous-rolling process was perfected that permitted plate glass to be produced more easily and cheaper. The same process was used to make patterned glass. Left unpolished, plate glass was translucent. To produce the familiar large clear sheets, plate glass was ground on both sides, then polished and cut to size.

Making plate glass was time consuming and costly, however, so floating was developed. Floating produces a glass that has the fire-polished finish and low cost of sheet glass and the flatness and low distortion properties of polished plate. In the floating process molten glass is deposited, or floated, on a bed of molten metal, usually tin. The process takes advantage of the physical law that the surfaces of a material of higher density floating on a material of lower density will become flat and parallel. In the process the glass is fire polished and annealed.

The floating process has been so successful that few, if any, manufacturers make any other kind of glass today. Blown, sheet, and plate glass will, however, be found in many older buildings, so they are important to our discussion here.

Sheet glass was produced in three thickness ranges. The thinnest, called picture glass, was not used in windows or doors. The next thickest, window glass, was made in two thicknesses. Single-strength (SS) window glass was 3/32 inch thick, double thickness (DS) 1/8 inch. Heavy sheet glass was 3/16 to 7/32 inch thick. Double-strength and heavy sheet glasses came in three grades: AA for high-grade work, A for superior glazing, B for general glazing.

Plate glass was divided into two thickness groupings. Regular plate was 1/8 to 1/4 inch thick, heavy plate 5/16 to 7/8 inch. Regular plate was available in silvering, mirror glazing, and glazing qualities. Heavy plate came only in glazing quality. Obscure plate was available polished on one side and rough on the other or rough on both sides.

Float glass is available in thicknesses ranging from 3/32 to 7/8 inch. Qualities include mirror, mirror select, glazing, and glazing select. Thinner float glass is available in single- and double-strength types with qualities similar to sheet glass, with its same designations but without the distortions inherent in sheet glass. Thin float glass can be used as replacement panes for broken sheet glass. Refer to ASTM Standard C 1036 for a detailed list of the available characteristics.

Sheet, plate, and float glass can all be tinted. The maximum readily available thickness of tinted glass, however, is one-half inch. The tints most commonly used are gray, dark gray, blue, blue-green, green, and bronze.

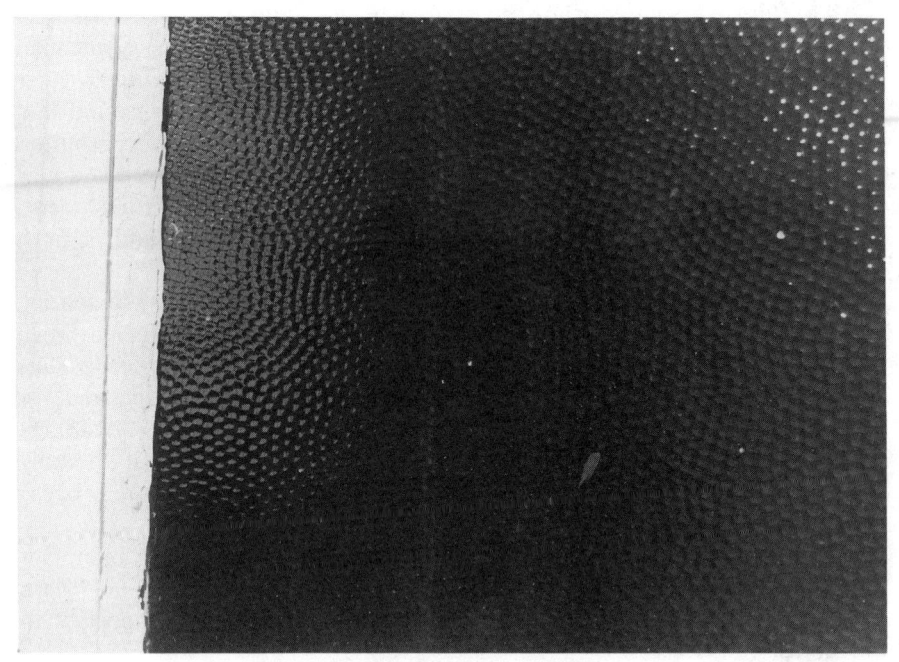

Figure 8-1 One of many available patterned glasses.

Tinted glass is used where it is desirable to reduce light or heat transmission. The degree of transmission differs with color, depth of tone, and manufacturer. Because of the differences between similar products, it is difficult to match a tinted glass except by using the same glass product from the same manufacturer.

Patterned glass is available in a variety of patterns (Fig. 8-1) and in differing degrees of light transmission and obscurity. Patterns may be geometric or random. Only the same proprietary patterned glass that was originally used will exactly match an existing patterned glass. Patterned glass may in addition be etched or sandblasted to increase its opacity, but such glass cannot be used where it will be subjected to stress. Some patterned glass is also tempered.

Wire glass is plate or float glass that has a wire mesh or grid suspended in it to reduce the likelihood of breakage. Wire glass may be flat or patterned and clear or translucent. Its normal thickness is one-fourth inch. It is available as a fire-resistive material complying with UL requirements, but not all wire glass is so classified. Wire glass may be either flat or patterned (Fig. 8-2). The pattern may be on one or both sides. Most mesh in wire

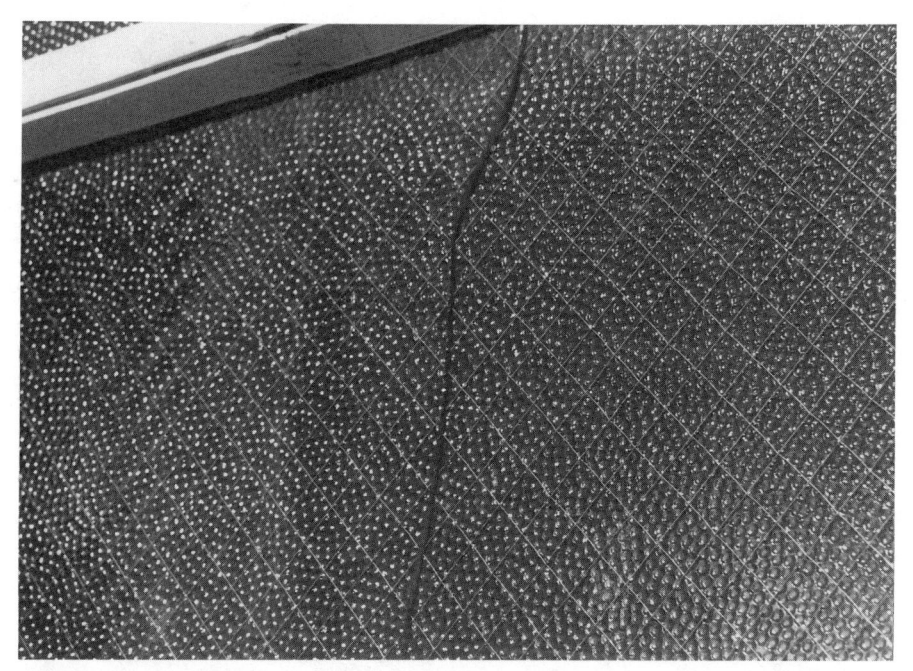

Figure 8-2 A patterned wire glass. The pane shown is broken, but is in no danger of falling from the frame.

glass is either diamond, square, or rectangular in shape, but other patterns may also be found in older buildings (Fig. 8-3).

Heat-Treated Glass. As mentioned earlier, all flat glass is subjected to the initial heat treatment known as annealing. Some glass is then further heat treated to alter its properties. Heat-treated glass is either tempered or heat strengthened. Tempered glass is also called fully tempered. Two manufacturing processes are used in heat treating glass: a vertical, or tong held, method and a horizontal, or roller hearth, process. The process used is usually left up to the manufacturer except when tong marks from using that method would be objectionable in the final work. When tong marks are present, it is usually required that they occur along the short edge of the sheet, unless one dimension of the glass exceeds 72 inches. When the horizontal method is used, parallel roller-wave distortions are generated in the glass. These are usually required to fall parallel with the bottom edge of the glass as it will be installed. Whichever the production method used, the distortions in heat-treated glass can sometimes be so large as to be objectionable. The only sure way to avoid having such distortions is to

Figure 8-3 These photos show two different types of wire glass.

select a particular proprietary product that is known to have distortion within acceptable limits.

Both clear and tinted flat primary glass one-eighth inch or thicker can be tempered; wire glass and patterned glass with a deep pattern cannot. Tempering is a heat-treating process that increases the strength of a glass by about 300 percent. It also alters the glass so that it breaks into small, rounded, relatively harmless pieces. Many national, state, and local codes and ordinances require that tempered glass be used in sliding glass doors, entrances, sidelights, low glazed panels in store fronts and curtain walls, and in other locations where people are likely to contact the glass. Tempered glass is the only kind of heat-treated glass that qualifies as a safety glass.

Heat strengthening is a similar process to tempering, but the glass is only partially tempered. This process makes the glass about twice as strong as annealed glass of the same type. Any glass that can be tempered can be heat strengthened, but its thickness is usually limited to either 1/4 or 5/16 inch. Spandrel glass is usually heat strengthened.

Coated Glasses. Glass can be coated with metal, metal oxide, ceramic materials, or a plastic sheet called an opacifier.

The category of coated glass called high-performance glass has on one side a film of metal or metal oxide. The film used on early coated glass was mainly intended to reduce the air-conditioning load in a building by reducing the amount of sunlight that entered. That glass was coated with a layer of highly reflective metal to make a material called reflective, or mirror, glass. Reflective glass is still produced today, of course, and continues to have its supporters. Unfortunately, reflective glass has three major disadvantages: 1) It severely reduces the amount of light transmitted, making occupants feel as though they are looking out through dark sunglasses; 2) In spite of the large reduction in light transmission, it does not greatly increase the insulating value of the glass; and 3) Its high reflectivity can create problems for its neighbors and even for traffic on the street when reflected bright sunlight blinds people and reflected heat drives up their cooling costs. Some cities now restrict the amount of reflectivity permitted in glass used on buildings.

Most such problems are solved when glass produced using a later version of the same coating technology is used. The new materials employ a different type of metallic coating that can be so thin as to be almost transparent. They produce only slight shading, permitting much light to enter, but reduce the emissivity of the glass. Emissivity refers to the radiant heat absorbed and then emitted by glass. The lower the emissivity, the lower is the amount of heat absorbed by the glass and then reradiated into the building. This form of high-performance glass is called low-emissivity, or Low-E, glass. Low-E glass can reflect as much as 90 percent of the radiant heat that strikes it.

Reflective and Low-E glass are both available in many colors and with many different performance characteristics. The only way to match one of them when it fails is to install another sheet of the same product that was originally installed.

Clear, tinted, and heat-treated plate and float glasses can all be coated. High-performance coatings are applied in one of two ways; pyrolytic deposition or vacuum sputtering.

In the pyrolytic deposition method, hot metal is applied to either the exterior (first) or interior (second) surface of hot glass. The glass is usually tinted. It can also be heat treated after the coating has been applied, but heat treatment may not be necessary when the coating is on the exterior surface of the glass. Safety-glass panels with a coating must of course be tempered. Products made by this method are sometimes called hard-coat. Some producers claim that hard-coat coatings can be exposed to contact with people, but others are not convinced.

Vacuum-sputtered coatings are applied to the interior (second) side of tinted, heat-treated glass. Such glass is usually tinted and either heat strengthened or tempered before the coating is applied. These coatings are sometimes called soft-coat. They cannot be used where they will be exposed to contact.

Spandrel glass is often coated using a ceramic material applied on the interior (second) side. It may also be coated with a high-performance coating that matches the coating used on the building's vision panels. Metal- or metal-oxide-coated spandrel glass is sometimes covered on the spandrel cavity (interior) side with a thin sheet of plastic called an opacifier. The plastic used depends on the construction in the spandrel cavity. When no insulation or other backing material will be applied directly to the spandrel glass, the opacifier is usually polyethylene. When such materials will be applied to the back of the glass, the opacifier is usually polyester.

Laminated Glass Products. Sheets of glass and interlayers of polyvinyl butyral plastic are laminated together under heat and pressure to produce glass products for many special purposes. Depending on the characteristics desired, two or more sheets of glass may be used. The sheets may be of any thickness of glass, and any type of primary, heat treated, or coated glass can be used. The interlayer is usually clear for clear glass and colored for tinted glass but may be tinted for use with tinted glass.

The primary use of laminated glass is as safety glass, but different combinations of glass type and thickness and interlayer thickness produce glass that is used for other purposes. The interlayer thickness for safety glass is usually 0.15 or 0.30 inch thick. The laminated glass used for sloped and other overhead glazing and for sound control has interlayers that are 0.30 or 0.60 inch thick. Burglar-resistant, bullet-resistant, and other security-type laminated glasses have 0.60- or 0.90-inch-thick interlayers.

The major advantage of laminated glass is that when it breaks the shards remain attached to the interlayer material and usually do not leave the opening.

Sealed Insulating-Glass Units. Insulating-glass units consist of two factory-assembled panes of glass with a sealed open space between them. The space is hermetically sealed, dehydrated, and filled with either air or a gas. The edges may be formed using glass or a spacer and sealants. When a spacer is used, the space contains a desiccant. Older insulating-glass units of the kind in which spacers are used were framed on all sides with metal channels. Later, these channels fell into disrepute and are no longer used.

Any type of primary, heat treated, coated, or laminated glass can be used in insulating-glass units. Glass types are usually selected according to their effect on the heat-transmission properties of the unit, but aesthetics and other factors also play a part. The properties of some types of glass may affect the selection of the other glasses used or other properties of the same light. For example, when an insulating-glass unit that will be subjected to high wind loads or thermal stress contains a reflective, Low-E, or heat-absorbing glass light, both lights should be tempered or heat strengthened.

Miscellaneous Glass Types and Their Uses. There are several glass types that are made from the other types. A relatively new one, called bent glass, includes any glass shape made from any type of glass. All the types mentioned earlier in this chapter—even laminated glass and insulating glass—can be bent.

Other types of special glass include art glass, framed and unframed mirrors, and so-called transparent mirrors. The latter are laminated glass products that appear to be mirrors from the lighted side but are transparent from the dark side.

Art glass is used in stained-glass installations with lead, zinc, brass, or other metal cames. It may be clear, translucent, flat, textured, colored, or painted.

8.2.2 Plastic Glazing Materials

Both acrylic plastic and polycarbonate sheets are used in glazing. They are tough, break-, shatter-, and crack-resistant thermoplastics. Compared to glass, plastic glazing materials have higher coefficients of expansion. They are lighter and more resistant to impact and breakage than glass, but are also more susceptible to scratching and abrasion, and are flammable.

Acrylics are harder and more impact resistant than polycarbonates and wear better, but polycarbonates can be used in higher temperatures than can acrylics. Both can be cold formed into curves. Polycarbonate sheets are sometimes used in bullet-resistant applications.

Both acrylic and polycarbonate glazing materials can be coated with an abrasion-resistant material that has a silicone base, to increase their resistance to weather and chemicals. Polycarbonate sheets can also be coated with a polymethyl methacrylate (acrylic) material to reduce the damage normally caused by sunlight and increase their service life.

8.2.3 Miscellaneous Glazing Materials

Glazing sealants, compounds, gaskets, and many other materials are necessary for proper glazing. Cleaners, primers, and sealers that will be used to prepare surfaces for the application of sealants, compounds, and gaskets should be of the type recommended by the sealant, compound, or gasket manufacturer.

Channel-glazing compounds may be either elastomeric or elastic oil- and resin-based glazing compounds formulated for channel glazing into metal or wood, as appropriate. They should be suitable for painting.

Face-glazing compounds should be elastic glazing compounds formulated for face-glazing into wood or metal, as appropriate, and be suitable for painting.

Glazing tape is polybutene or polyisobutylene-butyl tape. Its initial thickness should be 1/16 inch greater than the finished seal will be, to permit compression during application.

Other glazing compounds should be acrylic, single-component materials.

Putty should be oil-based and suitable for painting.

Glazing sealants should be selected for their compatibility with the other materials they will contact, including glass, aluminum and other non-glass panels, seals of insulating-glass units, laminated glass interlayers, and glazing channel substrates, under the conditions of installation and service that they will encounter, as demonstrated by testing and field experience. Sealants should comply with the recommendations of the sealant and glass manufacturers for the selection of glazing sealants that have performance requirements suitable for the applications to be encountered and the conditions at the time of installation. Glazing sealants for structural glazing and glass to glass and other butt glazing should be a chemically curing, one-part silicone type. Structural sealants should be of a type especially formulated for the purpose.

Glazing gaskets should be of the types and materials standard with the manufacturers of the windows, doors, store fronts, or curtain wall systems being glazed. They should be compatible with the surfaces and materials they will contact. Vinyl glazing gaskets are usually the channel type. Neoprene glazing gaskets can be extruded or molded, and should be of the type needed for the particular use and to produce the proper profile.

Setting blocks and spacers should be neoprene, EPDM, silicone, or rubber blocks or continuous extrusions, as required for compatibility with the glazing sealants to be used. For example, neoprene and EPDM cannot be used with silicone glazing sealants. Compatibility must be confirmed by testing, because not all formulations of the same material are compatible. Different silicone formulations, for example, are often incompatible.

Setting blocks and spacers should be of the size, shape, and hardness recommended by the glass and sealant manufacturers for the particular application. Standard shapes will not work in some applications. For example, in sloped glazing systems with structural silicone sealant glazing, the setting blocks must have a tapered edge to permit proper drainage. In all cases, edge blocks should limit the lateral movement (side walking) of the glass.

Compressible fillers (rods) should be a closed-cell or waterproof-jacketed rod stock of synthetic rubber or plastic foam that has been proven to be compatible with the sealants used, is flexible and resilient, and has adequate compression strength for the proper deflection.

Glazing clips should be of the type required by the conditions.

Special glazing compounds for art glass should be those recommended by the glass manufacturer or installer.

Mirror adhesive should be a type formulated for installing unframed mirrors by the spot-application method.

Lead and zinc cames for art glass should be H-shaped sections. Lead cames should be made from lead having approximately one-half of one percent copper content.

Solder for lead should be a 60/40 tin–lead solder.

Lead came weatherproofing compound should be a soft lead putty with a consistency similar to that of margarine. It should contain the proper amounts of linseed oil, kerosene, and lampblack.

8.3 Glazing in New Construction

Watertight and airtight installation of each glazing material is required. Each installation must withstand normal temperature changes, wind loading, and impact loading—for operating sashes and doors—without breakage of glass, failure of sealants or gaskets to remain watertight and airtight, deterioration of glazing materials, and other defects.

The installation should comply with the recommendations of the applicable industry standards and those of the manufacturers of the glass, plastic glazing materials, aluminum or other non-glass panels, window, door, store front, curtain wall system, sealants, compounds, gaskets, and other glazing materials used.

8.3.1 Delivery, Storage, and Handling

Glazing materials should be delivered to the installation site carefully packed, with each light bearing the manufacturer's label showing its strength, grade, thickness, type, and manufacturer. The labels can be omitted, however, when the manufacturer specifically recommends against their use or when they would damage the material.

Each piece of glass or plastic glazing material should be inspected immediately before it is installed. Pieces with observable damage or imperfections should not be used.

Glazing materials should be protected during delivery, storage, and handling in compliance with the manufacturer's instructions and as required to prevent edge damage to glass and plastic glazing materials and damage to glass, plastic, and other glazing materials from condensation and other moisture or temperature changes, direct exposure to sunlight, and other causes. A so-called rolling block should be used when rotating glass units, to prevent damage to corners. Glass should not be struck against metal framing. Suction cups should be used to shift glass units within openings. Glass should not be raised or drifted with a pry bar. Glass should be rotated so that flares or bevels along one horizontal edge that would occur in the vicinity of setting blocks are located at the top of the opening. Glass units should be removed from the project and disposed of when they have edge damage or other imperfections that would weaken the installed glass or impair its performance or appearance.

8.3.2 Preparation for Glazing

The location and conditions where glazing materials will be installed should be inspected and unsatisfactory conditions corrected before glazing is done. The inspection should verify proper design, application, and condition of framing and glazing channel surfaces, backing, and removable stops. It should also verify compliance with manufacturing and installation tolerances, including those for size, squareness, and offsets at corners; the presence and functioning of a weep and water-channel system; the existence of the minimum required face or edge clearances; and the effective sealing of joinery.

Surfaces to receive glass and plastic glazing materials or non-glass panels should be cleaned immediately before glazing. Coatings that are not firmly bonded to the substrate should be removed. Paint, lacquer, strippable coatings, and tape must be removed from metal surfaces where elastomeric sealants will be used. Metallic, ceramic, and other coatings and opacifiers must also be removed from surfaces to receive structural glazing sealants.

Ventilating sashes should be adjusted properly before they are glazed.

8.3.3 General Glazing Requirements

The glazier should review glass thickness, opening size, glazing method, elevation of the glazed opening above the exterior grade, and other factors affecting the structural performance of a glazing material, and verify the correctness of the material for its specific use. If thickness, edge retainage, or other factors are considered incorrect or the glazing material improperly used, the glazier should forward to the owner the manufacturer's recommendations before permitting the fabrication of new glazing materials or the reinstallation of removed glazing materials.

Glazing materials should not be installed under adverse job conditions or weather. The ambient and substrate temperatures should not be higher or lower than the limits permitted by the glazing material's manufacturer, and the substrates and glazing channels should not be wet from rain, frost, condensation, or other causes. Liquid sealants should be installed only when temperatures are within the lower or middle thirds of the range recommended by their manufacturer.

Glazing includes the installation of glazing beads for wood and metal doors, windows, and store fronts, and glazing beads and gaskets for aluminum doors, windows, store fronts, and curtain walls that are furnished by the manufacturers of those items, except, of course, when those items have been factory glazed.

Wood and metal doors and windows may be either factory preglazed or field glazed. Most windows are factory glazed.

The design of a glazing channel (Fig. 8-4) must provide the necessary minimum bite and edge clearance for the glazing material and allow for adequate sealant thicknesses, within reasonable tolerances. Some single-thickness glass can be cut in the field to adjust the edge clearance, but tinted, heat treated, laminated, and insulating glass cannot be altered in the field. In some designs the stops can be adjusted to provide the correct sealant thickness, but most metal doors, windows, and framing do not permit such relocation. Ultimately, the glazier is responsible for providing the correct glass size for each opening, within acceptable tolerances, but

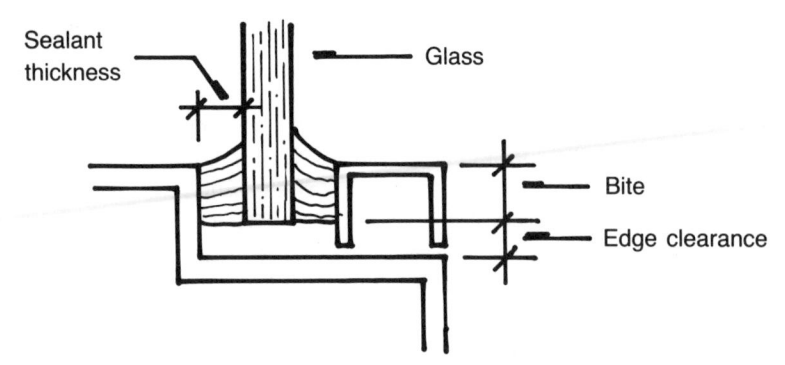

Figure 8-4 A typical glazing channel.

realistically many of today's glazing materials leave little latitude for adjustment. Usually, the design must be correct in the first place.

Glazing systems must be designed so that moisture does not accumulate in the glazing channels. The system and associated sills must be weeped outdoors.

Glazing materials must be compatible with the surfaces contacted.

Tinted, tempered, laminated safety, and insulating glass should be cut or manufactured to size at the factory, and should not be ground, nipped, cut, or fitted after fabrication. Such glass should not be subjected to springing, forcing, or twisting during setting or be permitted to strike against frames.

Glass and plastic should be fabricated to the sizes required, with edge clearances and tolerances complying with their manufacturers' recommendations. Thicknesses should be as recommended by the manufacturer for the particular application.

Glass clearance dimensions should be based on the type and thickness of the glass, as recommended in the Flat Glass Marketing Association's *Glazing Manual* or by the glass manufacturer.

Insulating glass should be installed with tinted glass toward the exterior and coated glass as the outer light. Opaque glass should be installed with the coated side toward the interior.

Glazing compounds for use with insulating glass, laminated safety glass, or transparent mirrors should not be thinned with chlorinated solvents (drycleaning fluids) or benzene-related compounds such as toluene. Neither should such materials be permitted to come in contact with those types of glass.

The edges of laminated safety glass and transparent mirrors should be sealed when and as recommended by their manufacturer.

A primer or sealant should be applied to joint surfaces when so recommended by the sealant manufacturer.

Primers should be applied to joint surfaces when required for adhesion of structural sealants, as determined by a preconstruction sealant–substrate test.

Setting blocks of the proper size should be installed in the sill rabbets beneath glass. They should be located one-quarter of the width of the glass from each corner, but usually no closer than six inches from a corner. The blocks should be set in a thin course of a sealant formulated for the purpose.

Except where gaskets or preshimmed tapes are used for glazing, glass larger than fifty united (width plus height) inches should be held in place by spacers of the proper size on both sides of the glass, at the proper spacing for the glass size. The spacers should bite at least 1/8 inch on the glass and should be of a thickness equal to the desirable sealant thickness for liquid compounds, but slightly less than the compressed thickness of tape sealants.

Edge blocking should comply with the requirements of the FGMA's *Glazing Manual*, except where the glass unit manufacturer's requirements are more stringent.

Glass units in each series should be set with uniformity of pattern, draw, bow, and similar characteristics.

Compressible filler rods or an equivalent back-up material, as recommended by the sealant and glass manufacturers, should be installed to prevent sealant from extruding into the glass-channel weep system and from adhering to the back surface of joints as well as to control the depth of the sealant for optimum performance.

Wet sealants should be forced into the glazing channel to eliminate voids and ensure complete "wetting" or bonding of the sealant to the glass and the channel surfaces.

The exposed surfaces of glazing liquids and compounds should be tooled to provide substantial "wash" away from the glass. Pressurized tapes and gaskets should protrude slightly out of the channel, to eliminate dirt and moisture pockets.

Excess glazing materials should be removed from glass, stops, and frames promptly after installation. Stains and discolorations should also be eliminated.

When wedge-shaped gaskets are driven into one side of a glazing channel to pressurize a sealant or gasket on the opposite side of the glass, adequate anchorage should be provided to ensure that the gasket will not "walk" out when the installation is subjected to movement. Gaskets should be anchored to their stops with matching ribs, a proven adhesive, or by embedding the gasket's tail in a cured heel bead of sealant.

The ends of wedge-shaped channel-glazing gaskets should be miter cut at the corners and the cuts bonded together using a sealant recommended by the gaskets' manufacturer so that the gaskets will not pull away from the corners and produce voids or leaks in the glazing system. Gaskets should not be joined except at their corners.

Glazing sealants and compounds should be cured in compliance with their manufacturers' instructions and recommendations, to obtain high early bond strength, internal cohesive strength, and surface durability.

Wire glass should be installed in accordance with NFPA 80. Square- and rectangular-mesh wire glass should be installed with the wires parallel with the stops, with the mesh running all in the same direction, and with the wires in line on opposite sides of mullions.

8.3.4 Specific Glazing Requirements

Glazing with Glass into Steel. Hollow metal and steel should not be glazed until abrasions have been touched up and the first field coat of paint has been applied and allowed to dry.

In steel windows, glass is often installed using face glazing. Glazing stops are more often used in hollow metal doors and frames.

When stops are used, shims and setting blocks should be provided and the glass bedded in channel-glazing compound, tape sealant, or gaskets. The glass should be held in place with metal glazing clips, channels, or angles. The glazing beads should then be installed with tape, gaskets, or a full ribbon of compound applied against the glass. A countersunk screw should be placed in each hole of the bead and tightened.

Glazing into Stainless Steel, Aluminum, and Copper Alloys. In stainless steel, aluminum, and copper-alloy doors and store fronts, and in aluminum windows and curtain walls, glass should be set using compressible gaskets and tape when those items have been furnished by the metal item manufacturer. The recommendations of the manufacturer and the applicable industry standards should be followed.

Openings in such metal items that have not been provided with glazing gaskets should be glazed using polyisobutylene-butyl tape against a fixed stop, with a heel bead around the perimeter of each light and an elastomeric sealant between the glass and a removable stop.

Glazed openings must comply with the performance requirements of the doors, windows, store fronts, and curtain walls. Refer to Chapters 4 through 7 for such requirements.

Glazing into Wood. Wood should not be glazed into until it has been primed or sealed and permitted to dry. The glazing surfaces must be clean, dry, and free from oil and dust.

Glass may be set in wood using wood beads (stops) and glazing tapes, liquid glazing compounds, or gaskets, or set without stops, by face glazing.

Where wood beads are used, they should be installed neatly and accurately.

Conventional Sloped Glazing. Sloped glazing is any glazing that is sloped more than fifteen degrees from the vertical. Conventional sloped glazing systems are constructed using metal framing members and a glazing material. The glazing material is usually set using gaskets and locked in place with a cap along each framing member. Refer to Chapter 7 for a discussion of sloped-glazing framing systems.

The selection of glazing material for sloped glazing has been controversial in the past, but in the last several years the following general consensus seems to have formed. Except for those normally associated with wind, snow, and other load resistances, there should be no restrictions on the glazing material when the area beneath the glass is inaccessible or when permanent protection is provided to prevent falling glass from reaching people below. A net, however, is not considered a permanent protection when the glass is a single pane of float glass. Some tests have shown that even tempered glass will break through a net. Single-thickness float glass is inappropriate in sloped glazing, but laminated glass with a minimum 0.03-inch-thick interlayer, wired glass, and plastic glazing materials may be used there. Tempered glass may be used up to ten feet above the floor if the slope does not exceed 30 degrees from the vertical. Insulating glass may be used, provided that the inner light is one of the acceptable glazing materials mentioned above. Wired glass is not appropriate in insulating glass, however.

The preceding restrictions apply to sloped glazing in buildings other than single-family residences. In houses, which are beyond the scope of this book, the requirements are less stringent.

Glass-to-Glass Joints. Two pieces of glass that abut end to end or at a corner without other support should be sealed together using beads of silicon sealant following the sealant manufacturer's recommendations. The surfaces of abutting glass panes in the same plane should finish flush. Corners should be square and plumb.

Butted glass joints that are backed up by a supporting mullion are called structural silicone-sealant glazing.

Structural Silicone-Sealant Glazing. A structural silicone-sealant glazing system is one in which one or more edges of the glass are bonded to metal or glass supports by a structural-grade silicone sealant. There are three basic types of structural silicone-sealant glazing systems: 1) those with silicone sealant supporting two sides of the glass; 2) those with structural silicone sealant supporting four sides of the glass; and 3) those with glass mullions, which are sometimes called fins. There are also variations within those types.

In two-sided and four-sided systems the glass is supported by metal—usually aluminum—mullions. At the silicone-supported joints the backup metal framing members are not usually visible on the exterior of the building. In a two-sided system the two sides that are not supported by silicone sealant are supported using more or less conventional channel-glazing methods. The silicone-supported joints may be either vertical or horizontal. In four-sided systems all four sides of the glass are bonded to the backup mullions by silicone sealant. In both systems butt joints in the glass are made using silicone sealants, but the butt-joint sealant may not always be a structural formulation.

Systems with glass mullions, called *fins,* also include metal—usually stainless steel plates—to anchor the glass panes to the mullions. The panes are sometimes held in place by metal plates supported by metal angles or other shapes that are in turn attached to the building's framing system. Except where horizontal glass fins occur, the silicone sealants in the horizontal joints between glass are not usually of the structural type. These joints are held together by metal clips.

The glass for use in structural silicone-sealant glazing systems must be carefully chosen. When proper precautions are taken, any of the glass types discussed in section 8.2.1 may be used. Insulating glass must be constructed using a structural silicone sealant as the secondary sealant within the glass unit. Refer to section 8.4 for an indication of the problems that can develop when the wrong glass is selected or the glass is improperly installed.

Glazing with Plastic. Each sash opening should be accurately measured and plastic sheets cut to fit. The size of each plastic sheet should be in accordance with the manufacturer's recommendations, to produce the proper bite and edge clearance.

Plastic sheets should be unmasked before they are installed but still be protected during and after installation. Excess sealant smears should be prevented by applying a paper-backed adhesive tape around the edges of the sheets, adjacent to the glazing rabbets.

The edges of each plastic sheet should be thoroughly cleaned immediately before it is installed using naphtha or another solvent recommended by the plastic sheet manufacturer. The sheets should be set using the sealant recommended by the manufacturer of the plastic.

Unframed Mirrors. Frameless mirrors should be installed with a combination of stainless steel or chrome-plated C-shaped clips and mirror adhesive, to provide a tamperproof installation.

Mirrors should come from the factory with a coat of moisture-resistant paint applied to their backs. Before they are installed they should be given an additional coat of the same paint and allowed to dry. Then about 25 percent of the back surface of the mirror should be covered with mirror mastic and the mirror should be set on the C-shaped clips and pressed against the substrate to ensure bond. A ventilation space approximately 1/8-inch thick should be left between the mirror and the substrate. This space should be left open at the top of the mirror.

The mirror should be given additional support at its bottom until the adhesive has set.

8.3.5 Cleaning and Protection

Labels should be left on single-thickness glass until it has been set and initially inspected, unless the glass manufacturer specifically recommends against using labels. The labels should be removed from insulating glass immediately after its installation and inspection, to prevent thermobreakage and staining.

Glass should be protected from breakage immediately upon installation, by the use of crossed streamers attached to the framing and held away from the glass. Markers should never be applied to the surfaces of glass or plastic glazing materials.

Glazing materials should be prevented from contacting contaminating substances resulting from construction operations. If, despite such precautions, contaminants come into contact with glazing materials, they should be removed immediately, using the methods recommended by the glazing materials' manufacturers.

Glazing materials should be removed and new materials installed when they are broken, chipped, cracked, abraded, scratched, or damaged in any other way during the construction period, whether the damage was due to natural causes, accidents, or vandalism.

Glazing materials should be kept reasonably clean during construction so that they will not be damaged by corrosive action or contribute by wash-off to the deterioration of glazing materials or other surfaces.

Glazing materials should be thoroughly washed and glass units polished on both faces. Labels, paint spots, and other defacements should be removed. Cleaning procedures and materials should be those recommended by the glazing materials' manufacturers.

8.4 Why Glazing Fails

Glazing materials fail for many reasons, most of which can be grouped into three categories. The first affects structural silicone-sealant glazing systems of the type in which the glass is supported directly from the structure. This category consists of problems with the surrounding or supporting construction. It includes excess structure movement and failed steel or concrete structures, which are discussed in sections 2.1 and 2.2. If the structure moves excessively or fails, the failure of a supported glass-wall system is inevitable.

The second category affects all other types of glazing systems. It has to do with the doors, windows, store fronts, and curtain wall systems in which the glazing materials are installed. When these items experience failure, the glazing materials in them will almost surely be affected. The extent of the effects is dictated by the extent of the damage to the supporting item. The types of failures that might occur are discussed in sections 4.11, 5.3, 6.8, and 7.3. When a glazing material has failed, the most likely cause is a problem with a supporting item, so those should be investigated first.

The third category includes causes of failure that are related directly to the glazing material or its installation. These causes include bad materials, improper design, inappropriate type selection, improper fabrication, improper installation, failure to protect glazing, failure of the immediate substrate, failure to properly maintain glazing, and natural aging. Those failure causes are discussed in this section.

8.4.1 Bad Materials

Improperly manufactured glass, glazing sealants or compounds, or related materials are a potential source of failure. An improperly formulated silicone sealant may not cure properly, for example. Occasionally a pane of glass may break due to inclusion of foreign materials. Material failures are, however, rarely the cause of problems with glazing.

Ripples or waves in sheet glass are not imperfections in the true sense of the word, because they are inherent in the material. Such appearance in sheet or float glass may be an imperfection in the material, but is more likely to have been caused by a design or installation error.

Glass that arrives at the installation site with rough edges or cracks and scratches along its edges should not be used. Such imperfections will often cause the glass to crack after it has been installed.

Delamination or discoloration in coated glass may be caused by improper coating application procedures, but are more likely to have been caused by improper handling.

Improperly fabricating insulating glass units may allow outside air to

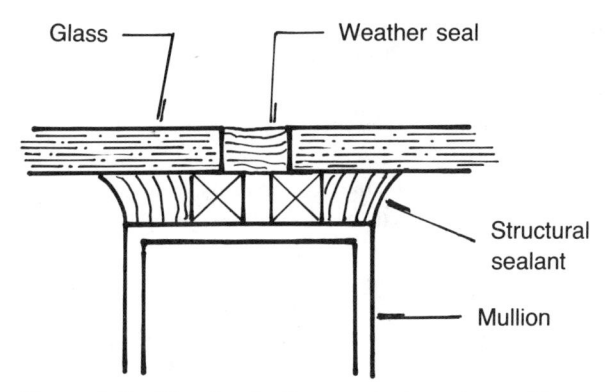

Figure 8-5 Structural silicone glazing at a mullion.

enter or the gas to leak out of the cavity between the glass panes. The result will be condensation within the cavity and discoloration of the coatings on the glass. Once the seal has failed, there is no corrective measure that can be taken short of refabricating the unit. In effect, this means that such failed units must be removed and new units installed.

Heat-treated glass will sometimes distort an unacceptable amount during the treatment process. This distortion becomes even more apparent if the glass is later coated. The only way to avoid this problem is to select specific products that are known to be free of excess distortion. Once a distorted pane has been installed, the only solutions are to live with the problem or remove the offensive pane and install a new one.

8.4.2 Improper Design

A serious possible design error is selecting a glass type that is too weak for the loads to be applied. This error includes using a glazing material that is too thin to support the loads or that deflects too much for the conditions to be encountered. Excess deflection can cause a glazing material to slip out of its glazing channel or break the glass.

A related error is installing a glazing material in a rabbet that is too shallow.

In a structural silicone-sealant glazing system, selecting a sealant that is too weak and using a metal mullion that is too narrow for the sealant selected are similar problems. In either case, insufficient bond will result in failure of the sealant. To help prevent this design error, some sealant producers recommend that only the sealant between glass and metal be calculated to resist shear. Then the sealant between glass and glass is considered only a weather seal (Fig. 8-5). Sometimes a different sealant is used to glue glass to mullions than is used in glass-to-glass joints.

Requiring that butt joints be used in tinted glass will often produce an unsatisfactory appearance. No two sheets of tinted glass will ever exactly match in color and shade.

Failing to allow for enough edge clearance or glazing material bite in the glazing channel can lead to several kinds of failures, including glass loss and bowed glazing panels.

Using glass that is inefficient thermally may result in cold spots and a poorly functioning heating and ventilating system. It is almost impossible to properly balance a mechanical system when the building's glazing permits excessive heat transmission.

Failing to include appropriate condensate drainage channels and weeps in the supporting framing for structural silicone-sealant glazing can result in delamination of laminated glass along the edges and to damaged insulating-glass units.

Improper design of spandrel cavities is a continuing source of problems in many curtain wall installations. Refer to section 7.3.1 for a discussion of this matter.

Sealants that are not adequate for the job may pick up dirt and become discolored.

Glass cracking is sometimes caused by differential expansion in a pane of glass. This condition is more likely to occur when the glass is not properly isolated from its frame or is glazed directly into concrete or masonry, but may also occur under other circumstances. Regardless of the condition that caused the problem, when the center of a glass pane heats up faster than the edges, the glass expands more at the center than at the edges, building up stresses in the glass. The glass may then break if there are even slight imperfections along its edges or if a notch or hole has been cut in it. To prevent such breakage the glass should be properly isolated from its surrounds, the edges of the panes must be strong and cleanly cut with no burrs, and there must be no notches or holes in the glass. Thermal cracking is more likely to happen when the glass is unevenly shaded by surrounding construction (Fig. 8-6). The worst case is when 25 percent or less of a pane is shaded and the shaded area includes more than 25 percent of the perimeter. Most major glass manufacturers' literature contains specific recommendations for acceptable shading patterns for particular types of glass.

The painting of signs or placing of large labels on glass can also cause thermal stresses that result in glass breakage.

Unaccounted-for thermal expansion and contraction can also cause cracking or glass loss when panes are abutted by other glass or fastened to metal using structural silicone sealants. Careful consideration must be given to closely matching the materials to be joined, providing the proper

8.4 Why Glazing Fails   333

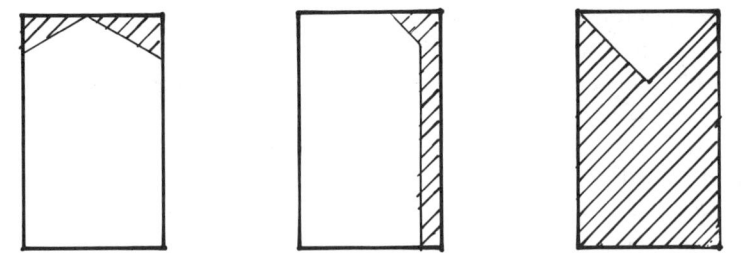

Figure 8-6 Examples of potentially harmful shading patterns.

width of joints, using sealants with the proper flexibility, and otherwise providing for differential thermal movement.

Heat traps on the interior side of a glass light can also cause cracking. Heat traps result from insufficient air circulation at the glass. This condition often occurs behind spandrel panels, but can also happen when the interior side of the light is exposed, if there is insufficient circulation across the glass. Deeply recessing the glass on the interior side without providing positive ventilation in the recess can cause this condition. Draperies, shades, and blinds that are placed less than 2 inches from the glass or less than 1-1/2 inches from the frame can also create a heat trap between the shading item and the glass. Positioning heat outlets beneath a glazed panel, as is often done, can contribute to heat-trap creation, especially if the outlet directs air into the space between the glass and a blind, shade, or drape.

Failing to back up spandrel-glass panels with an opacifier or other opaque material will result in the materials in the spandrel cavity becoming visible under some lighting conditions. Even reflective glass can be transparent when the light is right.

Unfortunately, solving the problem of visible materials in a spandrel cavity can itself become a problem, unless proper precautions are taken. For example, one way to prevent the material in the cavity from being visible is to use what is known as shadow-box construction. This technique consists of installing a layer of dark-colored rigid insulation behind the spandrel glass. If the shadow box does not contain a properly constructed vapor retarder, water will condense on the inner face of the spandrel glass and freeze there. The water and ice can cause deterioration or delamination of the opacifier or other coating on the spandrel panel. Even when deterioration does not occur, the water or ice can become visible from the exterior. A secondary problem is that the natural deterioration of some kinds of insulation, notably foams, produces volatile materials that can cause deterioration of the coatings on the spandrel panels.

Painted opacifiers were once used and may still be used today, but except in a few rare cases, their success rate has been low. The paint will usually deteriorate quite rapidly and crack or peel, often over the major portion of a panel. Plastic sheet opacifiers have a better record, but they can also fail. Those installed with a water-based adhesive are susceptible to loss of bond, especially in cold climates. Polyethylene or polyester-sheet opacifiers installed with a solvent-based adhesive have enjoyed more success. Nevertheless, some experts recommend against using an opacifier to solve the problem of the visibility of materials inside spandrel cavities, unless there is no other practicable solution.

8.4.3 Inappropriate Type Selection

Selecting insulating glass units having standard interior sealants for use in a structural silicone-sealant glazing system can lead to loss of the exterior glass light and possible failure of the entire system. The interior sealant in insulating glass units used in structural glazing must be structural silicone.

Some manufacturers recommend using heat-absorbing glass to help prevent breakage due to thermal stresses. Some suggest using laminated glass, others tempered glass, because it is several times stronger than annealed glass. Which type to use is controversial. Some experts say that using tempered glass where it will be subject to solar heat, especially in overhead locations, is bad, because that material has a tendency to break spontaneously under some conditions. The problem is caused by the inclusion of undetectable foreign materials that become embedded in the glass during manufacture. This condition cannot be truly called a defect in the glass, since it is only a problem when the glass is subject to high temperatures. Tempered glass is quite acceptable for locations where it will not be subjected to thermal loading.

The tempered glass supporters believe that no convincing argument has been presented that would demonstrate a valid reason to avoid tempered glass in overhead applications. They cite statistics showing that no injuries have resulted from tempered glass falling on people. They say that when laminated glass is subjected to large impact it will fall out too, because it will buckle.

We cannot resolve the controversy here, because both arguments have merit and neither type of glass has caused significant damage when used in overhead applications. In short, there is little evidence that either side is wrong. Different glasses seem best for different conditions. For example, tempered glass is more likely than laminated, heat strengthened, or even annealed glass to leave the opening when broken by flying roof aggregate or other debris in a windstorm, but it is much less likely to be broken by impact with people or large projectiles. Robert W. McKinley's 1986 article "Windloading of Glazed Curtain Walls" is a good discussion of the effects on different glasses of wind loads and wind-driven particles.

8.4.4 Improper Fabrication

Improperly fabricated glass can fail. For instance, when glass of a type that cannot be cut in the field has been made too large for its opening, it will not be able to expand. The result may be distorted or broken glass.

Improperly fabricated laminated glass may delaminate. The glass need not completely separate for the condition to be harmful. Partially separated sheets may show a mottled appearance. When a partially delaminated sheet breaks, the interlayer will not hold the shards in the opening, thus defeating a major reason for using laminated glass.

Improperly fabricated edge seals on insulating-glass units will result in condensation and fogging inside the cavity, with a resultant loss of insulating value.

Improperly fabricated gaskets will not seal properly.

Using damaged glazing materials, or damaging them during or after fabrication, must be avoided. Damaged materials should be discarded and new materials used.

8.4.5 Improper Installation

The primary cause of failure due to installation errors is not following the design and the recommendations of the manufacturer and recognized authorities. Failing to follow the design can mean using the wrong type of sealant, compound, or gasket for the condition. A related error is using the wrong accessories or ones made from the wrong materials. For example, neoprene and EPDM setting blocks and spacers are not compatible with silicone sealants and may cause the silicone to become discolored or lose its properties.

Improperly handling glass panes can result in chips and other damage along their edges. Such conditions may not be apparent at first, but they may later become cracks when thermal stresses occur in the glass. Such edge damage often occurs when a glass panel is rotated, or pitched, on its corner on a hard surface. To prevent such damage a device called a rolling block should be used to rotate glass.

Failing to remove oils and other fluids used in the cutting process before installing a glazing material can result in the glazing compound or sealant's not adhering to the glazing material.

Installing glazing materials onto improperly prepared substrates and adjacent construction or onto wet materials can cause many problems. Glazing compounds and sealants may not bond to a wet substrate. Water may deteriorate the plastic interlayer in laminated glass, damage the edge seals of insulating glass, or cause the coating on coated glass to delaminate from the glass.

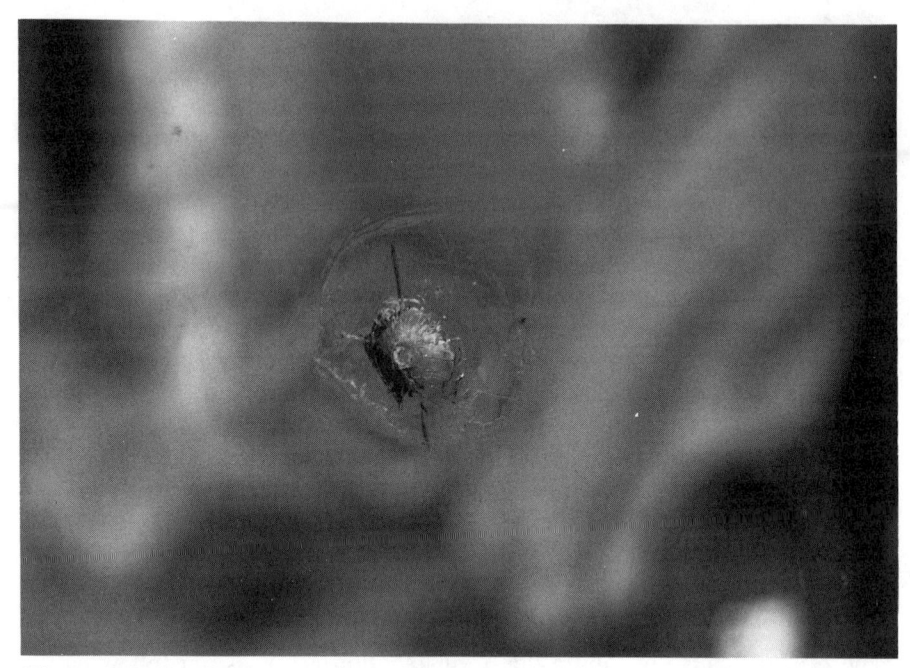

Figure 8-7 Vandalism was the culprit in the damage shown in this photo.

A related problem is permitting concrete or mortar dust to enter and remain in glazing channels. The lime from such materials will etch glass and can cause coatings to delaminate. Similarly, permitting water from concrete or liquids used to clean masonry to flow onto glass can etch and permanently discolor the glass.

Installing damaged glazing materials is a common error. While the damage may or may not be the fault of the glazier, installing damaged glazing material is.

Installing a glazing system in such a way that loads other than the weight of the glazing material are transmitted into it will surely cause the glazing material to fail.

Using old or the wrong types of compounds, sealants, or gaskets and improperly installing them are major causes of leaks. Gaskets must be butted tightly and be of the proper length to completely fill the spaces provided, without gaps of any kind. Sealant joints must be properly formed, using the appropriate sealants. Old gaskets and sealants will not function properly and will probably contribute to leaks.

Placing incompatible products in contact without compensating for their incompatibility will usually lead to problems. Some types of glazing com-

Figure 8-8 The glass pane shown penetrated in this photo has not yet gone through a complete heating–cooling cycle, but the sealant around the pipe has already begun to crack.

pounds and sealants will not bond to some types of coatings, for example. Some sealants will bond only when the coating is cleaned in a special way. Some types of sealants will cause the edges of laminated glass to delaminate. The thinning of glazing compounds using chlorinated solvents (dry-cleaning fluids) or benzene-related compounds such as toluene, or allowing such materials to come in contact with laminated glass, can cause deterioration of the plastic interlayer and subsequent failure of the unit. Using such materials in contact with insulating glass can damage the edge seals, to the extent that they no longer function properly.

8.4.6 Failure to Protect and Properly Maintain Glazing

Failing to maintain and reapply glazing compounds, sealants, and gaskets when necessary can result in leaks and eventual glass loss. To prevent this a regular maintenance schedule should be adopted and followed. Some experts recommend that a detailed inspection be made six months and again

Figure 8-9 Sooner or later, someone will bump the pipe connected to the hood shown in this photo and break the glass.

one year after a glazing system has been installed, then be repeated once a year thereafter.

The natural deterioration of liquid glazing compounds is a particular problem that must be dealt with regularly by removing old glazing compounds and reglazing the curtain walls. Similar problems may occur with plastic glazing gaskets and stops, even though they will probably last longer than will liquid glazing compounds.

Glazing sealants and compounds that are failing often show evidence of their impending failure that should not be ignored. Silicone glazing may leach oil before it fails, for example. The oil can often be seen running down the face of the glass and puddling on sills.

No amount of protection can prevent some types of damage. Vandalism, for example, is usually not preventable (Fig. 8-7).

8.4.7 Improper Modifications

Glass can be made to not perform properly or leak water or air when it is improperly modified after installation. The types of modifications shown in Figures 8-8 and 8-9, for example, will probably eventually lead to problems.

Figure 8-10 A temporary patch?

Some strange modifications, such as that shown in Figure 8-10, are not really harmful, but they are certainly unsightly. One would hope that they would eventually be corrected.

8.5 Glazing in Existing Construction

The requirements discussed in sections 8.1, 8.2, and 8.3 also apply when the glazing materials will be installed in existing construction. There are, however, some additional considerations that must be taken into account, many of which are addressed in this section.

Since the number of possible glazing conditions in existing buildings is large, it is not possible to cover the requirements for every possible one here. This discussion is therefore generic in concept and should be viewed in that light. These suggestions may not apply to all conditions, but their principles should be applicable to most. In the end there is no substitute for the application of professional judgment.

8.5.1 Standards, Controls, Warranties, Performance Requirements, Delivery, and Handling

The standards and controls in sections 1.6.1, 1.7.1, 8.1.1, and 8.1.2, the warranty requirements in section 8.1.3, and the performance requirements in section 8.1.4 all apply to glazing materials installed in existing construction.

In addition, existing glazing materials should be removed by the glazier who will install the new materials. Only experienced personnel should be permitted to do any of this work.

8.5.2 Glazing Materials and Accessories

With a few exceptions, the glazing materials and accessories used in existing buildings and those for new construction are essentially the same. Therefore, most of the requirements discussed in section 8.2 are applicable also to glazing in an existing building.

Unless a new material is desired, a glazing material removed from an opening can be reinstalled in the same opening, if it is in acceptable condition. New materials must be furnished, of course, when a removed glazing material is not fit for reinstallation or a different type of material is wanted.

New glazing materials that will be installed where an existing glazing material has been removed must usually be of the same thickness as the removed material, although the type of material may vary. A tinted glass may be used, for example, where the original glass was clear. A glazing material with a thickness different from that of the original may also be installed, of course, but some remodeling of the opening will then probably be necessary. It may not be possible to install radically different glazing materials in the same opening. For example, a door or window that was designed to receive single-thickness glass probably cannot be altered sufficiently to receive one-inch-thick insulating glass. It may well be possible, however, to modify a store front or curtain wall section to accommodate such large differences in glazing thickness. Sometimes all that is necessary is to use a different glazing bead. Some store front and curtain wall sections offer the option of using either thick or thin glass.

Using different material from that in the surrounding panes will often

Figure 8-11 The original patterned glass in this photo has been replaced with clear panes.

produce an unsightly condition. The use of clear glass where patterned glass was broken in Figure 8-11, for example, looks strange.

Sometimes a different material is used as a temporary patch (Fig. 8-12). Such materials should be removed as soon as possible and a proper repair made.

8.5.3 Preparation for Glazing

Just as preparations are necessary before glazing can be done in new construction, so is it necessary to make them before glazing openings in an existing structure. The preparations are similar in both cases, but existing conditions may require that additional procedures be performed.

The requirements for demolition and removal of existing items and for renovation work discussed in sections 1.6 and 1.7 are applicable to glazing. The requirements in this section are in addition to the ones given there.

As is true when the glazing and building structure are new, the location where glazing will take place in an existing building should be examined carefully by the glazier. Should conditions be encountered that differ sig-

Figure 8-12 The patch in this photo was made from two pieces of plastic glazing material and a pop rivet.

nificantly from those expected, the glazier should immediately notify the other concerned parties, especially the owner, so that alternative methods can be developed for handling the unexpected situation.

Glazing materials should be removed when their sealants, compounds, or gaskets have deteriorated or need to be replaced for other reasons; when the glazing material is broken, cracked, or otherwise damaged to the extent that it is not acceptable; and when a different type of glazing material is desired. When a new glazing material will be installed, the existing one must be completely removed and disposed of, along with associated glazing sealants, compounds, gaskets, and accessories.

Existing glazing materials that are not to be removed must be protected.

Damaged existing materials that are to remain in place, should be removed and new material installed.

Removed existing glazing materials that will be reinstalled should be protected from harm. Removed art (stained) glass and damaged cames must be delivered to the shop for reworking. Cames damaged beyond repair should be discarded and new ones provided.

8.5.4 Glazing

The recommendations in section 8.3 also apply to glazing in an existing building. When the doors, windows, store fronts, or curtain walls that are to receive the new glazing materials are new, section 8.3 applies completely. When existing glazed items are to receive new glazing materials, some additional requirements apply. The suggestions in this section are in addition to those in section 8.3.

During the course of work related to the removal of existing materials and new glazing, existing materials and items that are not to be removed and discarded should be protected from harm. Where such existing materials and items become damaged, they should be repaired to a condition at least equal to that existing before the damaging. Where making acceptable repairs is not possible, the damaged materials and items should be removed and new ones installed.

The methods and materials used for new glazing should be the same as those in the original, unless the glazing material's manufacturer recommends another method or the glazier or contractor will not warrant the installation if the existing method is used. If another method is proposed, the details of the proposed substitute methods and materials should be submitted to and approved by the owner before the installation is made.

Spandrel areas require special consideration. Refer to sections 7.3.1 and 7.5.4 for further discussions.

Art (stained) glass in lead, zinc, copper, or another metal cames should be repaired only in a remote specialty shop or a similarly equipped jobsite shop, and only by professional workers skilled in such work. Each panel should be inspected thoroughly. Repairs are needed when art-glass panels are not airtight; have damaged, cracked, stretched, or distorted cames; have damaged or deteriorated came weatherproofing, broken or unencased glass panels, deteriorated, distorted, cracked, broken, or otherwise damaged glass; or have broken or cracked soldering at joints. Cames that are not damaged should not be removed or disturbed unless doing so is necessary to install replacement glass. Where the damage is minor, minor adjustments should be made as necessary to make a sound installation.

Art glass with historic or other value beyond its normal worth is often reclaimed, even when it is cracked. The normal method is to add a came along the break, but if the light is protected, a high-performance adhesive can be used to glue the glass pieces together. Shattered glass, of course, must be discarded and new glass provided, regardless of the value of the glass.

In general, new art glass cames and glass should be installed and removed glass reinstalled using the same methods and materials as those used in the original installation, unless other acceptable methods and materials are recommended by the glazier.

More information about installing and repairing art glass in metal cames and a list of local distributors of caming materials can be obtained from the Chicago Metallic Corporation by calling them at (312) 563-4600.

8.6 Where to Get More Information

The AIA Service Corporation's *Masterspec* Basic Version sections marked with an [8] in the Bibliography contain requirements related to glazing into doors, windows, store fronts, and curtain walls.

Thomas A. Heinz's 1989 article "Use & Repair of Zinc Cames in Art-Glass Windows" contains basic information about the subject. It is a good starting point for someone with art-glass problems.

John A. Raeber's 1989 article "Exposed Metallic Coatings on Glass: A Cautionary Note" is an excellent description of some of the potential causes of stains on reflective glass when the coating is exposed. It also addresses some other types of glass damage caused by something other than problems with the glass itself. This article is recommended reading for anyone who has to deal with the staining of glass that has an exposed metallic coating.

Julie L. Sloan in her 1983 article "A Stained Glass Primer" offers some sound advice about dealing with problems related to art glass.

M. Stephanie Stubbs's 1986 article "Glued-on Glass" is an excellent introduction to structural silicone-sealant glazing systems.

Noel Valdes's 1988 article "Low-E Glass" is an excellent discussion of the purpose, production, and use of such glasses.

Refer to sections 4.14, 5.6, 6.11, and 7.6 for additional sources of information about doors, windows, store fronts, and curtain walls. Many of the additional sources of data listed in those sections also contain requirements for glazing into the items addressed in Chapter 4 through 7.

Refer also to other entries in the Bibliography that are marked with an [8].

APPENDIX

Data Sources

NOTE: The following list includes sources of data referenced in the text, included in the Bibliography, or both. **HP** following a source indicates that that source also contains data of interest to those concerned with historic preservation.

Adhesive and Sealant Council (ASC)
1627 K Street, N.W., Suite 1000
Washington, DC 20006
(202) 452-1500

Advisory Council on Historic
 Preservation
1100 Pennsylvania Avenue, Suite 809
Washington, DC 20004
(202) 786-0503 **HP**

AIA Professional Services Division
American Institute of Architects
1735 New York Avenue, N.W.
Washington, DC 20006
(202) 626-7300

Allied Building Metals Industries
211 East 43rd Street, #804
New York, NY 10017
(212) 697-5551

Aluminum Association (AA)
900 19th Street, N.W., Suite 300
Washington, DC 20006
(202) 862-5100

Aluminum Extruders Council
4300-L Lincoln Avenue
Rolling Meadows, IL 60008
(312) 359-8160

American Architectural Manufacturers Association (AAMA)
2700 River Road, Suite 118
Des Plaines, IL 60018
(312) 699-7310

American Council of Independent
 Laboratories (ACIL)
1725 K Street, N.W.
Washington, DC 20006
(202) 887-5872

American Galvanizers Association
1101 Connecticut Avenue, N.W.,
Suite 700
Washington, DC 20036
(202) 857-1119

American Institute of Architects
(AIA)
1735 New York Avenue, N.W.
Washington, DC 20006
(202) 626-7300

American Institute of Architects
Committee on Historic Resources
1735 New York Avenue, N.W.
Washington, DC 20006
(202) 626-7300 **HP**

American Institute of Steel Construction (AISC)
400 North Michigan Avenue
Chicago, IL 60611-4185
(312) 670-2400

American Institute of Timber Construction (AITC)
11818 S.E. Mill Plain Blvd., Suite 415
Vancouver, WA 98684
(206) 254-9132

American Iron and Steel Institute (AISI)
1133 15th Street, Suite 300
Washington, DC 20005
(202) 452-7100

American National Standards Institute (ANSI)
1430 Broadway
New York, NY 10018
(212) 354-3300

American Painting Contractor
American Paint Journal Company
2911 Washington Avenue
St. Louis, MO 63103

American Plywood Association
(APA)
P.O. Box 11700
Tacoma, WA 98411
(206) 565-6600

American Society of Architectural Hardware Consultants. *See* Door and Hardware Institute.

American Society for Metals
Metals Park, OH 44073
(216) 338-5151

American Welding Society (AWS)
550 N.W. LeJeune Road, N.W.
Box 351040
Miami, FL 33135
(305) 443-9353

American Wood-Preservers' Association (AWPA)
P.O. Box 849
Stevensville, MD 21666
(301) 643-4163

American Wood Preservers Bureau (AWPB)
P.O. Box 5283
Springfield, VA 22150
(703) 339-6660

Architectural Anodizers Council
1227 West Wrightwood Avenue
Chicago, IL 60614
(312) 871-2550

Architectural Technology
American Institute of Architects
1735 New York Avenue, N.W.
Washington, DC 20006
(202) 626-7300

Architectural Woodwork Institute
2310 South Walter Reed Drive
Arlington, VA 22206
(703) 671-9100

Architecture
American Institute of Architects
1735 New York Avenue, N.W.
Washington, DC 20006
(202) 626-7300

Associated Builders and Contractors
729 15th Street, N.W.
Washington, DC 20005
(202) 637-8800

Associated General Contractors of America
1957 E Street, N.W.
Washington, DC 20006
(202) 393-2040

Associated Laboratories
641 South Vermont Street
Palatine, IL 60067
(312) 358-7400

Association for Preservation Technology
Box 2487, Station D
Ottawa, Ont. K1P 5W6, Canada
(613) 238-1972 **HP**

ASTM
1916 Race Street
Philadelphia, PA 19103-1187
(215) 299-5585

Builders' Hardware Manufacturers Association
355 Lexington Avenue, 17th Floor
New York, NY 10017
(212) 661-4261

Building Design and Construction
Cahners Plaza
1350 East Touhy Avenue
P.O. Box 5080
Des Plaines, IL 60018
(312) 635-8800

Building Owners and Managers Association International
1250 I Street, N.W., Suite 200
Washington, DC 20005
(202) 289-7000

Color Association of the United States
409 W. 44th Street
New York, NY 10036
(212) 582-6884

Commercial Renovation (formerly *Commercial Remodeling*)
QR, Inc.
1209 Dundee Avenue, Suite 8
Elgin, IL 60120

Commercial Standard (U.S. Department of Commerce)
Government Printing Office
Washington, DC 20402
(202) 377-2000

Construction Industry Manufacturers Association
111 East Wisconsin Avenue, Suite 1700
Milwaukee, WI 53202
(414) 272-0943

Construction Research Council
1800 M Street, N.W., Suite 1040
Washington, DC 20036
(202) 785-3378

The Construction Specifier
Construction Specifications Institute
601 Madison Street
Alexandria, VA 22314-1791
(703) 684-0300

Consumer Product Safety Commission
5401 Westbard Avenue
Bethesda, MD 20816
(800) 638-2772

Copper and Brass Fabricators Council
1050 17th Street, N.W., #440
Washington, DC 20036
(202) 833-8575

Copper Development Association
    (CDA)
Greenwich Office Park 2
Box 1840
Greenwich, CT 06836
(203) 625-8210

Custom Roll Forming Institute
522 Westgate Tower
Cleveland, OH 44116
(216) 333-8848

Door and Hardware Institute (DHI)
7711 Old Springhouse Road
McLean, VA 22102-3474
(703) 556-3990

Environmental Protection Agency
    (EPA)
401 M Street, S.W.
Washington, DC 20460
(202) 382-2090

Exteriors
1 East 1st Street
Duluth, MN 55802
(218) 723-9200

Factory Mutual Engineering and Research Organization (FM)
1151 Boston–Providence Turnpike
Norwood, MA 02062
(617) 762-4300

Federal Housing Administration
    (FHA)
(U.S. Department of Housing and
    Urban Development)
451 7th Street, S.W., Room 915B
Washington, DC 20201
(202) 755-5210

Federal Specifications (U.S. General
    Services Administration)
Specifications Unit
7th and D Streets, S.W.
Washington, DC 20407
(202) 472-2205

Federation of Societies for Coatings
    Technology
1315 Walnut Street
Philadelphia, PA 19107
(215) 545-1506

Ferroalloys Association
1511 K Street, N.W., #833
Washington, DC 20005
(202) 393-0555

Flat Glass Marketing Association
    (FGMA)
White Lakes Professional Building
3310 Harrison Street
Topeka, KS 66611
(913) 266-7013

Forest Products Laboratory
U.S. Department of Agriculture
Gifford Pinchot Drive
P.O. Box 5130
Madison, WI 53705
(698) 264-5600

General Services Administration
    (GSA)
General Services Building
18th and F Streets, N.W.
Washington, DC 20405
(202) 472-1082

Glass Tempering Association
White Lakes Professional Building
3310 Harrison Street
Topeka, KS 66611
(913) 266-7064

Historic Preservation. See National
    Trust for Historic
    Preservation. **HP**

Insulated Steel Door Systems
    Institute
712 Lakewood Center North
14600 Detroit Avenue
Cleveland, OH 44107
(216) 226-7700

Insulating Glass Certification Council
 (IGCC)
Industrial Park, Route 11
Cortland, NY 13045
(607) 753-6711

International Lead Zinc Research
 Organization
P.O. Box 12036
Research Triangle Park
Raleigh, NC 27709-2036
(919) 351-4647

Iron Casting Society
455 State Street
Cast Metals Federal Building
Des Plaines, IL 60016
(312) 299-9160

Kawneer Company
Department C
Technology Park—Atlanta
555 Guthridge Court
Norcross, GA 30092
(404) 449-5555

Laminators Safety Glass Association
White Lakes Professional Building
3310 Harrison Street
Topeka, KS 66611
(913) 266-7014

Lead Industries Association
295 Madison Avenue
New York, NY 10017
(212) 578-4750

Library of Congress
1st Street, N.E.
Washington, DC 20540
(202) 287-5000 **HP**

Materials and Methods Standards
 Association
614 Monroe Street
Grand Haven, MI 49417
(616) 842-7844

McGraw-Hill Book Company
1221 Avenue of the Americas
New York, NY 10020
(212) 512-2000

Metal Architecture
7450 North Skokie Boulevard
Skokie, IL 60077
(312) 674-2200

Metal Building Manufacturers Association (MBMA)
1230 Keith Building
Cleveland, OH 44115
(216) 241-7333

Metal Construction Association
1133 15th Street, N.W., Suite 1000
Washington, DC 20005
(202) 429-9440

Metal Lath/Steel Framing Association (ML/SFA)
600 South Federal Street, Suite 400
Chicago, IL 60605
(312) 922-6222

Modern Metals
Delta Communications, Inc.
400 North Michigan Avenue
Chicago, IL 60611
(312) 222-2000

National Association of Architectural Metal Manufacturers (NAAMM)
600 South Federal Street, Suite 400
Chicago, IL 60605
(312) 922-6222

National Association of Corrosion
 Engineers
1440 South Creek
Houston, TX 77084
(713) 492-0535

National Bureau of Standards (NBS).
 See National Institute of Standards and Technology.

National Coil Coaters Association
1900 Arch Street
Philadelphia, PA 19103
(215) 564-3484

National Decorating Products
    Association
1050 North Lindbergh Boulevard
St. Louis, MO 63132
(314) 991-3470

National Electrical Manufacturers
    Association (NEMA)
2101 L Street, N.W., Suite 300
Washington, DC 20037
(202) 457-8400

National Fire Protection Association
    (NFPA)
Batterymarch Park
Quincy, MA 02169
(800) 344-3555

National Forest Products Association
    (NFPA)
1250 Connecticut Ave., N.W., Suite
    200
Washington, DC 20036
(202) 463-2700

National Glass Association
8200 Greensboro Drive
McLean, VA 22102
(703) 442-4890

National Institute of Standards and
    Technology (formerly National
    Bureau of Standards)
Gaithersburg, MD 20234
(301) 975-2000

National Institute of Standards and
    Technology (formerly National
    Bureau of Standards)
Center for Building Technology
Gaithersburg, MD 20234
(301) 975-5900

National Paint and Coatings Association (NPCA)
1500 Rhode Island Avenue, N.W.
Washington, DC 20005
(202) 462-6272

National Paint, Varnish and Lacquer
    Association (now National Paint
    and Coatings Association)

National Trust for Historic
    Preservation
1785 Massachusetts Avenue, N.W.
Washington, DC 20036
(202) 673-4000 **HP**

National Wood Window and Door
    Association (NWWDA) (formerly
    National Woodwork Manufacturers Association [NWMA])
1400 E. Touhy Avenue, #G54
Des Plaines, IL 60018
(312) 299-5200

Occupational Safety and Health Administration (OSHA) (U.S. Department of Labor)
Government Printing Office
Washington, DC 20402
(202) 523-6091

The Old-House Journal
435 Ninth Street
Brooklyn, NY 11215
(718) 788-1700 **HP**

Painting and Decorating Contractors
    of America
3913 Old Lee Highway, Suite 33B
Fairfax, VA 22030
(703) 359-0826

Painting and Decorating Contractors
    of America
Washington State Council
27606 Pacific Highway South
Kent, WA 98032
(206) 941-8823

Porcelain Enamel Institute
1101 Connecticut Avenue, N.W.,
    Suite 700
Washington, DC 20036
(202) 857-1134

The Preservation Press
National Trust for Historic
    Preservation
1785 Massachusetts Avenue, N.W.
Washington, DC 20036
(202) 673-4000 **HP**

Product Standards of NBS (U.S. Department of Commerce)
Government Printing Office
Washington, DC 20402
(202) 783-3238

PWC Magazine: Painting and Wallcovering Contractor (magazine of the Painting and Decorating Contractors of America)
130 West Lockwood Street
St. Louis, MO 63119
(314) 961-6644

Safety Glazing Certification Council
    (SGCC)
Industrial Park, Route 11
Cortland, NY 13045
(607) 753-6711

Screen Manufacturers Association
655 Irving Park at Lake Shore Drive
Park Place #201
Chicago, IL 60613-3198
(312) 525-2644

Sealant Engineering and Associated
    Lines (SEAL)
P.O. Box 24302
San Diego, CA 92124
(619) 569-7906

Sealant and Waterproofing Institute
3101 Broadway, #300
Kansas City, MO 64111-2416
(816) 561-8230

Sealed Insulating Glass Manufacturers Association (SIGMA)
111 East Wacker Drive
Chicago, IL 60601
(312) 644-6610

Sheet Metal and Air Conditioning
    Contractors National Association
    (SMACNA)
P.O. Box 70
Merrifield, VA 22116
(703) 790-9890

Society of the Plastics Industry
355 Lexington Avenue
New York, NY 10017
(212) 370-7340

Southern Pine Inspection Bureau
    (SPIB)
4709 Scenic Highway
Pensacola, FL 32504
(904) 434-2611

Steel Door Institute (SDI)
712 Lakewood Center North
14600 Detroit Avenue
Cleveland, OH 44107
(216) 226-7700

Steel Structures Painting Council
    (SSPC)
4400 Fifth Avenue
Pittsburgh, PA 15213
(412) 578-3327

Steel Window Institute (SWI)
1230 Keith Building
Cleveland, OH 44115-2180
(216) 241-7333

Technical Preservation Services
U.S. Department of the Interior,
    Preservation Assistance Division
National Park Service
P.O. Box 37127
Washington, DC 20013-7127
(202) 343-7394 **HP**

Truss Plate Institute (TPI)
583 D'Onofrio Drive, Suite 200
Madison, WI 53719
(608) 833-5900

Underwriters Laboratories (UL)
333 Pfingsten Road
Northbrook, IL 60062
(312) 272-8800

U.S. Department of Agriculture
14th Street and Independence Avenue, S.W.
Washington, DC 20250
(202) 447-8732

U.S. Department of Commerce
14th Street and Constitution Avenue, N.W.
Washington, DC 20230
(202) 377-2000

U.S. Department of Commerce
for PS (Product Standard of NBS)
Government Printing Office
Washington, DC 20402
(202) 783-3238

U.S. Department of the Interior
National Park Service
P.O. Box 37127
Washington, DC 20013-7127
(202) 343-7394 **HP**

U.S. General Services Administration
Historic Preservation Office
18th and F Streets, N.W.
Washington, DC 20405
(202) 655-4000 **HP**

U.S. General Services Administration
Specifications Unit
7th and D Streets, S.W.
Washington, DC 20407
(202) 472-2205/2140

U.S. Government Printing Office
North Capitol and H Streets, N.W.
Washington, DC 20405
(202) 275-3204

United States Gypsum Company
101 South Wacker Drive
Chicago, IL 60606
(312) 606-4000

Van Nostrand Reinhold
115 Fifth Avenue
New York, NY 10003
(212) 254-3232

West Coast Lumber Inspection Bureau (WCLIB)
P.O. Box 23145
Portland, OR 97223
(503) 639-0651

Western Wood Products Association (WWPA)
522 S.W. 5th Avenue, Yeon Building
Portland, OR 97204
(503) 224-3930

John Wiley & Sons
605 Third Avenue
New York, NY 10158
(212) 850-6000

Window and Door Specifier. Contact National Wood Window and Door Association

Woodwork Institute of California
P.O. Box 11428
Fresno, CA 93773
(209) 233-9035

Zinc Institute
295 Madison Avenue
New York, NY 10017
(212) 578-4750

Glossary

Terms that have been set in SMALL CAPS are defined elsewhere in this Glossary.

ACRYLIC BINDER. Acrylic resins are used to produce high-performance clear coatings for aluminum and copper alloys. The most common such coating is probably methacrylate lacquer. Most acrylics used in buildings are air setting types, although thermosetting types are also available. Incralac, developed by the International Copper Research Corporation, is an air-setting clear acrylic lacquer that is often used on copper alloys.

AGE HARDENING. Elevating the temperature of an aluminum alloy to just below its melting point, then rapidly quenching it in cold water.

AIR DRYING. When used in conjunction with an ORGANIC COATING, air drying means that the COATING will properly cure without the use of forced air. PAINT, for example, is always air drying.

ALKALINE DEGREASING. Immersing a metal in, or spraying it with, an alkaline solution. Sometimes mechanical brushing is also used. *See also* DEGREASING, VAPOR DEGREASING.

ALKYD BINDER. When most people today refer to oil paints they really mean one of the alkyds, which are actually oil-modified resins. Alkyd materials are available that will produce either a flat, a semigloss, or a high-gloss finish. Alkyds produce a strong, long-lasting enamel finish. Inexpensive clear alkyd coatings are also available for use on copper alloys, but they tend to yellow when exposed on the exterior, unless modified with melamine resins.

ALLIGATORING. Cracks in an applied coating that resemble an alligator's hide. Alligatoring is a severe form of CHECKING.

ALLOY. The material formed when two or more metals, or metals and nonmetallic substances, are joined together by being dissolved into one another while molten.

Aluminizing. Plating another metal with aluminum.

Annealing. When used with metals, the process of heating a metal to a temperature above its critical working temperature, then permitting it to cool slowly. Annealing makes a metal tougher (less brittle) and stronger, and relieves the strains imposed in it by cold working.

When used with glass, a controlled cooling process performed in an oven called a lehr. Flat glass must be annealed, because unannealed glass sheets do not contract at the same rate over their entire surface, which creates differential stress that results in breakage.

As fabricated finish. A finish imparted by the forming of aluminum or copper alloy items at the mill. This term does not apply to the finish after fabrication of the metal into a usable part or assembly. Synonymous with MILL FINISH as applied to ferrous metals.

Baking. The curing of a BAKING FINISH organic coating at temperatures between 250 and 600 degrees Fahrenheit.

Baking finish. An organic coating formulated or modified to cure only by force drying or baking.

Binder. Also called a film former. A non-volatile ingredient in an ORGANIC COATING that binds the solid particles together. Organic coatings are usually identified by their binder type. Binders include acrylics, alkyds, cellulose, elastomers, epoxies, fluorocarbon polymers, latex, nitrocellulose, oils, oil-alkyd combinations, oleoresins, phenolics, polyesters, rubbers, silicones, urethanes, vinyl chloride copolymers, and the various combinations of them, including alkyd-ureaformaldehyde, alkyd-melamine (alkydamine), and phenolic-alkyd, silicone-alkyd, vinyl-alkyd. Combination binders often have characteristics superior to those made with either component alone.

Blister. Small or large bubble in an ORGANIC COATING. May be caused by entrapped gases, or separation between coats, or of the coating from the substrate for some other reason.

Bonderizing. A hot phosphate treatment of a metal that leaves a crystalline zinc phosphate film on the surface.

Borrowed-light frame. A four-sided metal frame designed to be glazed.

Brake forming. Squeezing a piece of sheet metal between a punch and a die block to change it to a desired shape.

Brittle. Lacking in toughness; nonductile. *See also* TOUGH, DUCTILE.

Came. A soft metal bar used to divide, hold, and surround art (stained) glass.

Carburized metal. Metal impregnated with carbon.

Casehardening. A process in which steel is packed in charcoal and heated to impregnate the surface of the steel with carbon, making it harder than it would otherwise be.

Catalyzed. An organic coating to which an ingredient has been added to enhance performance.

Cellulose binder. Any of several materials, including cellulose acetate and cellulose ethers, derived from cellulose that are used as a binder in clear coatings. Used mostly in interior applications, because they tend to darken when exposed on the exterior.

Cementation. Forming an alloy between a coating metal and the coated metal. Often accomplished by covering the coated metal with a powder of the coating metal, then applying heat until the two metals combine (form an ALLOY).

Chalking. When the resin in an applied opaque ORGANIC COATING film disintegrates, a white powder called chalking is left on the surface. So-called chalking paints disintegrate at a controlled rate, leaving a light coat of chalking that is washed away by rain, along with collected soiling and dust. Excess chalking, however, destroys paint's ability to protect substrates.

Checking. Cracks in an applied coating that resemble ALLIGATORING but are much smaller and less noticeable. In its early stages also called CRAZING, crowfooting, or hairlining.

Chemical metal finishes. A finish produced by bringing a metal into contact with chemicals that change it chemically or physically. They include decorative finishes, but most so-called chemical finishes are actually cleaning procedures or treatments carried out to prepare a metal to receive other finishes.

Clear coating. *See* TRANSPARENT ORGANIC COATING.

Coating. One material applied over another to protect or decorate the coated material. Included are metallic coatings, PAINT, ORGANIC COATINGS other than paint, anodic coatings, vitreous coatings, and laminated coatings. All parts of a coating system are called a coating, collectively and individually.

Coating system. Some coatings consist of several layers (coats). These coats are called collectively a coating system. Each material used as a coat in a coating system is called a coating.

Coil coating. Sheet metal is rolled into large coils as it comes from the rolling mill. A coil coating is applied on the metal before it is cut from these coils into usable panel sizes.

Conversion varnish. A tough organic coating material that has excellent resistance to household chemicals.

Crazing. Fine cracks in an applied coating. In their early stages, CHECKING and ALLIGATORING are sometimes called crazing.

Critical working temperature. That temperature at which a metal can reorganize its molecular structure during working and at which the metal will automatically recover its plasticity after deformation.

CROWFOOTING. *See* CHECKING.
CURTAIN WALL. *See* GLAZED CURTAIN WALL.
DEGREASING. Removing oil and grease from metal. *See also* ALKALINE DEGREASING, VAPOR DEGREASING.
DESCALING. Removing mill scale from metal.
DIFFERENTIAL SEAL. A sealant joint between two spaces that have a pressure differential of more than one pound per square foot.
DRAWING. Pulling a metal bar or rod through a die, over a mandrel, or both, to produce a change in its shape.
DUCTILE. Capable of being drawn out into another shape or hammered or rolled thin, without fracturing and without returning to its original shape when the applied force is removed.
ELASTIC. Capable of returning to its original shape after being deformed. Metals are elastic to varying degrees. The elasticity of a given material may vary with its temperature.
ELASTIC DEFORMATION. Deformation caused by a STRESS small enough so that when the stress is removed the metal returns to its original shape. *See also* PLASTIC DEFORMATION.
ELASTIC LIMIT. The amount of STRESS that will cause a given material to permanently deform (set). *See also* MODULUS OF ELASTICITY.
ELASTOMER BINDER. Elastomers are synthetic rubber coatings such as neoprene and chlorosulfonated polyethylene that are highly resistant to chemicals and corrosion.
ENAMEL. Older enamels were essentially pigmented varnish. Modern enamels are AIR-DRYING or BAKING-FINISH ORGANIC COATINGS that are virtually indistinguishable from PAINT except for an additive or a difference in the formulation of the baking-finish types so that they require the addition of heat to make them cure properly.
EPOXY BINDER. Two-component epoxy-emulsion PAINTS are used where a high-performance, low-odor paint is required. Modified epoxy resins are also used in high-quality opaque coating systems. They include amine-cure epoxy resins; amine-catalyzed, cold-cured epoxy resins; epoxy-polymide resins; estrified epoxy resins; coal-tar epoxy resins; epoxy-zinc-rich coatings; epoxy esters; and epoxy emulsions. Clear epoxy coatings are sometimes used on copper alloys in interior applications. They tend to chalk and darken when used on the exterior.
EUTECTIC ALLOY. An alloy that is a mechanical mixture having a lower freezing point than any of the metals used to make the alloy. *See also* ALLOY.
EUTECTOID ALLOY. Similar to EUTECTIC ALLOY, except that this term is used when speaking of a SOLID SOLUTION rather than a liquid one, to which the term *eutectic* refers. *See also* ALLOY.

Explosive forming. Forming a metal into a desired shape by forcing it to conform to a die using shock waves generated by explosives.

Extender. A material used to impart some desirable quality to a COATING that the coating alone does not have. Extenders may make a coating flow more easily, for example. They do not, however, usually increase the hiding ability of a coating.

Extruded door or frame. A door or frame whose major components have been produced by extrusion.

Extrusion. Forcing a metal through a die; the metal product so produced.

Face glazing. The method of installing glass or another glazing material in which the glazing material is bedded in glazing putty or sealant that is exposed to view and weathering.

Failure. Every type of defect, from the slight crazing of a COATING to the total collapse of a structure, and everything in between.

Film former. *See* BINDER.

Finish hardware. Operating and security devices used on doors. This term replaces the older designation *builders' hardware*.

Flaking. In an applied coating, an advanced form of BLISTERING, ALLIGATORING, or other cracking in which small pieces of the coating fall off. *See also* SCALING.

Fluorocarbon polymer binder. COATINGS made with fluorocarbon polymers (also called just fluoropolymers) have high chemical, abrasion, and wear resistance. Frequently used today on aluminum. A common fluorocarbon polymer resin is Kynar 500. The better fluorocarbon polymer coatings contain at least 70 percent Kynar 500 resin or its equivalent.

Force drying. Curing a baking finish organic coating at temperatures below 200 degrees Fahrenheit.

Forging. Hammering, pressing, stamping, or otherwise forcing a metal to conform to a desired shape. Forging is usually done today using machines and dies.

Formed door or frame. A door or frame produced primarily from metal sheets and strips shaped by roller forming, brake forming, or a similar forming method.

Galvanize. The process of coating steel with zinc, either by the hot-dip process or electroplating.

Galvannealing. A heat treatment used on hot-dip galvanized sheets that removes spangles and leaves a gray, zinc-iron alloy surface.

Glazed curtain wall. A system of wood, metal, or glass framing elements that support vision and spandrel panels, and are used to enclose a building. Aluminum-framed glazed curtain wall is the only type specifically manufactured and sold for use as glazed curtain wall. Its components are designed and manufactured to fit together. Glazed curtain walls that use other metals, wood, or glass alone are made up from standard or specially fabricated components. Glazed curtain walls are sometimes called window walls. *See also* STORE FRONT.

HAIRLINING. *See* CHECKING.

HAMMERING. A type of forging in which a metal is formed into a desired shape using a hammer, usually a mechanical one. *See also* FORGING.

HARDNESS. When related to a metal, means resistant to abrasion, scratching, indentation, and PLASTIC DEFORMATION.

HYDROFORMING. Forming a metal into a desired shape by forcing it to conform to the shape of a die under hydraulic pressure. *See also* FORGING.

INORGANIC FINISH. Any finish that does not fit into the National Association of Architectural Metal Manufacturers (NAAMM) classification of "ORGANIC COATINGS."

KNOCKDOWN FRAME. A door or other frame produced and shipped in three or more parts for assembly at the installation site.

LACQUER. A clear (transparent), AIR-DRYING or BAKING FINISH ORGANIC COATING. Lacquer is often but not always based on modified cellulose resin.

LATEX BINDER. PAINTS with a latex film former dry by the evaporation of water. Both interior and exterior paints are available with latex binders, but factory-applied organic coatings do not usually have latex binders. Latex paints are used on doors and windows in nonresidential buildings, but not as often as are alkyd types. The latex in latex paints is either polyvinyl acetate, polyacrylic, or polystyrene-butadiene. Latex paints go on easily and dry quickly, with little odor. There are many colors available, and color retention is good.

MACHINING. Working a metal by milling, sawing, drilling, punching, shearing, tapping, or reaming it.

MALLEABLE. Capable of being hammered or rolled into a thinner shape without undergoing fracture.

MECHANICAL METAL FINISHES. Mechanical finishes include finishes left by the metal-production process and finishes made by mechanically altering the metal's surface by such actions as grinding, polishing, and rolling. They do not involve the use of chemicals or materials permanently applied over the metal. They are divided into mill (as fabricated) finishes and mechanical PROCESS FINISHES.

MILL FINISH. The finish imparted at the mill on ferrous metal products or aluminum. *See also* AS FABRICATED FINISH.

MODULUS OF ELASTICITY. The ratio of stresses lower than the ELASTIC LIMIT to strain.

NITRIDING. A steel-hardening process in which iron nitride is caused to form in the surface of the steel, making it harder than normal.

NITROCELLULOSE BINDER. AIR-DRYING clear nitrocellulose coatings are low cost and frequently used on copper alloys in interior locations. Usually modified with acrylic or alkyd resins. In exterior applications they usually require removal and reapplication at about one-year intervals.

OIL-ALKYD COMBINATION BINDER. When linseed oil is modified with alkyd resins, a binder results that improves a paint by reducing its drying time, making it harder, and reducing its tendency to fade. Oil alkyds are used as trim enamels for wood and metals and as primers on structural steel.

OIL BINDER. Paints with an oil binder were used extensively in the past and are still used occasionally today. The drying oil in oil-based paint is usually, but not always, linseed oil. Because paints with oil binders dry slowly, most current uses of oil-based paints are on exteriors of buildings. They provide high-quality applications.

OIL-CANNING. A wavy or oily appearance produced by undulations or deviations of flatness in a metal's surface.

OLEORESINOUS BINDER. Processing drying oils with hard resins produces an oleoresinous binder that is used as a varnish or mixing vehicle. Seldom used today, because they tend to yellow over time.

OPACITY. An organic coating's ability to cover the surface and hide what lies below the paint.

ORGANIC COATING. One of a group of coatings that includes AIR-DRYING and BAKING-FINISH opaque enamels and PAINTS, and transparent coatings, and consists of organic or resinous materials. Organic coatings for use on metals include those classified as such by the National Association of Architectural Metal Manufacturers (NAAMM). NAAMM divides organic coatings into four categories: PAINTS, ENAMELS, VARNISHES, and LACQUERS.

PAINT. An AIR-DRYING ORGANIC COATING that can be applied either in the shop or in the field, but usually in the field. By this definition, air-drying ENAMELS are PAINT but BAKING-FINISH enamels are not. Paint can be applied by hand or machine, in the factory, shop, or field, and may be either air-drying or a baking finish.

Opaque organic coatings that do not fit this definition of paint are called other organic coatings in the text. *See also* TRANSPARENT ORGANIC COATING.

PAINT SYSTEM. The several coats required to produce a complete paint coating are called collectively a paint system. Materials used as coats in paint systems are called paint. The various necessary coats include primers, emulsions, enamels, stains, sealers, fillers, and other applied materials used as prime, intermediate, or finish coats.

PHENOLIC BINDER. Products containing phenolic binders, an oleoresinous binder, are available as pigmented paint or COATINGS. Coatings are available in dark colors only. The finish may be either flat or high gloss.

Phenolics, which were among the first synthetic resins, can be used in wet environments.

PICKLING. The process of immersing metal in a diluted acid solution to remove mill scale and rust.

PIGMENT. The solid particles, usually a fine powder, in an ORGANIC COATING or stain that provide the material's color and ability to cover and fill. Pigments make organic coatings opaque and give some stains their color.

PLASTIC. Able to remain in a deformed shape after the deforming loads have been removed.

PLASTIC DEFORMATION. Deformation caused by a STRESS that exceeds the ELASTIC LIMIT of a material so that it does not return to its original shape when the stress is removed.

POLYESTER BINDER. Generally used in the construction industry to produce what are called "tile-like coatings," polyesters are also sometimes modified to produce metal coatings. Siliconized polyester, for example, is a high-performance coating for metals that some producers tout as competitive in performance with fluorocarbon polymers. Other polyester-resin coatings are used as industrial coatings on metals.

PRESSING. A type of forging in which a metal is forced into a desired shape in a hydraulic press. *See also* FORGING.

PRIMER. That coat of a paint or coating system that is applied to the bare substrate to prepare, or prime, it to receive succeeding coats. Most bare metal that will be painted in the field is given a prime coat in the shop.

PROCESS FINISH. All finishes on aluminum or copper alloys other than AS FABRICATED FINISHES.

PROFESSIONAL PAINT MATERIALS. Paint materials specifically formulated for application by professional painters. Often require thinning and the addition of materials not included in the paint as sold.

QUENCHING. A process by which a metal is hardened by submersing it in oil or water. Because quenching makes steel brittle, TEMPERING is necessary to eliminate the brittleness.

RETROFIT. To furnish with new parts or equipment not available at the time of manufacture. Usually means removing existing windows or doors and installing new ones, but can also mean upgrading existing windows or doors by changing the type of glazing, hardware, or some other component.

ROLLER COATING. *See* COIL COATING.

ROLLER FORMING, ROLLING. Passing a metal between a series of flat or shaped rollers to produce a desired shape.

RUBBER BINDER. While paints with latex binders are water-emulsion types, materials with rubber-based binders are solvent thinned. The resins used are actually synthetic rubber. Rubber-based materials are lacquer-type paints that dry quickly to form water- and chemical-resistant surfaces in areas where water is a problem, such as in showers, laundry rooms, and kitchens. Can also be used on exterior surfaces, but are hard to recoat because their strong solvents tend to lift previously applied material. Rubber-based coatings have characteristics similar to those for rubber-based paints.

SCALING. In an applied coating, an advanced form of FLAKING in which the pieces that fall off are large.

SHOP COAT. A coat of paint applied in the shop as a primer for the finish coats of paint that will be applied in the field. Protects the metal until the final coats are applied.

SILICONE BINDER. Clear silicone coatings have high resistance to heat, moisture, acids, salt spray, and alkalis. They are expensive and must be modified for exterior use, to prevent their darkening.

SOLIDS. All the non-volatile ingredients in a paint or other ORGANIC COATING system, including the BINDER (film former) and PIGMENTS.

SOLID SOLUTION. A metal alloy in which the metals are mutually soluble.

SOLVENT. When used in defining a paint or organic coating system's component, a volatile liquid used to dissolve the film former and PIGMENT.

SPINNING. A method of forming metal into a desired shape by clamping it to a block and spinning the block and metal while pressing it against a roller.

STAIN. A material applied as a liquid as all or part of an ORGANIC COATING system. Stains can be either opaque or transparent. Non-grain-raising (NGR) stains are dyes that contain no PIGMENTS. Wiping stains contain pigments.

STAMPING. Bending, shaping, cutting, indenting, embossing, coining, or otherwise forming metal in a press or using a mechanical hammer faced with a shaped die.

STANDARD LACQUER. Lacquer with a nitrocellulose binder and no additives.

STORE FRONT. A special kind of GLAZED CURTAIN WALL. Aluminum store front is specifically manufactured and sold for use as store front and is called store front by its manufacturers. Store front made of other metals, wood, or glass alone is not usually a manufactured system but is rather made up from standard or specially fabricated components. Store front is usually used in a more protected environment than other forms of glazed curtain wall, most often on ground floors and usually includes building entrances.

STRAIN. The change in cross-sectional area of a body produced by stress, as measured in inches per inch of length.

STRESS. The intensity of a mechanical force acting on a body. There are three types of stress: tensile, compressive, and shear. Stress is measured by the total force being applied divided by the area over which the force acts, as in pounds per square inch, for example.

STRETCH FORMING. Stretching a metal sheet or plate over one or more blocks to form it into a desired (usually irregular) shape.

STRONG. Capable of withstanding stress.

TELEGRAPHING. In an applied coating, the appearance of underlying defects or colorations through the applied coating.

TEMPERING. Raising the temperature of metal to a point below its CRITICAL WORKING TEMPERATURE, holding it there for a time, and quenching it to again cool it.

THINNER. A liquid used for thinning (reducing the viscosity of) a PAINT or ORGANIC COATING system's component. When the component is solvent based the thinner is a volatile material. When the component is water based, the thinner is usually water. The terms *thinner* and *solvent* are often used interchangeably.

TOUGH. Resistant to impact, when referring to metal.

TRANSMITTANCE. The amount of light that passes through a piece of glass, expressed as a percentage.

TRANSPARENT ORGANIC COATING. A liquid organic coating that can be applied by hand or machine in the shop, factory, or field. May be either AIR DRYING or a BAKING FINISH and leave a transparent film when cured. Sometimes called a clear coating.

URETHANE BINDER. Material with an oil-free urethane binder, used as a pigmented paint or COATING. Performs the same function as alkyds, but is more expensive. Usually used in interior applications, because it often loses its gloss and sometimes becomes chalky or yellow when exposed on the exterior. Clear urethane coatings are used on copper alloys. Older clear urethane coatings were likely to change color, but newer material probably will not. Clear urethanes are expensive but highly resistant to abrasion and chemicals.

VAPOR DEGREASING. Exposing a metal to chemical vapors to remove grease and oil. *See also* ALKALINE DEGREASING; DEGREASING.

VARNISH. A transparent organic finish composed of oils and resins. OLEORESINOUS varnishes cure by chemical reaction, spirit varnishes by evaporation.

VEHICLE. The resins that form a flexible film and bind a PIGMENT together are called an ORGANIC COATING's vehicle. The vehicle contains the coating's film former (binder).

VINYL CHLORIDE COPOLYMER BINDER. Coatings made with vinyl chloride copolymers are durable and resistant to moisture, oils, fats, alkalis, and acids.

WINDOW WALL. *See* GLAZED CURTAIN WALL.

WIPE COATED. Galvanized steel sheets wiped down while being withdrawn from the molten zinc, which removes spangle and leaves a thin zinc-iron alloy coating.

YIELD POINT. That level of stress at which a relatively large increase in strain occurs without a corresponding increase in stress. Applying stresses in excess of a metal's yield point will usually result in structural damage to it.

Bibliography

Most items here are followed by one or more numbers in brackets. Those numbers list the chapters in this book to which that bibliographical entry applies.

The **HP** designation following some entries here indicates that that entry has particular significance for historic preservation projects.

The sources for many of the entries, including addresses and telephone numbers, are listed in the Appendix.

Abate, Kenneth, 1985. Specifying Foam Core Architectural Metal Panel Systems. *The Construction Specifier*, 38(9)(September): 40–48. [7].

———. 1989. Metal Coatings, Fighting the Elements with Superior Paint Systems, Part I. *Metal Architecture*, 5(6)(June): 10, 69. [3].

———. 1989. Metal Coatings, Fighting the Elements with Superior Paint Systems, Part II. *Metal Architecture*, 5(7)(July): 40–41. [3].

AIA Service Corporation. *Masterspec*, Basic: Section 03310, Concrete Work, 5/87 ed. The American Institute of Architects. [2].

———. *Masterspec*, Basic: Section 03410, Structural Precast Concrete, 8/87 ed. The American Institute of Architects. [2].

———. *Masterspec*, Basic: Section 03450, Architectural Precast Concrete, 5/86 ed. The American Institute of Architects. [2].

———. *Masterspec*, Basic: Section 03470, Tilt-Up Concrete Construction, 5/87 ed. The American Institute of Architects. [2].

———. *Masterspec*, Basic: Section 04200, Unit Masonry, 5/85 ed. The American Institute of Architects. [2].

———. *Masterspec*, Basic: Section 04230, Reinforced Unit Masonry, 5/85 ed. The American Institute of Architects. [2].

———. *Masterspec,* Basic: Section 04405, Dimension Stone, 11/89 ed. The American Institute of Architects. [2].

———. *Masterspec,* Basic: Section 05120, Structural Steel, 8/86 ed. The American Institute of Architects. [2].

———. *Masterspec,* Basic: Section 05210, Steel Joists and Joist Girders, 8/86 ed. The American Institute of Architects. [2].

———. *Masterspec,* Basic: Section 05400, Cold-Formed Metal Framing, 5/89 ed. The American Institute of Architects. [2].

———. *Masterspec,* Basic: Section 06100, Rough Carpentry, 8/86 ed. The American Institute of Architects. [2].

———. *Masterspec,* Basic: Section 06130, Heavy Timber Construction, 5/88 ed. The American Institute of Architects. [2].

———. *Masterspec,* Basic: Section 06170, Structural Glue Laminated Units, 5/88 ed. The American Institute of Architects. [2].

———. *Masterspec,* Basic: Section 06192, Prefabricated Wood Trusses, 8/86 ed. The American Institute of Architects. [2].

———. *Masterspec,* Basic: Section 06200, Finish Carpentry, 2/83 ed. The American Institute of Architects. [5, 6].

———. *Masterspec,* Basic: Section 06401, Exterior Architectural Woodwork, 2/89 ed. The American Institute of Architects. [5, 6].

———. *Masterspec,* Basic: Section 06402, Interior Architectural Woodwork, 2/89 ed. The American Institute of Architects. [5, 6].

———. *Masterspec,* Basic: Section 08110, Steel Doors and Frames, 5/84 ed. The American Institute of Architects. [4, 8].

———. *Masterspec,* Basic: Section 08211, Flush Wood Doors, 2/87 ed. The American Institute of Architects. [5, 8].

———. *Masterspec,* Basic: Section 08212, Panel Wood Doors, 2/87 ed. The American Institute of Architects. [5, 8].

———. *Masterspec,* Basic: Section 08311, Aluminum Sliding Glass Doors, 8/88 ed. The American Institute of Architects. [6, 8].

———. *Masterspec,* Basic: Section 08312, Wood Sliding Glass Doors, 8/88 ed. The American Institute of Architects. [6, 8].

———. *Masterspec,* Basic: Section 08351, Folding Doors, 5/89 ed. The American Institute of Architects. [5].

———. *Masterspec,* Basic: Section 08385, Safety Glass Doors, 2/87 ed. The American Institute of Architects. [8].

———. *Masterspec,* Basic: Section 08410, Aluminum Entrances and Storefronts, 2/87 ed. The American Institute of Architects. [4, 8].

———. *Masterspec,* Basic: Section 08460, Automatic Entrance Doors, 2/87 ed. The American Institute of Architects. [4, 8].

———. *Masterspec,* Basic: Section 08470, Revolving Doors, 5/87 ed. The American Institute of Architects. [4, 8].

———. *Masterspec,* Basic: Section 08510, Steel Windows, 8/86 ed. The American Institute of Architects. [6, 8].

———. *Masterspec,* Basic: Section 08520, Aluminum Windows, 8/86 ed. The American Institute of Architects. [6, 8].

———. *Masterspec,* Basic: Section 08525, Aluminum Architectural Windows, 11/86 ed. The American Institute of Architects. [6, 8].

———. *Masterspec,* Basic: Section 08610, Wood Windows, 8/86 ed. The American Institute of Architects. [6, 8].

———. *Masterspec,* Basic: Section 08710, Finish Hardware, 5/84 ed. The American Institute of Architects. [4, 5].

———. *Masterspec,* Basic: Section 08800, Glass and Glazing, 5/86 ed. The American Institute of Architects. [8].

———. *Masterspec,* Basic: Section 08920, Glazed Aluminum Curtain Walls, 2/89 ed. The American Institute of Architects. [7, 8].

———. *Masterspec,* Basic: Section 09200, Lath and Plaster, 2/85 ed. The American Institute of Architects. [2].

———. *Masterspec,* Basic: Section 09250, Gypsum Drywall, 8/87 ed. The American Institute of Architects. [2].

———. *Masterspec,* Basic: Section 09800, Special Coatings, 2/88 ed. The American Institute of Architects. [3].

———. *Masterspec,* Basic: Section 09900, Painting, 11/88 ed. The American Institute of Architects. [3].

Allen, Edward. 1985. *Fundamentals of Building Construction: Materials and Methods.* New York: Wiley. [2, 3, 4, 5, 6, 7, 8].

American Architectural Manufacturers Association (AAMA).

———. AAMA TIR-A1-1975, Sound Control for Aluminum Curtain Walls and Windows. Des Plaines, IL: AAMA. [6, 7, 8].

———. AAMA TIR-A3-1975, Fire-Resistive Design Guidelines for Curtain Wall Assemblies. Des Plaines, IL: AAMA. [7, 8].

———. AAMA TIR-A4-1978, Recommended Glazing Guidelines for Reflective Insulating Glass. Des Plaines, IL: AAMA. [8].

———. AAMA TIR-A7-1983, Sloped Glazing Guidelines. Des Plaines, IL: AAMA. [7, 8].

———. ANSI/AAMA 101-1985, Voluntary Specifications for Aluminum Prime Windows and Sliding Glass Doors. Des Plaines, IL: AAMA. [6].

———. AAMA 501-1983, Methods of Test for Metal Curtain Walls. Des Plaines, IL: AAMA. [7, 8].

———. AAMA 603.8-1985, Voluntary Performance Requirements and Test Procedures for Pigmented Organic Coatings on Extruded Aluminum. Des Plaines, IL: AAMA. [3].

———. AAMA 604.2-1977, Voluntary Specification for Residential Color Anodic Finishes. Des Plaines, IL: AAMA. [3].

———. AAMA 605.2-1985, Voluntary Specification for High Performance Organic Coatings on Architectural Extrusions and Panels. Des Plaines, IL: AAMA. [3].

———. AAMA 606.1-1976, Voluntary Guide Specifications and Inspection Methods for Integral Color Anodic Finishes for Architectural Aluminum. Des Plaines, IL: AAMA. [3].

———. AAMA 607.1-1977, Voluntary Guide Specifications and Inspection Methods for Clear Anodic Finishes for Architectural Aluminum. Des Plaines, IL: AAMA. [3].

———. AAMA 608.1-1977, Voluntary Guide Specification and Inspection Methods for Electrolytically Deposited Color Anodic Finishes for Architectural Aluminum. Des Plaines, IL: AAMA. [3].

———. AAMA 609.1-1985, Voluntary Guide Specification for Cleaning and Maintenance of Architectural Anodized Aluminum. Des Plaines, IL: AAMA [3].

———. AAMA 610.1-1979, Voluntary Guide Specification for Cleaning and Maintenance of Painted Aluminum Extrusions and Curtain Wall Panels. Des Plaines, IL: AAMA. [3].

———. AAMA 701.2-1974, Voluntary Specifications for Pile Weather Strip. Des Plaines, IL: AAMA. [4, 5, 6].

———. AAMA 800-1986, Voluntary Specifications and Test Methods for Sealants. Des Plaines, IL: AAMA. [4, 5, 6, 7, 8].

———. AAMA 902-1987, Voluntary Specifications for Sash Balances. Des Plaines, IL: AAMA. [6].

———. AAMA 904-1987, Voluntary Specifications for Friction Hinges in Window Applications. Des Plaines, IL: AAMA. [6].

———. AAMA 906-1987, Voluntary Specifications for Sliding Glass Door Roller Assemblies. Des Plaines, IL: AAMA. [6].

———. AAMA 1002.10-1983, Voluntary Specifications for Aluminum Insulating Storm Products for Windows and Sliding Glass Doors. Des Plaines, IL: AAMA. [6].

———. AAMA 1102.7-1977, Voluntary Guide Specifications for Aluminum Storm Doors. Des Plaines, IL: AAMA. [4].

———. AAMA 1302, Voluntary Specifications for Forced-Entry Resistant Aluminum Prime Windows. Des Plaines, IL: AAMA. [6].

———. AAMA 1303, Voluntary Specifications for Forced-Entry Resistant Aluminum Sliding Glass Doors. Des Plaines, IL: AAMA. [6].

———. AAMA 1502.7-1981, Voluntary Test Methods for Condensation Resistance of Windows, Doors and Glazed Wall Sections. Des Plaines, IL: AAMA. [4, 6, 7, 8].

Bibliography

———. AAMA 1503.1-1988, *Voluntary Test Method for Thermal Transmittance and Condensation Resistance of Windows, Doors and Glazed Wall Sections*. Des Plaines, IL: AAMA. [4, 6, 7, 8].

———. AAMA 1504-1988, *Voluntary Standard for Thermal Transmittance and Condensation Resistance of Windows, Doors and Glazed Wall Sections*. Des Plaines, IL: AAMA. [4, 6, 7, 8].

———. AAMA GS-001. *Voluntary Guide Specifications for Aluminum Architectural Windows*. Des Plaines, IL: AAMA. [6].

———. 1979. *Aluminum Curtain Wall Design Guide Manual* (CW-I-9). Des Plaines, IL: AAMA. [7, 8].

———. 1989. *Metal Curtain Wall Manual* (CWM-1). Des Plaines, IL: AAMA. [7, 8].

———. *Aluminum Store Front and Entrance Manual* (SFM-1). Des Plaines, IL: AAMA. [3, 4, 8].

———. *Curtain Wall Manual*. Vol. 10 (CW-10), *Care and Handling of Architectural Aluminum from Shop to Site*. Des Plaines, IL: AAMA. [3].

———. *Curtain Wall Manual*. Vol. 11 (CW-11), *Design Windloads for Buildings and Barrier Layer Wind Tunnel Testing*. Des Plaines, IL: AAMA. [7, 8].

———. *Curtain Wall Manual*. Vol. 12 (CW-12), *Structural Properties of Glass*. Des Plaines, IL: AAMA. [8].

———. *Curtain Wall Manual*. Vol. 13 (CW-13), *Structural Sealant Glazing Systems*. Des Plaines, IL: AAMA. [8].

———. *Curtain Wall Manual* (CWM-1). *Installation of Aluminum Curtain Walls*. Des Plaines, IL: AAMA. [7, 8].

———. *Glass Design for Sloped Glazing* (SHDG-1). Des Plaines, IL: AAMA. [7, 8].

———. *Metal Curtain Wall, Window, Store Front and Entrance Guide Specifications Manual* (GSM-1). Des Plaines, IL: AAMA. [3, 4, 6, 7, 8].

———. *Window Selection Guide* (WSG-1). Des Plaines, IL: AAMA. [6].

American Institute of Steel Construction (AISC). 1989. *Manual of Steel Construction*. Chicago: AISC. [2].

———. 1989. *Code of Standard Practice for Steel Buildings and Bridges*. Chicago: AISC. (Also contained in *Manual of Steel Construction*, above.) [2].

———. 1989. *Specifications for the Design, Fabrication, and Erection of Structural Steel for Buildings*, with its *Commentary*. Chicago: AISC. (Also contained in *Manual of Steel Construction*, above.) [2].

———. 1989. *Specifications for Structural Joints Using ASTM A 325 or A 490 Bolts*. Chicago: AISC. (Also contained in *Manual of Steel Construction*, above.) [2].

American Institute of Timber Construction. *Timber Construction Standards*. Englewood, CO: American Institute of Timber Construction. [2].

———. *Timber Construction Manual*. Englewood, CO: American Institute of Timber Construction. [2].

Bibliography

American Iron and Steel Institute. 1975. *Stainless Steel Fasteners, Suggested Source List*. Washington, DC: AISI. [3, 4, 5, 6, 7].

———. 1976. *Stainless Steel Fasteners, A Systematic Approach to Their Selection*. Washington, DC: AISI. [3, 4, 5, 6, 7].

———. 1977. *Design Guidelines for the Selection and Use of Stainless Steel*. Washington, DC: AISI. [3, 4, 7].

———. 1979. *Welding of Stainless Steel and Other Joining Methods*. Washington, DC: AISI. [3, 4, 7].

American National Standards Institute (ANSI). Standard A39.1-1987. Safety Requirements for Window Cleaning. ANSI. [7].

———. Standard A58.1-1982. Minimum Design Loads, Buildings and Other Structures. ANSI. [2].

———. Standard A115.1-1982. Preparation for Mortise Locks for 1-3/8" and 1-3/4" Doors. ANSI. [4].

———. Standard A115.2-1982 (reaffirmed 1988). Preparation for Bored Locks for 1-3/8" and 1-3/4" Doors. ANSI. [4].

———. Standard A115.4-1982. Preparation for Lever Extension Flush Bolts. ANSI. [4].

———. Standard A115.5-1982. Preparation for 181 Series and 190 Series Deadlock Strikes. ANSI. [4].

———. Standard A115.6-1982. Preparation for Preassembled Locks. ANSI. [4].

———. Standard A115.7-1982. Preparation for Floor Closers—Light Duty, Center Hung, Single or Double Acting; Center Hung, Single or Double Acting; Offset Hung, Single Acting. ANSI. [4].

———. Standard A115.12-1982. Preparation for Offset Intermediate Pivots. ANSI. [4].

———. Standard A115.13-1982. Preparation for Tubular Deadlocks. ANSI. [4].

———. Standard A115.14-1982. Preparation for Open Back Strikes. ANSI. [4].

———. Standard A115.15-1985. Preparation for Bored or Cylindrical Locks in Prehung Insulated Steel Doors and Steel Frames. ANSI. [4].

———. Standard A115.16-1980. Preparation for Double-Type Locks in Prehung Insulated Steel Doors and Steel Frames. ANSI. [4].

———. Standard A115.17 (Reaffirmed 1988). Preparation for Double-Type Locks in Standard 1-3/8" and 1-3/4" Steel Doors and Steel Frames. ANSI. [4].

———. Standard A115.W1-1988. Mortise Lock Preparation for 1-3/4" Flush Doors. ANSI. [5].

———. Standard A115.W2-1988. Bored Lock Preparation for 1-3/4" Flush Doors. ANSI. [5].

———. Standard A115.W3-1988. Bored Lock Preparation for 1-3/8" Flush Doors. ANSI. [5].

———. Standard A115.W4-1988. Preassembled Lock Preparation for 1-3/4" Flush Doors. ANSI. [5].

———. Standard A115.W5-1988. Lever Extension Flush Bolt Preparation for 1-3/4" Flush Doors. ANSI. [5].

———. Standard A115.W6-1988. Specifications for Double-Type Lock Preparation for 1-3/4" Flush Doors. ANSI. [5].

———. Standard A115.W7-1988. Specifications for Push/Pull Latch Preparation for 1-3/4" Flush Doors. ANSI. [5].

———. Standard A115.W8-1988. Specifications for Bored Deadlock Preparation for 1-3/4" Flush Doors. ANSI. [5].

———. Standard A115.W9-1988. Specifications for Round Flush Pull Preparation for 1-3/4" Flush Doors. ANSI. [5].

———. Standard A117.1-1986. American National Standards for Buildings and Facilities—Providing Accessibility and Usability for Physically Handicapped People. ANSI. [4, 5].

———. Standard A151.1-1980. Physical Endurance for Steel Doors and Hardware Reinforcings. ANSI. [4].

———. Standard A156.1-1988. Butts and Hinges. ANSI. [4, 5].

———. Standard A156.2-1983. Bored and Preassembled Locks and Latches. ANSI. [4, 5].

———. Standard A156.3-1984. Exit Devices. ANSI. [4, 5].

———. Standard A156.4-1986. Door Controls—Closers. ANSI. [4, 5].

———. Standard A156.5-1984. Auxiliary Locks and Associated Products. ANSI. [4, 5].

———. Standard A156.6-1986. Architectural Door Trim. ANSI. [4, 5].

———. Standard A156.7-1988. Templet Hinge Dimensions. ANSI. [4, 5].

———. Standard A156.8-1988. Door Controls—Overhead Holders. ANSI. [4, 5].

———. Standard A156.10-1985. Power Operated Pedestrian Doors. ANSI. [4].

———. Standard A156.12-1986. Interconnected Locks and Latches. ANSI. [4, 5].

———. Standard A156.13-1987. Mortise Locks and Latches. ANSI. [4, 5].

———. Standard A156.14-1985. Sliding/Folding Door Hardware. ANSI. [4, 5, 6].

———. Standard A156.15-1986. Life Safety Closer/Holder/Release Devices. ANSI. [4, 5].

———. Standard A156.16-1989. Auxiliary Hardware. ANSI. [4, 5].

———. Standard A156.17-1987. Self Closing Hinges and Pivots. ANSI. [4, 5].

———. Standard A156.18-1987. Materials and Finishes. ANSI. [4, 5].

———. Standard A156.19-1984. Power Assisted and Low Energy Power Operated Doors. [4].

———. Standard A156.21-1989. Thresholds. ANSI. [4, 5].

———. Standard A208.1-1979. Particleboard, Mat-Formed Wood. ANSI. [5].

———. Standard Z97.1-1984. Safety Glazing Materials Used in Buildings. ANSI. [8].

American National Standards Institute/Steel Door Institute. 1983. ANSI/SDI—119-83. Performance and Test Procedures for Steel Door Frames and Frame Anchors. ANSI/SDI. [4].

———. 1982. ANSI/SDI—123.1-82. Nomenclature for Steel Doors and Steel Door Frames. ANSI/SDI. [4].

———. 1980. ANSI/SDI—151.1-80. Test Procedure and Acceptance Criteria for Physical Endurance of Steel Doors and Hardware Reinforcings. ANSI/SDI. [4].

———. 1980. ANSI/SDI—224.1-80. Test Procedure and Acceptance Criteria for Prime Painted Steel Surfaces of Steel Doors and Frames. ANSI/SDI. [4].

American National Standards Institute/Screen Manufacturers Association. 1976. ANSI/SMA—1004-1976. Specifications for Aluminum Tubular Frame Screens for Windows. ANSI/SMA. [6].

———. 1976. ANSI/SMA—2005-1976. Specifications for Aluminum Sliding Doors. ANSI/SMA. [6].

American Paint Contractor. 1988. Direct to Rust Coatings. *American Paint Contractor,* June, 65(6): 8–19. [3].

American Society for Metals. 1982. *Metals Handbook. 9th ed. Vol. 1: Properties and Selection of Metals.* Metals Park, OH: American Society for Metals. [3].

———. 1982. *Metals Handbook. 9th ed. Vol. 5: Surface Cleaning, Finishing, and Coating.* Metals Park, OH: American Society for Metals. [3].

American Welding Society. 1983. *AWS D1.1. Structural Welding Code—Steel.* Miami: AWS. [4, 6, 7].

———. 1983. *AWS D1.2. Structural Welding Code—Aluminum.* Miami: AWS. [4, 6, 7].

American Wood-Preservers' Association. *Book of Standards.* Springfield, VA: American Wood Preservers Association. [5, 6].

Architectural Graphic Standards. See Ramsey/Sleeper.

Architectural Technology. 1986. Technical Tips: Paints and Coatings Primer. *Architectural Technology,* July/August: 64–65. [3].

Architectural Woodwork Institute. 1988. *Architectural Woodwork Quality Standards, Guide Specifications and Quality Certification Program,* 5th ed. Arlington, VA: AWI. [5, 6].

Associated Laboratories, Inc. *Certified Products Directory—Fenestration Products.* Palatine, IL: ALI. [8].

ASTM. Standard A 6, Standard Specification for General Requirements for Rolled Steel Plates, Shapes, Sheet Piling, and Bars for Structural Use. ASTM. [3].

372   Bibliography

———. Standard A 27, Standard Specification for Carbon-Steel Castings for General Application. ASTM. [3].

———. Standard A 36, Standard Specification for Structural Steel. ASTM. [3].

———. Standard A 108, Standard Specification for Steel Bars, Carbon, Cold-Finished, Standard Quality. ASTM. [3].

———. Standard A 123, Specification for Zinc (Hot-Galvanized) Coatings on Products Fabricated from Rolled, Pressed, and Forged Steel Shapes, Plates, Bars, and Strip. ASTM. [3].

———. Standard A 153, Specification for Zinc Coating (Hot-Dip) on Iron and Steel Hardware. ASTM. [3].

———. Standard A 165, Standard Specification for Electrodeposited Coatings of Cadmium on Steel. ASTM. [3].

———. Standard A 167, Specification for Stainless and Heat-Resisting Chromium-Nickel Steel Plate, Sheet, and Strip. ASTM. [3].

———. Standard A 269, Specification for Seamless and Welded Austenitic Stainless Steel Tubing for General Service. ASTM. [3].

———. Standard A 276, Specification for Stainless and Heat-Resisting Steel Bars and Shapes. ASTM. [3].

———. Standard A 283, Standard Specification for Low and Intermediate Tensile Strength Carbon Steel Plates, Shapes and Bars. ASTM. [3].

———. Standard A 307, Specification for Carbon Steel Externally Threaded Standard Fasteners. ASTM. [3, 4, 5, 6, 7].

———. Standard A 312, Specification for Seamless and Welded Austenitic Stainless Steel Pipe. ASTM. [3].

———. Standard A 366, Specification for Steel, Sheet, Carbon, Cold-Rolled, Commercial Quality. ASTM. [3].

———. Standard A 386, Specification for Zinc Coating (Hot-Dip) on Assembled Steel Products. ASTM. [3].

———. Standard A 424, Specification for Steel Sheet for Porcelain Enameling. ASTM. [3].

———. Standard A 446, Specification for Sheet Steel, Zinc-Coated (Galvanized) by the Hot-Dip Process, Structural (Physical) Quality. ASTM. [3].

———. Standard A 463, Standard Specification for Sheet Steel, Cold-Rolled Aluminum-Coated Type 1. ASTM. [3].

———. Standard A 525, Specification for General Requirements for Steel Sheet Zinc-Coated (Galvanized) by the Hot-Dip Process. ASTM. [3].

———. Standard A 526, Specification for Steel Sheet Zinc-Coated (Galvanized) by the Hot-Dip Process, Commercial Quality. ASTM. [3].

———. Standard A 527, Specification for Steel Sheet Zinc-Coated (Galvanized) by the Hot-Dip Process, Lock-Forming Quality. ASTM. [3].

———. Standard A 528, Specification for Steel Sheet Zinc-Coated (Galvanized) by the Hot-Dip Process, Drawing Quality. ASTM. [3].

———. Standard A 554, Specification for Welded Stainless Steel Mechanical Tubing. ASTM. [3].

———. Standard A 568, Standard Specification for General Requirements for Steel, Carbon and High-Strength Low-Alloy Hot-Rolled Sheet and Cold-Rolled Sheet. ASTM. [3].

———. Standard A 569, Specification for Steel, Carbon (0.15 Maximum, Percent), Hot-Rolled Sheet and Strip, Commercial Quality. ASTM. [3].

———. Standard A 570, Specification for Hot-Rolled Carbon Steel Sheet and Strip, Structural Quality. ASTM. [3].

———. Standard A 572, Specification for High-Strength Low-Alloy Columbium-Vanadium Steels of Structural Quality. ASTM. [3].

———. Standard A 575, Standard Specifications for Steel Bars, Carbon, Merchant Quality, M-Grades. ASTM. [3].

———. Standard A 591, Specification for Steel Sheet, Cold-Rolled Electrolytic Zinc-Coated. ASTM. [3].

———. Standard A 607, Standard Specification for Steel Sheet and Strip, Hot-Rolled and Cold-Rolled, High-Strength, Low-Alloy Columbium and or Vanadium. ASTM. [3].

———. Standard A 611, Specification for Steel, Cold-Rolled Sheet, Carbon, Structural. ASTM. [3].

———. Standard A 663, Standard Specification for Steel Bars, Carbon, Merchant Quality, Mechanical Qualities. ASTM. [3].

———. Standard A 675, Standard Specification for Steel Bars, Carbon, Hot-Wrought, Special Quality, Mechanical Properties. ASTM. [3].

———. Standard A 743, Standard Specification for Castings, Iron-Chromium-Nickel, and Nickel-Base, Corrosion-Resistant, for General Application. ASTM. [3].

———. Standard A 780, Practice for Repair of Damaged Hot-Dip Galvanized Coatings. ASTM. [3].

———. Standard A 792, Standard Specification for Steel Sheet, Aluminum–Zinc Alloy-Coated by the Hot-Dip Process. ASTM. [3].

———. Standard A 875, Standard Specification for Steel Sheet, Zinc 5 Percent, Aluminum-Misch Metal Alloy, Coated by the Hot-Dip Process. ASTM. [3].

———. Standard B 26, Specification for Aluminum-Alloy Sand Castings. ASTM. [3].

———. Standard B 62, Specification for Composition Bronze or Ounce Metal Castings. ASTM. [3].

———. Standard B 108, Specification for Aluminum-Alloy Permanent Mold Castings. ASTM. [3].

------. Standard B 177, Recommended Practice for Chromium Electroplating on Steel for Engineering Use. ASTM. [3].

------. Standard B 209, Specification for Aluminum and Aluminum Alloy Sheet and Plate. ASTM. [3].

------. Standard B 210, Specification for Aluminum-Alloy Drawn Seamless Tubes. ASTM. [3].

------. Standard B 211, Specification for Aluminum-Alloy Bars, Rods, and Wire. ASTM. [3].

------. Standard B 221, Specification for Aluminum-Alloy Extruded Bars, Rods, Wire, Shapes, and Tubes. ASTM. [3].

------. Standard B 247, Specification for Aluminum-Alloy Die and Hand Forgings. ASTM. [3].

------. Standard B 253, Recommended Practice for Preparation of and Electroplating on Aluminum Alloys by the Zincate Process. ASTM. [3].

------. Standard B 254, Recommended Practice for Preparation and Electroplating on Stainless Steel. ASTM. [3].

------. Standard B 271, Specification for Copper-Base Alloy Centrifugal Castings. ASTM. [3].

------. Standard B 320, Recommended Practice for Preparation of Iron Castings for Electroplating. ASTM. [3].

------. Standard B 370, Specification for Copper Sheet and Strip for Building Construction. ASTM. [3].

------. Standard B 429, Specification for Aluminum-Alloy Extruded Structural Pipe and Tube. ASTM. [3].

------. Standard B 455, Specification for Copper-Zinc-Lead Alloy (Leaded Brass) Extruded Shapes. ASTM. [3].

------. Standard B 456, Specification for Electrodeposited Coatings of Copper Plus Nickel Plus Chromium and Nickel Plus Chromium. ASTM. [3].

------. Standard B 483, Specification for Aluminum and Aluminum-Alloy Extruded Drawn Tubes for General Purpose Applications. ASTM. [3].

------. Standard B 584, Specification for Copper Alloy Sand Castings. ASTM. [3].

------. Standard B 650, Specification for Electrodeposited Engineering Chromium Coatings on Ferrous Substrates. ASTM. [3].

------. Standard C-236, Test Method for Steady-State Thermal Performance of Building Assemblies by Means of a Guarded Hot Box. ASTM. [2, 4].

------. Standard C 282, Test Method for Acid Resistance of Porcelain Enamels (Citric Acid Spot Test). ASTM. [3].

------. Standard C 283, Test Method for Resistance of Porcelain Enamel Utensils to Boiling Acid. ASTM. [3].

------. Standard C 286, Definition of Terms Relating to Porcelain Enamel and Ceramic-Metal Systems. ASTM. [3].

Bibliography

———. Standard C 313, Test Method for Adherence of Porcelain Enamel and Ceramic Coatings to Sheet Metal. ASTM. [3].

———. Standard C 346, Test Method for 45-deg Specular Gloss of Ceramic Materials. ASTM. [3].

———. Standard C 448, Test Method for Abrasion Resistance of Porcelain Enamel. ASTM. [3].

———. Standard C 481, Test Method for Laboratory Aging of Sandwich Construction. ASTM. [7].

———. Standard C 509, Specification for Cellular Elastomeric Preformed Gasket and Sealing Material. ASTM. [6].

———. Standard C 538, Test Method for Color Retention of Red, Orange, and Yellow Porcelain Enamels. ASTM. [3].

———. Standard C 540, Test Method for Image Gloss of Porcelain Enamel Surfaces. ASTM. [3].

———. Standard C 542, Specification for Lockstrip Gaskets. ASTM. [7, 8].

———. Standard C 578, Specification for Preformed Cellular Polystyrene Thermal Insulation. ASTM. [3].

———. Standard C 645, Specification for Non-Load (Axial) Bearing Steel Studs, Runners (Track), and Rigid Furring Channels for Screw Application of Gypsum Board. ASTM. [2].

———. Standard C 703, Test Method for Spalling Resistance of Porcelain Enameled Aluminum. ASTM. [3].

———. Standard C 716, Specification for Installing Lockstrip Gaskets and Infill Glazing Materials. ASTM. [7, 8].

———. Standard C 920, Specification for Elastomeric Joint Sealants. ASTM. [4, 5, 6, 7, 8].

———. Standard C 955, Specification for Load-Bearing (Transverse and Axial) Steel Studs, Runners (Track), and Bracing or Bridging, for Screw Application of Gypsum Board and Metal Plaster Bases. ASTM. [2].

———. Standard C 964, Standard Guide for Lockstrip Glazing. ASTM. [7, 8].

———. Standard C 976, Method of Test for Thermal Performance of Building Assemblies by Means of a Calibrated Hot Box. ASTM. [6].

———. Standard C 1007, Installation of Load-Bearing (Transverse and Axial) Steel Studs and Accessories. ASTM. [2].

———. Standard C 1036, Standard Specification for Flat Glass. ASTM. [8].

———. Standard C 1048, Specification for Heat Treated Flat Glass—Kind HS, Kind FT Coated and Uncoated Glass. ASTM. [8].

———. Standard D 16, Standard Definitions of Terms Relating to Paint, Varnish, Lacquer, and Related Products. ASTM. [3].

———. Standard D 659, Evaluating Degree of Chalking of Exterior Paints. ASTM. [3].

———. Standard D 660, Evaluating Degree of Checking of Exterior Paints. ASTM. [3].

———. Standard D 661, Evaluating Degree of Cracking of Exterior Paints. ASTM. [3].

———. Standard D 662, Evaluating Degree of Erosion of Exterior Paints. ASTM. [3].

———. Standard D 1187. Test Method for Asphalt Emulsions for Use as Protective Coatings for Metal. ASTM. [3].

———. Standard D 1653, Test Method for Moisture Vapor Permeability of Organic Coating Films. ASTM. [3].

———. Standard D 1730, Recommended Practices for Preparation of Aluminum and Aluminum-Alloy Surfaces for Painting. ASTM. [3].

———. Standard D 1731, Recommended Practices for Preparation of Hot-Dip Aluminum Surfaces for Painting. ASTM. [3].

———. Standard D 1784, Specification for Rigid Poly(Vinyl Chloride) (PVC) Compounds and Chlorinated Poly(Vinyl Chloride) (CPVC) Compounds. ASTM. [3].

———. Standard D 2000, Classification System for Rubber Products in Automotive Applications. ASTM. [6, 7, 8].

———. Standard D 2092, Recommended Practices for Preparation of Zinc-Coated Steel Surfaces for Painting. ASTM. [3].

———. Standard D 2287, Specification for Nonrigid Vinyl Chloride Polymer and Copolymer Molding and Extrusion Compounds. ASTM. [6, 7, 8].

———. Standard D 2833, Standard Index of Methods for Testing Architectural Paints and Coatings. ASTM. [3].

———. Standard D 3276, Standard Guide for Painting Inspectors (Metal Substrates). ASTM. [3].

———. Standard D 3794, Practice for Testing Coil Coatings. ASTM. [3].

———. Standard D 3927, Standard Guide for State and Institutional Purchasing of Paint. ASTM. [3].

———. Standard E 84, Test Method for Surface Burning Characteristics of Building Materials. ASTM. [3].

———. Standard E 90, Laboratory Measurement of Airborne-Sound Transmission Loss of Building Partitions. ASTM. [4, 5].

———. Standard E 97, Test Method for 45-deg, 0-deg Directional Reflectance Factor of Opaque Specimens by Broad-Band Filter Reflectometry. ASTM. [3].

———. Standard E 152, Fire Tests of Door Assemblies. [4, 5].

———. Standard E 163, Fire Tests of Window Assemblies. ASTM. [6, 8].

———. Standard E 283, Test Method for Rate of Air Leakage Through Exterior Windows, Curtain Walls, and Doors. ASTM. [4, 6, 7].

———. Standard E 330, Test Method for Structural Performance of Exterior Windows, Curtain Walls, and Doors by Uniform Static Air Pressure Differential. ASTM. [4, 6, 7].

———. Standard E 331, Test Method for Water Penetration of Exterior Windows, Curtain Walls, and Doors by Uniform Static Air Pressure Differential. ASTM. [4, 6, 7].

———. Standard E 413, Classification for Determination of Sound Transmission Class. ASTM. [4, 5, 6].

———. Standard E 547, Test Method for Water Penetration of Exterior Windows, Curtain Walls, and Doors by Cyclic Air Pressure Differential. ASTM. [4, 6, 7].

———. Standard E 699, Practice for Criteria for Evaluation of Agencies Involved in Testing, Quality Assurance, and Evaluating Building Components in Accordance with Test Methods Promulgated by ASTM Committee E-6. ASTM. [7].

———. Standard E 773, Test Method for Seal Durability of Sealed Insulating Glass Units. ASTM. [8].

———. Standard E 774, Standard Specification for Sealed Insulating Glass Units. ASTM. [8].

———. Standard E 783, Field Measurement of Air Leakage Through Installed Exterior Windows and Doors. ASTM. [4, 6].

———. Standard E 987, Standard Tests for Deglazing Force of Fenestration Products. ASTM. [6].

———. Standard E 1233, Test Method for Structural Performance of Exterior Windows, Curtain Walls, and Doors by Cyclic Air Pressure Differential. ASTM. [7].

———. Standard F 588, Standard Test Methods for Resistance of Window Assemblies to Forced Entry, Excluding Glazing. ASTM. [6].

———. Standard F 842, Standard Test Methods for Measurement of Forced Entry Resistance of Horizontal Sliding Door Assemblies. ASTM. [6].

Bakhalov, G. T. and A. V. Turkovakaya. 1965. *Corrosion and Protection of Metals.* Elmsford, NY: Pergamon Press. [3].

Banov, Able. 1973. *Paints and Coatings Handbook for Contractors, Architects and Builders.* Farmington, MI: Structures Publishing Co. [3].

Bartlett, Thomas L. 1983. Cleaning with Corncobs. *The Construction Specifier,* Feb. 36(2): 6–7. [3].

Batcheler, Penelope Hartshorne. 1968. Paint Color Research and Restoration. Technical Leaflet 15. Nashville, TN: American Association for State and Local History. **HP.**

Bauer, Mike. 1987. Letterbox: Too Much to the Imagination? *The Construction Specifier,* July, 40(7): 7–8. [3].

Belles, Donald. 1987. Preventing Flame Spread on Exterior Walls. *Exteriors,* 5(2)(Summer): 28–38. [7].

——— and Jesse Beitel. 1988. Fire Performance of Curtainwalls Questioned. *Exteriors*, 6(2)(Summer): 44–50. [7].

Bennett, C. R. 1987. Paints and Coatings: Getting Beneath the Surface. *The Construction Specifier*, Feb. 40(2): 36–41. [3].

Benny, James C. 1988. Ensuring that Wood Doors and Hardware Are Compatible. *Window and Door Specifier*, 1(1)(Summer): 69–70. [5].

Bessmer, Stan. 1979. Edge-Sealed Insulating Glass. *The Construction Specifier*, 32(6)(June): 50–53. [8].

Black, David R. Dealing with Peeling Paint. *North Carolina Preservation*, Dec. 1986–Feb. 1987, 66: 16–17. [3].

Bocchi, Greg. 1986. Powder Coatings, A New Technology Takes Off. *The Construction Specifier*, 39(9)(Sept.): 102–105. [3].

———. 1988. Powder Coatings Making Inroads in Metal Construction Market. *Metal Architecture*, April, 4(4): 8–9. [3].

Bock, Gordon. 1988. Glass Notes. *The Old-House Journal*, 16(4)(July/Aug.): 35–43. [5, 6, 8].

———. 1989. The Sash Window Balancing Act. *The Old-House Journal*, 17(5)(Sept./Oct.): 31–40. [6].

———. 1989. Stripping Paint from Windows. *The Old-House Journal*, 17(5)(Oct.): 39–40. [3].

Bordenaro, Michael. 1989. Specifying and Applying Wet Sealants. *Building Design and Construction*, 30(5)(April): 72–74. [4, 5, 6, 7, 8].

Bower, Norman F. 1985. Insurance by the Gallon. *The Construction Specifier*, 38(4)(April): 96–99. [3].

Brewer, Wilfred B. 1986. Avoiding Sealant Failure with Preformed Expanded Foam. *The Construction Specifier*, 39(12)(Dec.): 23–24. [4, 5, 6, 7].

British Iron and Steel Research Association (BISRA). 1977. *How to Prevent Rusting*. London: BISRA. [3].

Brunnell, Gene. 1977. *Built to Last: A Handbook on Recycling Old Buildings*. Washington, DC: Preservation Press. **HP**.

Building Design and Construction. 1987. Survey Reveals Advantages of Wood Doors and Windows. *Building Design and Construction*, 28(6)(June): 157–66. [5, 6].

———. 1988. Focus on Metals in Building Construction. *Building Design and Construction*, 29(6)(June): 67. [3, 4, 6, 7].

———. 1989. Roundtable: New Horizons in Curtain Wall Design and Construction. *Building Design and Construction*, 30(1)(Jan.): 67–73. [7, 8].

Canadian Heritage. 1985. Take It All Off? Advice on When and How to Strip Interior Paintwork. *Canadian Heritage*, Dec. 1985–Jan. 1986: 44–47. [3].

Carter, Roy. 1988. The Ins and Outs of Access Control Systems. *The Construction Specifier*, 41(1)(Jan.): 84–87. [4, 5].

Cassidy, Victor M. 1989. How Windows Get Better. *Modern Metals*, 45(1)(Feb.): 32–46. [6].

Catani, Mario J. 1985. Protection of Embedded Steel in Masonry. *The Construction Specifier*, 38(1)(Jan.): 62–68. [3, 4, 5, 6, 7].

Chase, Sarah B. 1984. Home Work: The ABC's of House Painting. *Historic Preservation*, 36(4)(Aug.): 12–14. [3].

City Limits. 1986. Exterior Paints. *City Limits*, Oct., 11(8): 15. [3].

Commerce Publishing Corp. 1988. *The Woodbook*. Seattle: Commerce Publishing Corp. [2].

Commercial Renovation. 1987. Performance Standards for the Wood Door Industry. *Commercial Renovation*, 9(4)(Aug.): 50. [5].

———. 1987. Product Focus: Door and Hardware Innovations. *Commercial Renovation*, 9(4)(Aug.): 52–53. [4, 5].

———. 1988. New Technology Makes Window Efficiency Shine. *Commercial Renovation*, 10(3)(June): 48–53. [6].

———. 1988. Door Security Hinges on Hardware. *Commercial Renovation*, 10(5)(Oct.): 52–56. [4, 5].

Construction Specifications Institute (CSI). 1988. CSI Monograph 07M411, Precoated Metal Building Panels. Alexandria, VA: Construction Specifications Institute. [3].

———. CSI Monograph 08M710, Finish Hardware. Alexandria, VA: Construction Specifications Institute. [4, 5].

———. 1988. *Specguide* 09900, Painting. Alexandria, VA: Construction Specifications Institute. [3].

Construction Specifier, The. 1979. CS Builders Hardware Conference, Part 1. *The Construction Specifier*, 32(6)(June): 39–49. [4, 5].

———. 1979. CS Builders Hardware Conference, Part 2. *The Construction Specifier*, 32(7)(July): 47–53. [4, 5].

———. 1979. CS Builders Hardware Conference, Part 3. *The Construction Specifier*, 32(9)(Sept.): 55–65. [4, 5].

———. 1983. Twenty-Four Karat Glass. *The Construction Specifier*, 36(3)(March): 28–32. [8].

———. 1983. Rethinking Stained Glass. *The Construction Specifier*, 36(5)(June): 56–62. [8].

———. 1985. Painting to Protect. *The Construction Specifier*, 38(4)(April): 92–94, 123. [3].

Consumer Product Safety Commission. 16 CFR Part 1201. Safety Standards for Architectural Glazing Materials. Bethesda, MD: CPSC. [8].

Copper Development Association, Inc. *Building Expansion Joints*. New York: Copper Development Association. [2].

———. *Clear Film-Coated Copper for Decorative Applications.* New York: Copper Development Association. [3].

———. *Clear Organic Finishes for Copper and Copper Alloys.* New York: Copper Development Association. [3].

———. *Copper, Brass, Bronze, Design Handbook, Architectural Applications.* New York: Copper Development Association. [3].

———. 1980. *Copper, Brass, Bronze Design Handbook, Sheet Copper Applications.* New York: Copper Development Association. [3].

———. *Design Guide: Forgings.* New York: Copper Development Association. [3].

———. *How to Apply Statuary and Patina Finishes.* New York: Copper Development Association. [3].

———. *Joining Copper-Nickel Alloys.* New York: Copper Development Association. [3].

———. *Joining Copper-Tin Alloys (Phosphor Bronzes).* New York: Copper Development Association. [3].

———. *Mechanical Properties of Soldered Copper.* New York: Copper Development Association. [3, 4].

———. *Properties of Clear Organic Coatings on Copper and Copper Alloys.* New York: Copper Development Association. [3].

———. *Sheet Copper Fundamentals.* New York: Copper Development Association. [3].

———. *Soldering and Brazing Copper Tube.* New York: Copper Development Association. [3].

———. *Standard Designations for Copper and Copper Alloys.* New York: Copper Development Association. [3].

———. *Standards Handbook, Part 2—Alloy Data.* New York: Copper Development Association. [3].

———. *Standards Handbook, Part 3—Terminology.* New York: Copper Development Association. [3].

———. *Standards Handbook, Part 7—Cast Products.* New York: Copper Development Association. [3].

———. *Taking Care of the Metal in Your Building.* New York: Copper Development Association. [3].

———. *Welding Handbook.* New York: Copper Development Association. [3, 4].

Cox, Sarah and Billy Edwards. 1987. Keys to Specifying the Right Locking System. *The Construction Specifier,* 40(1)(Jan.): 25–26. [4, 5].

Dahlen, Richard R. and Brian K. Hill. 1986. Sophisticated Window Films. *The Construction Specifier,* 39(6)(June): 60–67. [8].

Dean, Sheldon W. and T. S. Lee, eds. 1988. *Degradation of Metals in the Atmosphere*. Philadelphia: ASTM. [3].

Door and Hardware Institute. 1975. *Recommended Locations for Builder's Hardware for Standard Steel Doors and Frames*. DHI. [4].

———. 1976. *Hardware Reinforcements on Steel Doors and Frames*. DHI. [4].

———. 1979. *Processing Hardware for Custom Aluminum Doors*. DHI. [4].

———. 1979. *Recommended Locations for Builder's Hardware for Custom Steel Doors and Frames*. DHI. [4].

———. 1980. DHI-WDHS-1, *Templet Book Criteria for Wood Doors*. DHI. [5].

———. 1980. DHI-WDHS-2, *Recommended Templet Fasteners for Wood Doors*. DHI. [5].

———. 1980. DHI-WDHS-3, *Recommended Hardware Locations for Wood Flush Doors*. DHI. [5].

———. 1985. *Basic Architectural Hardware*. DHI. [4, 5].

———. 1986. *Installation Guide for Doors and Hardware*. DHI. [4, 5].

Easter, R. Lee. 1985. Can Glass and Hurricanes Mix? *The Construction Specifier*, 38(5)(May): 46–50. [8].

Elswick, Dan. 1987. Preparing Historic Woodwork for Repainting, Part 2—Thermal and Chemical Cleaning. *The New South Carolina State Gazette*, 19(3)(Spring): 4–5. [3] **HP**.

Evans, Lyon D. 1985. Aluminum Windows and Sliding Glass Doors: The New Specs. *The Construction Specifier*, 38(9)(Sept.): 56–63. [6].

———. 1986. Design Techniques for Glazed Curtainwall Retrofit. *Exteriors*, 4(3)(Autumn): 56–60. [7, 8].

Evans, U. R. 1972. *The Rusting of Iron: Causes and Control, Studies in Chemistry No. 7*. London: Edward Arnold, Ltd. [3].

———. 1972. *The Corrosion and Oxidation of Metals: Scientific Principles and Practical Applications*. London: Edward Arnold, Ltd. [3].

Exteriors. 1986. Skyline: Window Standards. *Exteriors*, 4(2)(Summer): 20. [6].

———. 1986. Skyline: Glazing Technics. *Exteriors*, 4(4)(Winter): 24–25. [8].

———. 1987. Racetrack Grandstand Retrofit Is Off and Running. *Exteriors*, 5(2)(Summer): 60. [7].

———. 1987. Skyline: Switchable Coatings. *Exteriors*, 5(3)(Autumn): 42–44. [3].

———. 1987. Skyline: Curtainwall Insulation. *Exteriors*, 5(3)(Autumn): 44–46. [7].

———. 1987. Skyline: Low E Glass Use Rising. *Exteriors*, 5(4)(Winter): 18–19. [8].

———. 1988. Portfolio: Replacement Windows Help School Save History and Money. *Exteriors*, 6(1)(Spring): 70. [6].

———. 1988. Skyline: Glass Strength Debated. *Exteriors*, 6(3)(Autumn): 18–19. [8].

———. 1988. Skyline: Curtainwall Issues. *Exteriors*, 6(4)(Winter): 75. [7].

Federal Specifications. *See* U.S. General Services Administration Specifications Unit.

Fisher, Charles E. 1986. Rehabilitating Historic Windows. *The Construction Specifier,* 39(7)(July): 68–83. [6].

Fishman, Herbert B. 1986. Architectural Copper Work. *The Construction Specifier,* 39(11)(Nov.): 60–66. [3].

Flat Glass Marketing Association. *Glazing Manual.* Topeka: Flat Glass Marketing Association. [8].

———. *Sealant Manual.* Topeka: Flat Glass Marketing Association. [8].

Forest Products Laboratory. *Handbook No. 72—Wood Handbook.* Washington, DC: U.S. Department of Agriculture. [2].

Francis, Geoffrey V. 1987. A Practical Approach to Sealant and Joint Design. *Exteriors,* 5(1)(Spring): 64–71. [4, 5, 6, 7].

———. 1988. New Developments in Structural Glazing. *Exteriors,* 6(1)(Spring). 50–55. [8].

———. 1989. Innovation in Structural Sealant Glazing. *The Construction Specifier,* 42(3)(Mar.): 54–59. [8].

Freund, Eric C. and Gary L. Olsen. 1985. Renovating Commercial Structures: A Primer. *The Construction Specifier,* 38(7)(July): 36–47. [2, 3, 4, 5, 6, 7].

Fulton, Frank. 1988. High Performance Window Design. *The Construction Specifier,* 41(8)(Aug.): 80–84. [6].

George, Louis F. 1988. Specifying Hardware for Schools. *The Construction Specifier,* 41(8)(Aug.): 52–59. [4, 5].

———. 1989. Specifying Hardware in Housing for the Elderly. *The Construction Specifier,* 42(2)(Feb.): 62–66. [4, 5].

———. 1989. Specifying Systems for Access Control. *The Construction Specifier,* 42(6)(June): 52–58. [4, 5].

Glass Tempering Association. 1976. 66-9-20 Revision #3. *Engineering Standards Manual, Section 4, Specifications for Ceramic Enameled Spandrel Glass Fully Tempered or Heat Strengthened.* ASTM. [8].

Goldsmith, Aaron. 1984. Glass: The Invisible Hazard. *Commercial Renovation,* 6(6)(Dec.): 66–67. [8].

Gregerson, John. 1988. Metal Panels Have Designs on New Markets. *Building Design and Construction,* 29(6)(June): 68–71. [7].

———. 1989. New Glass Technologies Solve Site Challenges. *Building Design and Construction,* 30(3)(Mar.): 74–77. [8].

———. 1989. Metal Retrofits Raise Face Value of Older Buildings. *Building Design and Construction,* 30(7)(June): 58–61. [7].

Griffiths, Howard. 1987. Glass: Colors of the City. *The Construction Specifier,* 40(8)(Aug.): 100–107. [8].

Grossen, John F. 1985. New Openings for Plastic Laminate Doors. *The Construction Specifier*, 38(1): 34–39. [5].

Grossi, Anthony F. 1984. A Brief on Low E Glass. *The Construction Specifier*, 37(9)(Sept.): 22–23. [8].

Gypsum Association. 1985. *Using Gypsum Board for Walls and Ceilings*. (GA-201-85). Evanston, IL: Gypsum Association. [2].

———. 1986. *Recommended Specifications: Recommendations for Installation of Steel Fire Door Frames in Steel Stud–Gypsum Board Fire-Rated Partitions*. (GA-219-86). Evanston, IL: Gypsum Association. [2, 4].

Hardingham, David. 1980. Preparing for Painting. *The Old-House Journal* (Oct.): 133–36. [3].

Harvey, John. 1972. *Conservation of Buildings*. London: Baker. **HP**.

Heerwagen, Judith H. 1987. Windowscapes—The Psychology of View. *The Construction Specifier*, 40(8)(Aug.): 31–32. [6].

Heitmann, L. J. 1986. The Trouble with Shadow Boxes. *Exteriors*, 4(2)(Summer): 10. [7, 8].

———. 1986. A Closer Look at Structural Glazing. *Exteriors*, 4(4)(Winter): 12. [8].

———. 1987. The Rain Screen Principle Can Work for You. *Exteriors*, 5(1)(Spring): 12. [7].

———. 1987. The Trouble with Curtainwall. *Exteriors*, 5(2)(Summer): 12. [7].

———. 1987. Proper Value Engineering for Curtainwall. *Exteriors*, 5(3)(Autumn): 12. [7].

———. 1987. Safety First in Overhead Glazing. *Exteriors*, 5(4)(Winter): 12. [8].

———. 1988. On Designing a Spandrel Cavity. *Exteriors*, 6(2)(Summer): 12. [7, 8].

———. 1988. A Question of Compatibility. *Exteriors*, 6(3)(Autumn): 12. [7, 8].

———. 1988. The Role of the Consultant. *Exteriors*, 6(4)(Winter): 12. [1].

Heinz, Thomas A. 1989. Use and Repair of Zinc Cames in Art-Glass Windows. *The Old-House Journal*, 17(5)(Sept./Oct.): 35–38. [8].

Hornbostel, Caleb. 1978. *Construction Materials, Types, Uses, and Applications*. New York: Wiley. [3].

Howell, J. Scott. 1987. Architectural Cast Iron: Design and Restoration. *The Construction Specifier*, 40(7)(July): 70–74. [3].

Hudnut, Richard. 1989. The Hardware Metals. *The Construction Specifier*, 41(8)(Aug.): 31–33. [3, 4, 5].

Insall, Donald W. 1972. *The Care of Old Buildings Today: A Practical Guide*. London: Architectural Press. **HP**.

Insulating Glass Certification Council. *Certified Products Directory—Sealed Insulating Glass*. Cortland, NY: IGCC. [8].

Johnson, Stephen. 1988. Improvement in Domestic Steel Quality Translating into Better Metal Paneling. *Metal Architecture*, 4(9)(Sept.): 5, 85. [7].

Johnston, Bob. 1986. Quality Control Critical with Structural Silicone. *Exteriors,* 4(2)(Summer): 44–50. [8].

Jones, Larry. 1984. Painting Galvanized Metal. *The Old-House Journal,* Jan.–Feb. 12(1): 10–11. [3].

———. 1984. Don't Overlook the Heat Plate. *The Old-House Journal,* Jan.–Feb., 12(1): 12–13. [3].

Kawneer Co., Inc. 1984. *Architectural Finishes.* Norcross, GA: Kawneer. [3].

Kelley, Stephen J. and Dennis K. Johnson. 1989. Looking Out, Window Replacement and Quality Assurance. *The Construction Specifier,* 42(3)(Mar.): 36–42. [6].

Kincaid, Mary. 1982. What Paint Experts Say. *The Construction Specifier,* July, 35(5): 54–61. [3].

———. 1983. Facets of Glass. *The Construction Specifier,* 36(3)(Mar.): 34–40. [8].

Koller, Alice. 1981. Hot-Dip Galvanizing: How and When to Use It. *The Construction Specifier,* 34(8)(Sept.): 47–51. [3, 4, 6, 7].

Labine, Clem. 1982. Restoring Clear Finishes. *The Old-House Journal,* Nov, 10(11): 221, 238–41. [3].

Laminators Safety Glass Association. 1987. *Standards Manual.* Topeka: Laminators Safety Glass Association. [8].

LaQue, F. E. and H. R. Copson, eds. 1963. *Corrosion Resistance of Metals and Alloys,* 2nd ed. New York: Van Nostrand Reinhold. [3].

Latona, Raymond W., Thomas A. Schwartz, and Glen R. Bell. 1988. New Standard Permits More Realistic Curtain Wall Testing. *Building Design and Construction,* 29(11)(Nov.): 42–46. [7, 8].

Leidheiser, Henry, Jr. 1971. *The Corrosion of Copper, Tin, and Their Alloys.* New York: Wiley. [3].

Lesniak, Joseph G. and Judith P. Guy. 1985. The Architectural Hardware Consultant. *The Construction Specifier,* 38(6)(June): 34. [1, 4, 5].

Lofgren, Michael. 1983. Making Building Security "Airtight." *The Construction Specifier,* 36(11)(Nov.): 68–69. [4, 5].

Lowes, Robert. 1988. Abrasive Blasting. *PWC Magazine,* May/June, 50(3): 26–30. [3].

MacDonald, John E. 1978. The Effects of Light on Wood and Wood Finishing Systems. *The Construction Specifier,* 31(3)(Mar.): 23–26. [3].

———. 1978. Architectural Flush Wood Doors. *The Construction Specifier,* 31(3)(Mar.): 28–32. [5].

Mahowald, Dave. 1988. Specifying Paint Coatings for Harsh Environments. *The Construction Specifier,* Oct., 41(10): 13–16. [3].

Maisel, Murry. 1983. Custom Hollow Metal Doors, Problems and Solutions. *The Construction Specifier,* 36(1)(Jan.): 74–77. [4].

Marinelli, Janet. 1988. Architectural Glass and the Evolution of the Store Front. *The Old-House Journal*, 16(4)(July/Aug.): 34–42. [5, 8].

Martens, Charles R. and Sherwin-Williams Co. 1974. *Technology of Paints and Lacquers*. New York: Robert E. Krieger Publishing Co. [3].

Maruca, Mary. 1984. 10 Most Common Restoration Blunders. *Historic Preservation*, Oct., 36(5): 13–17. **HP.**

Mason, Donald. 1983. Selecting an Access Control System. *The Construction Specifier*, 36(11)(Nov.): 71–73. [4, 5].

McAuliffe, William B. 1985. Finish Hardware, A Complex Specification. *The Construction Specifier*, 38(6)(June): 30–37. [4, 5].

McDonald, Timothy B. 1987. Technical Tips: Coatings That Protect Against the Corrosion of Steel. *Architecture*, (July): 101–102. [3].

McInerney, William D. 1987. Selecting the Proper Hinge and Pivot. *The Construction Specifier*, 40(3)(Mar.): 116–122. [4, 5].

McKinley, Robert W. 1986. Saving Energy with Low E Glazing. *Exteriors*, 4(1)(Mar.): 50–58. [8].

———. 1986. Windloading of Glazed Curtain Walls. *Exteriors*, 4(4)(Winter): 32–36. [7, 8].

———. 1989. Insulating Glass: Designing for the 1990's. *Exteriors*, 7(1)(Spring): 12–16. [8].

Metal Architecture. 1987. Building Repainting System Duplicates Coil Coatings in Appearance, Life Expectancy. *Metal Architecture*, Nov., 3(11): 24. [3].

———. 1987. Guide to Metal Wall and Roof Panels. *Metal Architecture*, 3(12)(Dec.): 20, 22–23, 28–31, 38. [7].

———. 1988. Guide to Pre-Insulated Panels. *Metal Architecture*, April, 4(4): 40–41, 43–46. [7].

———. 1988. 1988 Guide to Architectural Coil Coatings. *Metal Architecture*, Nov., 4(11): 22–23. [3].

Metal Lath/Steel Framing Association. 1985. *Lightweight Steel Framing Systems Manual*, 2nd ed. Chicago: Metal Lath/Steel Framing Association. [2].

———. 1986. *Specifications for Metal Lathing and Furring*. Chicago: Metal Lath/Steel Framing Association. [2].

———. *Technical Bulletin No. 18: Fire Rated Metal Lath/Steel Stud Exterior*. Chicago: Metal Lath/Steel Framing Association. [2].

Miller, Michael H. 1984. Selecting the Right Sealant. *The Construction Specifier*, 37(1)(Jan.): 72–76. [4, 5, 6, 7, 8].

Minor, Joseph E. 1987. Accommodating Wind Forces in Glazing Design. *The Construction Specifier*, 40(8)(Aug.): 25–26. [8].

Modern Metals. 1989. Presses on Roll Form Lines: How to Choose the Best. *Modern Metals*, Feb., 45(1): 10–20. [3].

———. 1989. Finishing Forum: Coil Anodized Sheet Survives with Flying Colors. *Modern Metals*, Feb., 45(1): 22–28. [3].

Moit, Dan. 1988. Coatings for Metals: Preplan the Selection. *Metal Architecture*, Sept., 4(9): 8. [3].

Monnich, Joni. 1983. Restorer's Notebook: Beware Brass-Cleaner Damage. *The Old-House Journal*, Oct., 11(8): 180. [3].

Montella, Ralph. 1985. *Plastics in Architecture: A Guide to Acrylic and Polycarbonate*. New York: Marcel Dekker. [8].

Moormann, Ambrose F., Jr. 1982. Paint and the Prudent Specifier. *The Construction Specifier*, 35(5)(July): 69–71. [3].

———. 1983. Paint and the Prudent Specifier: Working Hard to Look Natural. *The Construction Specifier*, 36(1)(Jan.): 84–87. [3].

Munger, Charles G. 1984. *Corrosion Prevention by Protective Coatings*. Houston: National Association of Corrosion Engineers. [3].

Myers, John H. 1981. Preservation Briefs: #9, *The Repair of Historic Wooden Windows*. Washington, DC: Technical Preservation Services Division, U.S. Department of the Interior. [6].

National Association of Architectural Metal Manufacturers (NAAMM). 1988. *Metal Finishes Manual for Architectural and Metal Products*. Chicago: NAAMM. [3].

National Decorating Products Association. 1988. *Paint Problem Solver*. St. Louis: National Decorating Products Association. Also available from Painting and Decorating Contractors of America. [3].

National Electrical Manufacturers Association. NEMA LD 3-1985. *High Pressure Decorative Laminates*. Washington, DC: NEMA. [3, 5].

National Fire Protection Association. 1986. *NFPA 80: Fire Doors and Windows*. Quincy, MA: NFPA. [4, 5, 6].

———. 1988. *NFPA 101: Life Safety Code*. Quincy, MA: NFPA. [4, 5, 6, 7, 8].

———. 1985. *NFPA 105: Smoke and Draft Control Door Assemblies*. Quincy, MA: NFPA. [4, 5].

National Forest Products Association. *Manual for House Framing*. Washington, DC: National Forest Products Association. [2].

———. *National Design Specifications for Wood Construction*. Washington, DC: National Forest Products Association. [2].

———. *Span Tables for Joists and Rafters*. Washington, DC: National Forest Products Association. [2].

National Glass Association. 1987. *Guide to the Federal Glazing Laws and the Model Safety Glazing Code*. McLean, VA: National Glass Association. [8].

———. *Glass Standards: A Collection of National and Voluntary Standards Pertaining to Glass and Glazing*. McLean, VA: National Glass Association. [8].

———. *Glazing Reference Guide*. McLean, VA: National Glass Association. [8].

National Trust for Historic Preservation. 1985. *All About Old Buildings—The Whole Preservation Catalog*. Washington, DC: Preservation Press. An extensive reference containing the names and addresses of many organizations active in the historic preservation field and lists of publications' sources. Anyone facing a preservation problem should obtain this catalog as soon as possible. It will save much time in finding the right organization or data source. **HP.**

National Wood Window and Door Association. *Care and Finishing of Wood Doors*. Park Ridge, IL: NWWDA. [5].

———. *How to Store, Handle, Finish and Maintain Wood Doors*. Park Ridge, IL: NWWDA. [5].

———. ANSI/NWWDA I.S. 1-1986 Series, Industry Standard for Wood Flush Doors. Park Ridge, IL: NWWDA. [5].

———. NWWDA I.S. 2-1988, Industry Standard for Wood Window Units. Park Ridge, IL: NWWDA. [3, 6].

———. NWWDA I.S. 3-1988, Industry Standard for Sliding Patio Doors. Park Ridge, IL: NWWDA. [3, 6].

———. NWWDA I.S. 4-1981, Industry Standard for Water-Repellent Preservative Non-Pressure Treatment for Millwork. Park Ridge, IL: NWWDA. [3, 5, 6].

———. ANSI/NWWDA I.S.6-1986, Industry Standard for Wood Stile and Rail Doors. Park Ridge, IL: NWWDA. [5].

Nicastro, David H. 1988. Uncovering the Reasons for Curtainwall Failure. *Exteriors*, 6(2)(Summer): 26–32. [7, 8].

———. 1989. Parameters for Comparing High-Performance Sealants. *The Construction Specifier*, 42(4)(Apr.): 122–25. [4, 5, 6, 7, 8].

Old-House Journal, The. 1982. Stripping Paint. *The Old-House Journal* (Dec.): 249–52. [3]. **HP.**

———. 1983. 48 Paint Stripping Tips. *The Old-House Journal*, 11(2)(Mar.): 44–45. [3]. **HP.**

———. 1983. Our Opinion of 'Peel Away.' *The Old-House Journal*, 11(4)(May): 80. [3]. **HP.**

———. 1983. Ask OHJ: Paint on Paint. *The Old-House Journal*, 11(9)(Nov.): 202. [3]. **HP.**

———. 1985. Stripping Clinic. *The Old-House Journal*, 13(10)(Dec.): 212B. [3]. **HP.**

———. 1987. Exterior Painting: Problems and Solutions. *The Old-House Journal* (Sept.–Oct.): 35–39. [3]. **HP.**

———. 1988. Commercial Paint Stripping. *The Old-House Journal*, 16(4) (July/Aug.): 29–33. [3]. **HP.**

Olin, Harold B., John L. Schmidt, and Walter H. Lewis. 1983. *Construction Principles, Materials and Methods*, 5th ed. Chicago: U.S. League of Savings Institutions. [2, 3, 4, 5, 6, 7, 8].

Olson, Christopher. 1987. Heightening Impact with Structural Glazing. *Building Design and Construction*, 28(3)(Mar.): 132–36. [8].

Painting and Decorating Contractors of America. 1975. *Painting and Decorating Craftsman's Manual and Textbook,* 5th ed. Falls Church, VA: Painting and Decorating Contractors of America. [3].

———. 1982. *Painting and Decorating Encyclopedia.* Falls Church, VA: Painting and Decorating Contractors of America. [3].

———. 1984. *Painting and Wallcovering: A Century of Excellence.* Falls Church, VA: Painting and Decorating Contractors of America. [3].

———. 1986. *Architectural Specification Manual, Painting, Repainting, Wallcovering and Gypsum Wallboard Finishing,* 3rd ed. Kent, WA: Specifications Services, Washington State Council of the PDCA. [3].

———. 1988. *Hazardous Waste Handbook.* Falls Church, VA: Painting and Decorating Contractors of America. [3].

———. 1988. *The Master Painters Glossary.* Falls Church, VA: Painting and Decorating Contractors of America. [3].

———. 1988. *PDCA Estimating Guide.* Falls Church, VA: Painting and Decorating Contractors of America. [3].

Petersen, Maurice R. 1984. Finishes on Metals: A View from the Field—Part 1. *The Construction Specifier,* 37(12)(Dec.): 36–39. [3].

———. 1985. Finishes on Metals: A View from the Field—Part 2. *The Construction Specifier,* 38(1)(Jan.): 70–73. [3].

Peterson, Charles O., Jr. 1984. Structural Silicone. *The Construction Specifier,* 37(9)(Sept.): 56–63. [8].

Phillips, Morgan, and Judith Selwyn. 1978. Epoxies for Wood Repairs in Historic Buildings. Washington, DC: Technical Preservation Service, U.S. Department of the Interior. [5, 6].

Poore, Patricia. 1983. The Trouble with Pocket Doors. *The Old-House Journal* 11(4)(May): 79, 90–92. [5].

———. 1983. Repairing Top Hung Pocket Doors. *The Old-House Journal* 11(6)(July): 115, 128–29. [5].

———. 1985. Stripping Paint from Exterior Wood. *The Old-House Journal* 13(10)(Dec.): 207–11. [3].

Porcelain Enamel Institute (PEI). 1970. *Guide to Designing with Architectural Porcelain Enamel on Steel.* PEI. [3, 7].

———. 1969. PEI: ALS-105(69). Recommended Specification for Architectural Porcelain Enamel on Aluminum for Exterior Use. PEI. [3].

———. 1970. PEI: S-100(65). Specification for Architectural Porcelain Enamel on Steel for Exterior Use. PEI. [3].

———. 1973. The Weatherability of Porcelain Enamel. PEI. [3].

———. 1986. Color Guide for Architectural Porcelain Enamel. PEI. [3].

———. Bulletin T-2, Test for Resistance of Porcelain Enamel to Abrasion. PEI. [3].

Bibliography

———. Bulletin T-20, Image Gloss Test. PEI. [3].

———. Bulletin T-21, Test for Acid Resistance of Porcelain Enamels. PEI. [3].

———. Bulletin T-22, Cupric Sulfate Test for Color Retention. PEI. [3].

———. Bulletin T-51, Antimony Trichloride Spall Test for Porcelain Enameled Aluminum. PEI. [3].

Quigg, Paul and William Leavitt. 1988. New Test Verifies Rock Wool's Fire Performance. *Exteriors*, 6(4)(Winter): 86–88. [7].

Raeber, John A. 1985. Selecting Door Hardware. *The Construction Specifier*, 38(6)(June): 38–47. [4, 5].

———. 1988. Selecting Door Contract Hardware. *The Construction Specifier*, 41(8)(Aug.): 62–71. [4, 5].

———. 1989. Exposed Metallic Coatings on Glass: A Cautionary Note. *The Construction Specifier*, 42(6)(June): 31–32. [8].

———. 1989. Selecting Door-Hanging Hardware. *The Construction Specifier*, 42(6)(June): 36–42. [4, 5].

Ramsey/Sleeper and the AIA Committee on Architectural Graphic Standards. 1981. *Architectural Graphic Standards*, 7th ed. New York: Wiley. [2, 3, 4, 5, 6, 7, 8].

Reader's Digest Association. 1973. *Reader's Digest Complete Do-it-yourself Manual*. Pleasantville, NY: The Reader's Digest Association. [4, 5, 6, 8].

Reynolds, James. 1986. Specifying Coated Glazing Material. *The Construction Specifier*, 39(6)(June): 46–59. [8].

Rich, Jack C. 1947, 1988. *The Materials and Methods of Sculpture*. New York: Dover. [3].

Rush, Richard. 1987. Refining Window Energy Performance. *Building Design and Construction*, 28(12)(Dec.): 148–53. [6, 8].

Russo, Michael. 1987. Improving the Art of Glass Design. *Exteriors* 5(3)(Autumn): 6. [8].

———. 1989. Modified Bitumens, Low-E Glass Hot in '89. *Exteriors*, 7(1)(Spring): 18–23. [8].

Safety Glazing Certification Council. *Certified Products Directory—Safety Glazing Materials Used in Buildings*. Cortland, NY: SGCC. [8].

Sanford, A. G. 1987. Stress Tests and Safety Factors in Structural Glazing. *The Construction Specifier*, 40(6)(June): 31–37. [8].

Scassellati, Rudy R. 1988. Architectural Flush Doors. *The Construction Specifier*, 41(4)(Apr.): 112–20. [5].

Scharfe, Thomas R. 1988. Meeting the Challenges of Sloped Glazing. *Building Design and Construction*, 29(3)(Mar.): 140–47. [7, 8].

———. 1988. New Metal Coating Technologies Enhance Design Opportunities. *Building Design and Construction*, 29(6)(June): 86–89. [3].

Schulthesis, Joseph A. 1985. Guidelines for Sloped Glazing. *The Construction Specifier*, 38(3)(Mar.): 90–95. [7, 8].

———. 1986. Sloped Glazing with Structural Silicone. *The Construction Specifier*, 39(6)(June): 36–45. [7, 8].

Scott, Gerald. 1965. *Atmospheric Oxidation and Antioxidants*. New York: Elsevier. [3].

Screen Manufacturers Association (SMA). 1976. ANSI/SMA 1004-1976, Specifications for Aluminum Tubular Frame Screens for Windows. Chicago: SMA. [6].

———. 1976. ANSI/SMA 2005-1976, Specifications for Aluminum Sliding Screen Doors. Chicago: SMA. [6].

Sealant Engineering and Associated Lines. 1987. SEAL Guide Specification for Sealants and Caulking. San Diego: SEAL. [4, 5, 6, 7, 8].

Sherwin-Williams Co. 1988. Painting and Coating Systems for Specifiers and Applicators. Sherwin-Williams. [3].

Sittnick, Ralph. 1985. The Evolution of the Electromagnetic Lock. *The Construction Specifier*, 38(6)(June): 48–49. [4, 5].

Sloan, Julie L. 1983. A Stained Glass Primer. *Historic Preservation*, 35(6)(Nov.–Dec.): 14–17. [8].

Smith, Charles F., Jr. 1989. Specifying Weatherstripping for Energy, Smoke, and Sound Control. *The Construction Specifier*, 42(6)(June): 44–51. [4, 5].

Southern Pine Inspection Bureau. *Standard Grading Rules for Southern Pine Lumber*. Pensacola, FL: Southern Pine Inspection Bureau. [2].

Stanbrough, Jerry. 1986. Energy Efficiency: The Curtainwall as Filter. *Exteriors*, 4(2)(Summer): 34–39. [7, 8].

Stanwood, Les. 1983. Acid Rain: A Cloudy Issue. *The Construction Specifier*, 36(11)(Nov.): 74–79. [3].

Steel Door Institute. 1985. SDI-100-85, Recommended Specifications for Standard Steel Doors and Frames. SDI. [4].

———. 1982. SDI-105-82, Recommended Erection Instructions for Steel Frames. SDI. [4].

———. SDI-106, Recommended Standard Door Type Nomenclature. SDI. [4].

———. 1984. SDI-107-84, Hardware on Steel Doors. SDI. [4].

———. 1983. SDI-108-83, Recommended Selection and Usage Guide for Standard Steel Doors. SDI. [4].

———. 1975. SDI-109-75, Hardware for Standard Steel Doors and Frames. SDI. [4].

———. 1984. SDI-110-84, Standard Steel Doors and Frames for Masonry Construction. SDI. [4].

———. SDI-111, Recommended Standard Details for Steel Doors and Frames. SDI. [4].

———. 1979. SDI-112-79, Test Procedure and Acceptance Criteria for Apparent Thermal Performance of Steel Door and Frame Assemblies. SDI. [4].

———. 1979. SDI-114-79, Test Procedure and Acceptance Criteria for Acoustical Performance of Steel Door and Frame Assemblies. SDI. [4].

———. 1979. SDI-116-79, Standard Test Procedure and Acceptance Criteria for Rate of Air Flow through Closed Steel Door and Frame Assemblies. SDI. [4].

———. SDI-117, Manufacturing Tolerances for Standard Steel Doors and Frames. SDI. [4].

———. 1976. SDI-118-76, Basic Fire Door Requirements. SDI. [4].

Steel Structures Painting Council (SSPC). 1983. *Steel Structures Painting Manual*, Vol. 1, *Good Paint Practice*, 2nd ed. Pittsburgh: SSPC. [3].

———. 1983. *Steel Structures Painting Manual*, Vol. 2, *Systems and Specifications*, 2nd ed. Pittsburgh: SSPC. [3].

Stubbs, M. Stephanie. 1986. Glued-on Glass. *Architectural Technology* (May–June): 46–51. [8].

Tatum, Rita. 1984. The Return of Wood Windows. *The Construction Specifier*, 37(9)(Sept.): 51–55. [6].

———. 1987. The Evolution of Energy-Saving Windows. *The Construction Specifier*, 40(8)(Aug.): 33–34. [6, 8].

Taylor, Paul. 1983. Glass Skyscrapers Vulnerable. *The Construction Specifier*, 36(11)(Nov.): 10–12. [7, 8].

Technologies Media Corp. 1989. Architectural Hardware. *Building Design and Construction*, 30(12)(Oct.): 115–43. [4, 5].

Thomsett, Michael C. 1985. Should You Hire a Consultant? *The Construction Specifier*, 38(10)(Oct.): 23–24. [1].

Ting, Raymond. 1986. Metal Panel Behavior in Exterior Wall Design. *Exteriors*, Autumn: 65–69. [7].

———. 1987. Ensuring a Trouble-Free Curtainwall Renovation. *Exteriors*, 5(3)(Autumn): 64–66. [7].

———. 1988. Designing a Leak-Free Curtainwall System. *Exteriors*, 6(3)(Autumn): 30–34. [7].

———. 1989. Performance Parameters of Composite Foam Panels. *Metal Architecture*. 5(9)(Sept.): 7–8, 73–74. [7].

Trechsel, Heinz. 1988. Specifying an Energy Efficient Thermal Window. *Exteriors*, 6(3)(Autumn): 36–40. [6, 8].

Truss Plate Institute. *Design Specifications for Light Metal Plate Connected Wood Trusses*. Madison, WI: Truss Plate Institute. [2].

Umlauf, Elyse. 1989. Specifying the Right Wood Windows and Doors. *Building Design and Construction*, 30(7)(June): 100–103. [5, 6].

United States Gypsum Co. 1972. *Red Book: Lathing and Plastering Handbook*, 28th ed. Chicago: United States Gypsum Co. [2].

392 Bibliography

———. 1987. *Gypsum Construction Handbook*, 3rd ed. Chicago: United States Gypsum Co. [2].

U.S. Department of the Army. *Technical Manual TM 5-801-2. Historic Preservation Maintenance Procedures*. Washington, DC: Department of the Army. **HP**.

———. 1980. *Painting: New Construction and Maintenance*, EM 1110-2-3400. Washington, DC: Superintendent of Documents, U.S. Government Printing Office. [3].

———. 1982. Corps of Engineers Guide Specifications, Military Construction, CEGS-08110, Steel Doors and Frames. Office of the Chief of Engineers, Department of the Army. [4].

———. 1981. Corps of Engineers Guide Specifications, Military Construction, CEGS-09910, Painting, General. Office of the Chief of Engineers, Department of the Army. [3].

———. 1980. Corps of Engineers Guide Specifications, Military Construction, CEGS-11701, Casework, Metal and Wood. Office of the Chief of Engineers, Department of the Army. [5, 6].

U.S. Departments of the Army and the Air Force. 1965. *Building Construction Materials and Practices, Caulking and Sealing*. Washington, DC: U.S. Government Printing Office. [4, 5, 6, 7].

U.S. Departments of Army, the Navy, and the Air Force. 1969. *Technical Manual TM 5-618, Paints and Protective Coatings*. Washington, DC: U.S. Government Printing Office. [3].

U.S. Department of Commerce. *PS 1—U.S. Product Standard for Construction and Industrial Plywood*. Washington, DC: U.S. Department of Commerce. [2].

———. *PS 20—American Softwood Lumber Standard*. Washington, DC: U.S. Department of Commerce. [3].

———. 1968. *Organic Coatings BSS 7*. Washington, DC: U.S. Department of Commerce, National Bureau of Standards. [3].

U.S. Department of Defense. 1970. Military Specification MIL-P-21035. Paint, High Zinc Dust Content, Galvanizing Repair. DOD. [3].

U.S. Department of the Navy, Naval Facilities Engineering Command. 1983 (Mar.). Guide Specifications Section 06100, Rough Carpentry. Department of the Navy. [2].

———. 1984 (Sept.). Guide Specifications Section 06200, Finish Carpentry. Department of the Navy. [5, 6].

———. 1981 (Aug.). Guide Specifications Section 07920, Sealants and Caulkings. Department of the Navy. [4, 5, 6, 7].

———. 1986 (Feb.). Guide Specifications Section 08110, Steel Doors and Frames. Department of the Navy. [4].

———. 1984 (Feb.). Guide Specifications Section 08120, Aluminum Doors and Frames. Department of the Navy. [4].

———. 1980 (Feb.). Guide Specifications Section 08210, Wood Doors. Department of the Navy. [5].

———. 1982 (June). Guide Specifications Section 08520, Aluminum Windows. Department of the Navy. [6].

———. 1980 (Apr.). Guide Specifications Section 08610, Wood Windows. Department of the Navy. [6].

———. 1985 (Mar.). Guide Specifications Section 08710, Finish Hardware. Department of the Navy. [4, 5].

———. 1983 (Dec.). Guide Specifications Section 08800, Glazing. Department of the Navy. [8].

———. 1984 (Feb.). Guide Specifications Section 09100, Metal Support Systems. Department of the Navy. [2].

———. 1986 (Feb.). Guide Specifications Section 09910C, Painting of Buildings (Field Painting). Department of the Navy. [3].

———. 1983 (Jan.). Guide Specifications Section 10201, Metal Wall and Door Louvers. Department of the Navy. [4, 5].

U.S. General Services Administration. 1971. (Feb.). Public Building Services Guide Specification, Section 3-0990, Painting and Finishing, Renovation, Repair and Improvement. U.S. General Services Administration. [3].

———. 1973. (Oct.). Public Building Services Guide Specification, Section 4-0990.01, Painting and Finishing. U.S. General Services Administration. [3].

U.S. General Services Administration Specifications Unit. Federal Specification L-S-125, Screening, Insect, Nonmetallic. Washington, DC: GSA Specifications Unit. [6].

———. Federal Specification RR-W-365, Wire Fabric (Insect Screening). Washington, DC: GSA Specifications Unit. [6].

———. Federal Specification TT-C-535, Coating, Epoxy, Two Component, for Interior Use on Metal, Wood, Wallboard, Painted Surfaces, Concrete, and Masonry. Washington, DC: GSA Specifications Unit. [3].

———. Federal Specification TT-C-542E, Coating, Polyurethane, Oil-free, Moisture Curing. Washington, DC: GSA Specifications Unit. [3].

———. Federal Specification TT-E-489G, Enamel, Alkyd, Gloss (For Exterior and Interior Surfaces). Washington, DC: GSA Specifications Unit. [3].

———. Federal Specification TT-E-496, Enamel, Heat-resisting, (400 Deg F), Black. Washington, DC: GSA Specifications Unit. [3].

———. Federal Specification TT-E-505A, Enamel, Odorless, Alkyd, Interior, High Gloss, White and Light Tints. Washington, DC: GSA Specifications Unit. [3].

———. Federal Specification TT-E-506K, Enamel, Alkyd, Gloss, Tints and White (For Exterior and Interior Surfaces). Washington, DC: GSA Specifications Unit. [3].

———. Federal Specification TT-E-508C, Enamel, Interior, Semigloss, Tints and White. Washington, DC: GSA Specifications Unit. [3].

———. Federal Specification TT-E-509B, Enamel, Odorless, Alkyd, Interior, Semigloss, White and Tints. Washington, DC: GSA Specifications Unit. [3].

———. Federal Specification TT-E-527C, Enamel, Lusterless. Washington, DC: GSA Specifications Unit. [3].

———. Federal Specification TT-E-543A, Enamel, Interior, Undercoat, Tints and White. Washington, DC: GSA Specifications Unit. [3].

———. Federal Specification TT-E-545, Enamel, Odorless, Alkyd, Interior, Undercoat, Flat, Tints and White. Washington, DC: GSA Specifications Unit. [3].

———. Federal Specification TT-E-1593B, Enamel, Silicone Alkyd Copolymer, Gloss (for Exterior and Interior Use). Washington, DC: GSA Specifications Unit. [3].

———. Federal Specification TT-F-322D, Filler Two-component Type: For dents, Cracks, Small-holes, and Blow Holes. Washington, DC: GSA Specifications Unit. [3].

———. Federal Specification TT-F-336E, Filler, Wood, Paste. Washington, DC: GSA Specifications Unit. [3].

———. Federal Specification TT-F-340C, Filler, Wood, Plastic. Washington, DC: GSA Specifications Unit. [3].

———. Federal Specification TT-L-58E, Lacquer, Spraying, Clear and Pigmented for Interior Use. Washington, DC: GSA Specifications Unit. [3].

———. Federal Specification TT-L-190D, Linseed Oil, Boiled (For Use in Organic Coatings). Washington, DC: GSA Specifications Unit. [3].

———. Federal Specification TT-L-201A, Linseed Oil, Heat Polymerized. Washington, DC: GSA Specifications Unit. [3].

———. Federal Specification TT-P-25E, Primer Coating, Exterior (Undercoat for Wood, Ready-mixed, White and Tints). Washington, DC: GSA Specifications Unit. [3].

———. Federal Specification TT-P-28E, Paint, Aluminum, Heat Resisting. Washington, DC: GSA Specifications Unit. [3].

———. Federal Specification TT-P-29J, Paint, Latex Base, Interior, Flat, White and Tints. Washington, DC: GSA Specifications Unit. [3].

———. Federal Specification TT-P-30E, Paint, Alkyd, Odorless, Interior, Flat White and Tints. Washington, DC: GSA Specifications Unit. [3].

———. Federal Specification TT-P-37D, Paint, Alkyd Resin; Exterior Trim, Deep Colors. Washington, DC: GSA Specifications Unit. [3].

———. Federal Specification TT-P-47F, Paint, Oil, Nonpenetrating-flat, Ready-mixed Tints and White (for Interior Use). Shakes and Rough Siding. Washington, DC: GSA Specifications Unit. [3].

———. Federal Specification TT-P-52D, Paint, Oil (Alkyd Oil), Wood Shakes and Rough Siding. Washington, DC: GSA Specifications Unit. [3].

———. Federal Specification TT-P-55B, Paint, Polyvinyl Acetate Emulsion, Exterior. Washington, DC: GSA Specifications Unit. [3].

———. Federal Specification TT-P-81E, Paint, Oil, Alkyd, Ready Mixed Exterior, Medium Shades. Washington, DC: GSA Specifications Unit. [3].

———. Federal Specification TT-P-86G, Paint, Red-lead-base, Ready-mixed. Washington, DC: GSA Specifications Unit. [3].

———. Federal Specification TT-P-615D, Primer Coating, Basic Lead Chromate, Ready Mixed. Washington, DC: GSA Specifications Unit. [3].

———. Federal Specification TT-P-636D, Primer Coating, Alkyd, Wood and Ferrous Metal. Washington, DC: GSA Specifications Unit. [3].

———. Federal Specification TT-P-641G, Primer Coating; Zinc Dust–Zinc Oxide (For Galvanized Surfaces). Washington, DC: GSA Specifications Unit. [3].

———. Federal Specification TT-P-645A, Primer, Paint, Zinc Chromate, Alkyd Type. Washington, DC: GSA Specifications Unit. [3].

———. Federal Specification TT-P-650C, Primer Coating, Latex Base, Interior, White (For Gypsum Wallboard). Washington, DC: GSA Specifications Unit. [3].

———. Federal Specification TT-P-664C, Primer Coating, Synthetic, Rust-inhibiting, Lacquer-resisting. Washington, DC: GSA Specifications Unit. [3].

———. Federal Specification TT-P-791A, Putty, Pure-Linseed-Oil (For) Wood-sash-glazing. Washington, DC: GSA Specifications Unit. [3].

———. Federal Specification TT-P-1511A, Paint, Latex-base, Gloss and Semi-gloss, Tints and White (for Interior Use). Washington, DC: GSA Specifications Unit. [3].

———. Federal Specification TT-S-176E, Sealer, Surface, Varnish Type, Floor, Wood and Cork. Washington, DC: GSA Specifications Unit. [3].

———. Federal Specification TT-S-300A, Shellac, Cut. Washington, DC: GSA Specifications Unit. [3].

———. Federal Specification TT-S-708A, Stain, Oil; Semi-transparent, Wood, Exterior. Washington, DC: GSA Specifications Unit. [3].

———. Federal Specification TT-S-711C, Stain, Oil Type, Wood, Interior. Washington, DC: GSA Specifications Unit. [3].

———. Federal Specification TT-T-291F, Thinner, Paint, Mineral Spirits, Regular and Odorless. Washington, DC: GSA Specifications Unit. [3].

———. Federal Specification TT-V-86C, Varnish, Oil, Rubbing (For Metal and Wood Furniture). Washington, DC: GSA Specifications Unit. [3].

Valdes, Noel. 1988. Low E Glass. *The Construction Specifier,* 41(8)(Aug.): 71–78. [8].

Vild, Donald J. 1986. Glass and the Building Codes. *Exteriors,* 4(1)(Mar.): 12. [8].

———. 1986. Glass: The Energy Saver. *Exteriors,* 4(2)(Summer): 12. [8].

———. 1986. The Case for Sloped Glazing. *Exteriors,* 4(3)(Autumn): 10. [7, 8].

———. 1986. Fully Tempered vs. Heat-Strengthened Glass. *Exteriors,* 4(4)(Winter): 10. [8].

———. 1987. Why Glass Goes Bad. *Exteriors*, 5(1)(Spring): 10. [8].

———. 1987. The Overhead Glazing Controversy. *Exteriors*, 5(3)(Autumn): 10. [7, 8].

———. 1987. Who Selects the Glass. *Exteriors*, 5(4)(Winter): 10. [8].

———. 1988. Proper Engineering of Structural Silicone. *Exteriors*, 6(1)(Spring): 10. [8].

———. 1988. What's Wrong with Glass Spandrels? *Exteriors*, 6(2)(Summer): 10. [7, 8].

———. 1988. How Much Glass Will Break? *Exteriors*, 6(3)(Autumn): 10. [8].

———. 1988. Some Precautions When Specifying Low-E Glass. *Exteriors*, 6(4)(Winter): 7. [8].

———. 1988. Clearing up Building Code, Glass Selection Confusion. *Exteriors*, 6(4)(Winter): 76–80. [8].

———. 1989. More Problems with Glass Spandrels. *Exteriors*, 7(1)(Spring): 7. [7, 8].

Weaver, Martin E. 1988. Acid Rain. *The Construction Specifier*, 41(7)(July): 54–62. [3].

———. 1989. Fighting Rust, Part I: A Backgrounder. *The Construction Specifier*, 42(5)(May): 143–45. [3].

———. 1989. Fighting Rust, Part II: Remedies. *The Construction Specifier*, 42(6)(June): 129–30. [3].

———. 1989. Caring for Bronze. *The Construction Specifier*, 42(7)(July): 58–66. [3].

Weeks, Kay D. and David W. Look. 1982. Exterior Paint Problems on Historic Woodwork. Washington, DC: Technical Preservation Services, U.S. Department of the Interior, Preservation Brief No. 10. **HP**.

Weismantel, Guy E., ed. 1981. *Paint Handbook*. New York: McGraw-Hill. [3].

West Coast Lumber Inspection Bureau. *Standard Grading Rules for West Coast Lumber*. Portland, OR: West Coast Lumber Inspection Bureau. [2].

Western Lath, Plaster and Drywall Contractor's Association. 1988. *Plaster/Metal Framing System/Lath Manual*. New York: McGraw-Hill. [2, 3].

Western Wood Products Association (WWPA). *A-2, Lumber Specifications Information*. Portland, OR: WWPA. [2].

———. *Grade Stamp Manual*. Portland, OR: WWPA. [2].

———. *Grading Rules for Western Lumber*. Portland, OR: WWPA. [2].

———. *Western Woods Use Book*. Portland, OR: WWPA. [2].

———. *Wood Frame Design*. Portland, OR: WWPA. [2].

Wherry, Allen P. 1986. Standard Steel Doors for Customized Design. *The Construction Specifier*, 39(7)(July): 106–111. [4].

Wilson, Forrest. 1984. *Building Materials Evaluation Handbook*. New York: Van Nostrand Reinhold. [3].

Wilson, Randy J. 1987. Sloped Glazing and the Rain Screen Principle. *The Construction Specifier*, 40(8)(Aug.): 94–98. [7, 8].

Wright, Gordon. 1987. Trends in Specifying Architectural Coatings. *Building Design and Construction*, 28(6)(June): 188–92. [3].

———. 1988. Curtain Walls Drawing More Critical Attention. *Building Design and Construction*, 29(3)(Mar.): 162–65. [7, 8].

———. 1989. Storefront Designs Accommodate Multiple Requirements. *Building Design and Construction*, 30(1)(Jan.): 60–62. [4].

Wysocki, Robert J. 1988. A Treasure in Stained Glass. *The Construction Specifier*, 41(8)(Aug.): 49–50. [8].

Zinc Institute. *Zinc Coatings for Corrosion Protection*. New York: Zinc Institute. May not be available, since the Zinc Institute is no longer in operation (mid-1989).

Zingeser, Joel P. 1988. Applauding Glass. *The Construction Specifier*, 41(8)(Aug.): 47–48. [8].

Index

Acrylic plastic sheets, 319-320
Adhesive, mirror glass, 321
AIA Service Corporation, 40, 190, 223, 268, 303, 344
Air infiltration, 235, 277
Air leaks, 264, 298-299
Alclad, 66
Alterations, 25-28
  controls, 25-26
  making of, 27-28
  materials for, 26-27
Aluminized steel, 65-66
Aluminizing, 63
Aluminum, 49-52
  alloy designations, 50
  anodic coatings, 59-60, 113
  as fabricated finish, 58
  chrome plated, 66
  clad wood windows, 45
  cleaning, 117
  doors and frames, 148-149, 151, 161-165, 167, 169, 173
  finishes for, 57-60, 80
  finish hardware, 167, 169
  glazing into, 326
  heat treatment of, 50-51
  louvers, 165
  mechanical and chemical cleaning of, 57-60
  porcelain enamel on, 68
  products, 51-52
  sliding glass doors, 233-236
  temper designations, 50
  windows, 233-236
Aluminum Association (AA), 50, 57, 59, 60, 148, 272
American Architectural Manufacturers Association (AAMA), 69, 137, 148, 173, 190, 233-235, 270, 271, 272, 303
American Iron and Steel Institute (AISI), 49, 148
American National Standards Institute (ANSI), 148, 193, 228, 272
American Society for Metals, 138
Anchors, metal frame, 169-170
Annealing, 315
Anodic coatings, 58, 59-60
Anodizing. *See* Anodic coatings
Architects
  as help for building owners or managers, 7-8
  other architects as help for, 15
  pre-work on-site examination by, 18-19
  product manufacturers as help for, 15-16
  professional help for, 15-17
  specialty consultants as help for, 16-17
Architectural Hardware Consultants (AHC), 14, 167, 184, 196, 218

Index

Architectural Woodwork Institute (AWI), 43, 138, 193, 224, 237, 247, 268
Art glass, 319, 321, 343
Associated Laboratories, Inc. (ALI), 307
Automatic operators, 168, 207
Awning windows, 230, 235, 239

Baking-finish coatings, 80, 113, 136
Bent glass, 319
Black, as rolled finish, 55
Bow/bay windows, 239
Brass, 52, 66
Bronze, 52
Buffed finishes, 58
Builder Hardware Manufacturers Association (BHMA), 193
Building contractors. *See* Contractors, building
Building owners/managers
   architects and engineers as help for, 7–8
   general contractors as help for, 8–9
   manufacturers as help for, 9–11
   pre-work on-site examination by, 18
   professional help for, 5–6
   specialty consultants as help for, 12–15
   specialty contractors as help for, 11–12

Cadmium, 54, 66
Carbon steel. *See* Steel, carbon
Casehardening, 48
Casement windows, 230, 235, 239
Certified Door Consultants (CDC), 14, 167
Channel-glazing compounds, 320
Chemical cleaning and finishes
   for aluminum, 57–59
   for copper alloys, 61–62
   repair of finishes, 113
   for steel, 56, 57
Chemical finishes. *See* Chemical cleaning and finishes
Chromium, 54, 66
Cladding, 264
Clad wood sliding glass doors, 45
Clad wood windows, 45
Clear coatings, 80
Clear float glass, 312
Clips, glazing, 321
Closet folding doors, 197
Coated glass, 309, 310, 311, 317–318, 330, 335
Coatings. *See also* specific types, e.g., Anodic coatings; Laminated coatings
   factory, 113
   flash, 57
   inorganic

     laminated, 68–69
     on metal, 63–69
     repair of, 113
     vitreous, 67–68
   metallic, 63
   organic, 58, 69–86
     applying, 82–86, 136–137
     blistering, 130
     chalking, 128
     checking and alligatoring, 129
     cracking, peeling, scaling, 130–131
     crazing, 129
     factory-applied, 79–80
     factory primers for metal, 81
     failure to dry, 131
     field touch-up, 82
     improper finish application, 95–98
     manufacturers and products, 70–71
     materials and manufacturers for refinishing, 121–122
     naturally aged, 131
     opaque, 72
     paints (*see* Paints)
     preparation for refinishing, 122–135
     preparation of metal surfaces, 79
     preparation of wood surfaces, 78–79
     preparing site for field application, 81–82
     removal, 132–135
     repair of, 113
     runs and sags, 131
     shop primers for wood, 80–81
     soft, gummy, 129
     stains and discolorations, 127–128
     standards, 69
     systems, 75–76, 122
     telegraphing or shadowing, 128
     transparent, 72–75, 131, 132, 136
     wash-off, 131
     wrinkles, 129
   plastic, 68–69, 113
   powder, 77
   resinous, 58
   transparent, 72–75, 113, 131, 132, 136
Column-cover-and-spandrel curtain wall systems, 272
Combination windows, 232, 236, 239
Compounds, glazing, 320, 336, 338
Compressible fillers (rods), 321
Concrete structures, 31–32
Condensation, 277, 281
Contractors, building
   as help for building owners and managers, 8–9
   pre-work on-site examination, 19–20
   professional help for, 17
Conversion coatings, 57, 62

Index

Copper alloys, 52–54
  cleaning, 118
  conversion coatings for, 62
  designations, 53
  doors and frames, 148–149, 151, 160, 167, 169
  finishes for, 61–62
  finish hardware, 167, 169
  glazing into, 326
  mechanical and chemical cleaning, 61–62
  non-etched cleaned, 62
  products, 53–54
Copper Development Association (CDA), 53, 116, 118, 139, 148, 191, 229
Copper plating, 66
Corrosion, 99–103, 111
  chemical attack, 100–101
  errors that lead to, 102–103
  galvanic, 101–102
  oxidation, 100
  structural, 102
  weathering and exposure, 100
Curtain wall. *See also* Glazed curtain walls
  consultants, 12
  definition, 271

Damage consultants, 12–13
Deflection, 30, 32, 33
Demolition, 20–25
  controls, 20–21
  disposition of removed materials, 24–25
  performance of work, 22–24
  protection of persons and property, 21–22
Differential seals, 281
Directional textured mechanical finishes, 58
Door and Hardware Institute (DHI), 14, 207–208
Door consultants, 12
Doors and frames. *See also* specific materials, e.g., Aluminum, doors and frames; specific types of doors and frames, e.g., Sliding glass doors
  adjustments, cleaning, protection, 174, 211
  controls for, 149–152, 194–196, 221
  finish hardware, 167–169
  frame supports and anchors, 169–170
  louvers, 165–167, 173–174, 190, 197–198, 201–202, 211, 223
  metal, 147–191
  standards for, 148–149, 187–188, 193–194
  wood, 192–224

Double-hung windows, 230, 235, 239
Dutch doors, 155

Edge blocking, 325
Electroplated and metallic coatings, 58, 63
Emergencies, 3, 6–7
Engineers
  as help for building owners or managers, 7–8
  other engineers as help for, 15
  pre-work on-site examination by, 18–19
  product manufacturers as help for, 15–16
  professional help for, 15–17
  specialty consultants as help for, 16–17
Entranceways, 162–163, 165
Excess structure movement, 30–31
Expansion/contraction, 30, 32, 33
Extruded aluminum doors, 161–165

Face-glazing compounds, 320
Factory coatings, 113
Failure
  concrete, stone, or masonry walls or partitions as cause of, 37–39
  excess structure movement as cause of, 30–31
  of glazed curtain walls, 287–294
  of glazing, 330–339
  of immediate substrate, 99
  to maintain applied finishes, 107–108
  of metal doors and frames, 175–182
  metal wall or partition framing as cause of, 35–36
  miscellaneous elements as cause of, 39–40
  to protect materials and finishes, 106–107
  steel or concrete structures as cause of, 31–32
  types and conditions, 3
  underlying metal finishes, 105
  of windows and sliding glass doors, 250–256
  of wood doors and frames, 211–215
  wood structure, wall, or partition framing as cause of, 32–35
Film laminate, 77
Finishes. *See also* specific finishes
  for aluminum, 57–60, 80
  black, as rolled, 55
  chemical, 56, 57, 62, 113
  for copper alloys, 61–62
  for glazed curtain walls, 280
  mechanical, 55–57, 61–62, 113
  metal, 54–55, 69–110
  opaque, 44
  patterned, 57, 58, 61

polished, 57
sliding glass door, 233–234, 238
steel window, 229–230
transparent, 44
wood (*see* Wood, finishes)
wood window, 45, 238
Finish hardware
  for metal swinging doors, 167–169
  for sliding glass doors, 240–241
  for windows, 232, 236, 240–241
  for wood doors, 196, 206–208
Fins (glass mullions), 328
Fire-resistance-rated wire glass, 306
Fixed windows, 232, 236, 239
Flash coating, 57
Flat Glass Marketing Association (FGMA), 306, 324
Float glass, 312, 313
Fluoropolymer finishes, 113
Flush-panel doors, 152–156, 161
Flush wood doors, 198–199
Folding doors, 200
Forensic consultants. *See* Damage consultants
Forest Products Laboratory, 41
Fracture, 103–105, 111
  brittle, 104
  ductile, 104
  errors that lead to, 104–105
  fatigue, 104
  stress-corrosion, 104
Frames. *See* Doors and frames; specific frame types
Fully tempered glass, 315, 317

Galvanizing, 63–65, 112
Gaskets, glazing, 321, 325, 335, 336, 338
General building contractors. *See* Contractors, building
Glass. *See also* Glazing; specific types of glass, e.g., Tinted glass
  broken, 335
  cracking, 332, 335
  definition, 311
  glass-to-glass joints, 327
  mullions, 328
Glazed curtain walls, 270–304
  adjustments, cleaning, protection, 287
  air/water leaks, 298–299
  allowable deflection, 278
  bent, twisted, deformed components, 296–297
  controls, 273–276
  design, 274, 281–283, 288–291
  in existing construction, 300–303
  fabrication, 281–283, 291
  failure of, 287–294
  finishes, 280

  general requirements, 271–285
  improper design, 288–291
  improper fabrication, 291
  improper maintenance, 293–294
  installation, 285–287, 292–293, 301–303
  loose, sagging, out-of-line components, 297
  materials, 279–280, 301
  miscellaneous components and accessories, 283–285
  natural aging, 294
  panels, 282–283, 299–300
  performance requirements, 276–278
  preconstruction testing, 274
  rain-screen principle, 282
  repairing, 294–300
  samples, 275
  shop drawings, 274–275
  standards, 272–273
  structural strength, 277
  systems, 271–272
  warranties, 276
Glazing, 305–344
  accessories, 306, 340–341
  controls, 306–309, 340
  for curtain walls, 283 (*see also* Glazed curtain walls)
  definition, 305
  in existing construction, 339–344
  failures, 330–339
  field construction mock-ups, 308
  general requirements, 305–311
  improper design, 331–334
  improper fabrication, 335
  improper installation, 335–337
  improper modifications, 338–339
  inappropriate selection, 334
  installation in metal doors and frames, 173–174, 190
  installation in wood doors and frames, 211, 223
  materials, 311–321, 340–341 (*see also* specific glass types)
    bad, 330–331
    definition, 305
    glass types and quality, 311–319
    miscellaneous, 320–321
    plastic, 319–320, 328
  in new construction, 321–329
    cleaning and protection, 329
    delivery, storage, handling, 322
    design of channel, 323
    general requirements, 323–326
    preparation for, 322–323
    specific requirements, 326–329
  performance requirements, 310–311, 340
  preconstruction meeting, 308
  protection and maintenance, 337–338

Glazing *(cont.)*
  standards, 306, 340
  warranties, 309–310, 340
  for windows and sliding glass doors, 228, 265, 267
Grilles, 284
Gypsum Association, 41

Handicapped-person hardware, 168, 206
Hardware consultants, 12
Hazardous materials, 134–135
Heat-absorbing glass, 334
Heat-treated glass, 306, 315, 331
High-performance glass, 317
Hinged-access windows, 232
Horizontal sliding windows, 236, 239
Hot-dip process, 63, 64

Impact loading, 310
Inorganic coatings. *See* Coatings
Insulating glass, 309, 310, 324, 330, 335
Insulating Glass Certification Council (IGCC), 307
Iron, 66

Jalousie windows, 236
Joining errors, 291

Knockdown frames, 159–160

Lacquer, 74, 80
Laminated coatings, 58, 68–69, 77, 113
Laminated glass, 310, 318–319, 324, 335
Lead, 321
Lead-lined doors, 155
Lead-lined frames, 160
Linseed oil, 73
Lintels, 158
Liquid glazing compounds, 338
Louvers, 165–167, 173–174, 190, 197–198, 201–202, 211, 223, 284
Low-emissivity glass, 317–318

Managers. *See* Building owners/managers
Manufacturers
  as help for architects and engineers, 15–16
  as help for building contractors, 17
  as help for building owners and managers, 9–11
  of organic coatings, 70–71, 121–122
Masterspec, 40, 190, 223, 268, 303, 344
Mechanical cleaning and finishes
  for aluminum, 57–60
  for copper alloys, 61–62
  repair of finishes, 113
  for steel, 55–57

Mechanical finishes. *See* Mechanical cleaning and finishes
Metal
  cleaning and repairing, 110–119
  doors and frames, 147–191
    in existing construction, 186–190
    failure of, 175–182
    improper design, 175–177
    improper fabrication, 179–180
    inappropriate selection, 177–179
    installation, 180–182, 188–190
    natural aging, 182
    repairing, 182–186
    unforeseen trauma, 182
  factory-applied organic coatings, 80
  finishes, 54–55, 69–110
    bad, 88
    cleaning and repairing, 110–119
    damaged, 111–114
    factory primers for, 81
    failure of immediate substrate, 99
    failure to maintain applied, 107–108
    failure to protect, 106–107
    improper application, 95–98
    improper preparation, 91–95
    inappropriate, 89–91
    natural aging, 108–110
    preparation for refinishing, 126–127
    refinishing in field, 119–121
    transparent organic coatings, 72
    underlying failure, 105
    unrefinishable in field, 119
  in glazed curtain walls, 279–280
  inorganic coatings on, 63–69
  materials, 47, 86–110
    bad, 87
    cleaning and repairing, 110–119
    corrosion, 99–103, 111
    dents and holes, 114
    fracture, 103–105, 111
    inappropriate, 88–89
  preparation of surfaces for organic coatings, 79
  primers for galvanized, 76
Metal curtain walls. *See also* Glazed curtain walls
  classifications, 271
  definition, 271
Metal/Lath Steel Framing Association, 41
Metallic coatings, 63
Methacrylate lacquer, 80
Mill finish, 55, 57
Mirror glass, 310, 317, 319, 321, 324, 329

National Association of Architectural Metal Manufacturers (NAAMM),

47, 58, 61, 62, 69, 71, 140, 148, 233, 272
National Council on Radiation Protection and Measurement (NCRPM), 155
National Decorating Products Association, 140
National Fire Protection Association, 193, 229
National Forest Products Association, 41
National Wood Window and Door Association (NWWDA), 46, 208, 210, 237, 247
Nickel, 54, 66
Nitriding, 48
Non-directional textured mechanical finishes, 58

Oil-canning effect, 299
Opacifiers, 290, 318, 333
Opaque coatings, 72, 112
Opaque finishes, 44
Organic coatings. *See* Coatings
Oxidation, 100

Packaged entrances, 165
Painting and Decorating Contractors of America (PDCA), 141
Paints, 113, 131
  applying, 135–136
  removal, 133–134
  special, 77
Panel curtain wall systems, 272
Panning. *See* Retrofit casings
Partition framing, 32–35
Patches, 27–28
Patina (verde antique) finish, 62
Patterned finishes, 57, 58
Patterned glass, 312, 314
Plastic coatings, 68–69, 113
Plastic facings, 44
Plate glass, 313
Polished finishes, 57
Polycarbonate sheets, 319–320
Polyurethane, 74–75
Porcelain enamel
  on aluminum, 68
  cleaning, 118
  repair of, 114
  on steel, 67–68
Porcelain Enamel Institute (PEI), 67, 142
Powder coatings, 77
Pre-work on-site examination, 18–20
Primary glass, 306, 312–315
Professional help, 4. *See also* specific professions, e.g., Architects
  for building contractors, 17
  for building owners and managers, 5–6

Projected windows, 230, 236
Putty, 320
Pyrolitic coating deposition, 318

Quenching, 48

Rain-screen principle, 282
Reflective glass, 317
Resinous coatings, 58
Retrofit casings, 265–266, 267
Reversible steel windows, 230
Revolving doors, 174–175
Rolling block, 335

Safety glass, 306, 310, 318–319
Safety Glazing Certification Council (SGCC), 308
Samples, 195, 196, 226, 275
Screens, 244–246
Sealants, glazing, 320, 325, 327–328, 331, 332, 334, 336, 338
Sealed insulating glass, 306, 319
Sealed Insulating Glass Manufacturers Association (SIGMA), 306
Security-type hardware, 207
Selective demolition. *See* Demolition
Setting blocks, 321, 325
Settlement, 30, 32, 33
Shadow-box construction, 333
Sheet glass, 312, 313
Shellac, 74
Shop drawings, 194, 226, 247, 274–275
Silicone glazing sealants, 306, 311, 327, 331, 332, 334
Single-hung windows, 235
Sliding bypass doors, 200
Sliding glass doors, 45
  aluminum, 233–236
    grades and performance classes, 234–235
    hardware, 236
    materials, products, finishes, 233–234
  controls, 226–228, 265
  in existing construction, 264–267
  fabrication, 246–247, 253–254
  failure of, 250–256
  frames and trim, 243
  glazing, 228, 265, 267
  improper design, 250–251
  inappropriate selection, 252–253
  installation, 248–250, 254–255, 266–267
  loose, sagging, sticking, out-of-line, 262–263
  miscellaneous components and accessories, 241–246, 265–266
  natural aging, 255
  repairing, 256–264

Sliding glass doors *(cont.)*
  thermal barriers, 241–242
  tracks, 243
  weather stripping, 242
  wood
    hardware, 240–241
    materials, products, finishes, 238
    performance requirements, 238–239
    standards, 237
Sliding pocket doors, 200
Sloped glazing, 272, 327
Soda-lime glass, 311
Solder, 321
Sound Transmission Classification (STC), 278
Spacers, 321
Spandrel cavities, 289, 303, 332, 333, 343
Spandrel glass, 318
Specialty consultants
  as help for architects and engineers, 16–17
  as help for building contractors, 17
  as help for building owners and managers, 12–13
Specialty contractors, 11–12
Spraying, 63
Stained glass. *See* Art glass
Stainless steel. *See* Steel, stainless
Standard intermediate windows, 230
Standards
  glazed curtain wall, 272–273, 301
  glazing, 306, 340
  industry, 15–16
  metal door and frame, 148–149, 187–188
  organic coating, 69
  sliding glass door, 237, 265
  window, 228–229, 265
  wood door and frame, 193–194
Statuary bronze, 62
Steel
  failed, 31–32
  glazing, with glass into, 326
  porcelain enamel on, 67–68
  windows, 228–232
Steel, carbon, 66
  aluminized, 65–66, 76
  cadmium-plated, 66
  chemical finishes for and cleaning of, 56
  doors and frames, 148–150, 152–160, 167, 169
  galvanized, 63–65, 76
  heat treatment of, 48
  hollow-metal doors and frames, 47
  louvers, 165
  mechanical finishes for and cleaning of, 55–56
  products, 48–49

Steel, stainless, 66
  chemical finishes for, 57
  cleaning, 57, 117
  doors and frames, 47, 148–151, 160, 167, 169
  glazing into, 326
  heat treatment of, 48
  louvers, 165
  mechanical finishes for and cleaning of, 56–57
  products, 48–49
Steel Door Institute, 152, 160, 171
Steel Structures Painting Council (SSPC), 55, 77, 126, 142
Steel Window Institute (SWI), 228, 230
Stick glazed curtain wall systems, 271–272
Stile-and-rail doors, 161–162, 199
Store fronts, 160, 162–164, 173, 203–204
Storm water removal, 281
Storm windows, 244
Strippable coatings, 72, 293
Structural glazing sealants, 311, 320, 327–328, 331, 332, 334
Supports, metal frame, 169–170
Swinging doors, 161–162, 167–169, 173

Tape, glazing, 320, 325
Tempered glass, 306, 315, 317, 324, 334
Tempering, 48
Thermal barriers, 241–242, 283–284
Thermal breaks, 281, 291
Thermal expansion/contraction, 30, 277
Tinted glass, 311, 312, 313, 324, 331
Top-hinged
  in-swinging windows, 236
  out-swinging windows, 232
Transparent coatings, 72–75, 113, 131, 132, 136
Transparent finishes, 44
Traumatic events, 31, 33

Unified Numbering System (UNS), 53
Unit-and-mullion curtain wall systems, 272
Unit curtain wall systems, 272
United States Gypsum Company, 41

Vacuum sputtering, 318
Varnish, 73–74
Veneers, 44
Venetian blinds, 244
Vertically pivoted windows, 236
Vibration, 31, 32, 33
Vinyl clad wood windows, 45
Vitreous coatings, 58, 67–68

Walls
    concrete, stone, or masonry, 37–39
    glazed curtain (*see* Glazed curtain walls)
    metal, 35–36
    wood, 32–35
Warranties, 6, 9, 195, 208, 227, 276, 309–310, 340
Water-infiltration test, 235
Water leaks, 264, 277, 298–299
Weathering and exposure, 100
Weather stripping, 242
Welded buck and trim frames, 159
Western Lath, Plaster, and Drywall Contractors Association, 41
Wet sealants, 325
Wind loading, 310
Window consultants, 12
Windows. *See also* specific types
    aluminum, 233–236
        grades and performance classes, 234–235
        hardware, 236
        materials, products, finishes, 233–234
        types, 235–236
    controls, 226–228, 265
    in existing construction, 264–267
    fabrication, 246–247, 253–254
    failure of, 250–256
    frames and trim, 243
    glazing, 228, 265, 267
    improper design, 250–251
    inappropriate selection, 252–253
    installation, 248–250, 254–255, 266–267
    loose, sagging, sticking, out-of-line, 262–263
    metal bent, twisted, deformed, 261
    miscellaneous components and accessories, 241–246, 265–266
    mullions, 243
    muntins, 244
    natural aging, 255
    repairing, 256–264
    sills and stools, 243
    steel, 226–232
        grades, 230
        hardware, 232
        materials, products, finishes, 229–230
        standards, 228–229
        types, 230–232
    thermal barriers, 241–242
    tracks, 243
    weather stripping, 242
    wood
        broken or deteriorated components, 261–262
        hardware, 240–241
        materials, products, finishes, 45, 238

    performance requirements, 238–239
    standards, 237
    types, 239
Wind pressure, variable, 30
Wire glass, 306, 312, 314, 316, 326
Wood, 43–46
    cleaning and repairing, 110–119
    doors and frames, 192–224
        controls, 194–196, 221
        in existing construction, 221–223
        fabrication and operation of, 193–208
        failure of, 211–215
        finish hardware, 206–208
        flush panels, 200
        frame and trim materials, 45–46
        hardware, 196, 206–208
        improper design, 212
        improper fabrication, 214
        inappropriate selection, 212–214
        installation, 208–211, 214–215, 222–223
        loose, sagging, binding, out-of-line, 219–221
        louvers, 197–198, 201–202, 223
        materials and finishes for new, 197
        natural aging, 215
        repairing, 215–221
        samples, 195, 196
        site-and-rail, 199
        special, 200
        standards, 193–194, 221
        warped or twisted, 219
    factory-applied organic coatings, 79
    finishes, 69–86, 86–110
        bad, 88
        cleaning and repairing, 110–119
        damaged, 111–114
        failure of immediate substrate, 99
        failure to maintain applied, 107–108
        failure to protect, 106–107
        improper application, 95–98
        improper preparation, 91–95
        inappropriate, 89, 90–91
        lacquer, 74
        linseed oil, 73
        natural aging, 108–110
        polyurethane, 74–75
        preparation for refinishing, 125–126
        preservatives and sealers, 75
        refinishing in field, 119–121
        shellac, 74
        shop primers for, 80–81
        stain, 73
        transparent organic coatings, 72–75
        unrefinishable in field, 119
        varnish, 73–74
    glazing into, 326–327

Wood *(cont.)*
  materials, 86–110
    bad, 87
    cleaning and repairing, 110–119
    dented or punctured, 114–115
    inappropriate, 88–89
    panel, 44
    sliding glass door, 45
  preparation of surfaces for organic coatings, 78–79
  preservative treatment, 46
  stain, 73
  structure, 32–36
  walls, 32–35
  windows, 45, 237–241
Woodwork Institute of California (WIC), 193

Zinc, 54, 321
Zincating, 66
Zinc Institute, 145